Inventions
and
Inventors

Inventions and Inventors

Volume 2

Laser — Yellow fever vaccine

Index

459 – 936

edited by

Roger Smith

SALEM PRESS, INC.
Pasadena, California   Hackensack, New Jersey

Essays originally appeared in *Twentieth Century: Great Events* (1992, 1996), *Twentieth Century: Great Scientific Achievements* (1994), and *Great Events from History II: Business and Commerce Series* (1994). New material has been added.

∞ The paper used in these volumes conforms to the American National Standard for Permanence of Paper for Printed Library Materials, Z39.48-1992 (R1997).

Library of Congress Cataloging-in-Publication Data
Inventions and inventors / edited by Roger Smith
    p.cm. — (Magill's choice)
    Includes bibliographical reference and index
    ISBN 1-58765-016-9 (set : alk. paper) — ISBN 1-58765-017-7 (vol 1 : alk. paper) — ISBN 1-58765-018-5 (vol 2. : alk. paper)
1. Inventions—History—20th century—Encyclopedias. 2. Inventors—Biography—Encyclopedias. I. Smith, Roger, 1953-   .
II. Series.

T20 .I59 2001
609—dc21   2001049412

TABLE OF CONTENTS

Inventions
and
Inventors

Laser

THE INVENTION: Taking its name from the acronym for *l*ight *a*mplification by the *s*timulated *e*mission of *r*adiation, a laser is a beam of electromagnetic radiation that is monochromatic, highly directional, and coherent. Lasers have found multiple applications in electronics, medicine, and other fields.

THE PEOPLE BEHIND THE INVENTION:

Theodore Harold Maiman (1927-   ), an American physicist

Charles Hard Townes (1915-   ), an American physicist who was a cowinner of the 1964 Nobel Prize in Physics

Arthur L. Schawlow (1921-1999), an American physicist, cowinner of the 1981 Nobel Prize in Physics

Mary Spaeth (1938-   ), the American inventor of the tunable laser

Coherent Light

Laser beams differ from other forms of electromagnetic radiation in being consisting of a single wavelength, being highly directional, and having waves whose crests and troughs are aligned. A laser beam launched from Earth has produced a spot a few kilometers wide on the Moon, nearly 400,000 kilometers away. Ordinary light would have spread much more and produced a spot several times wider than the Moon. Laser light can also be concentrated so as to yield an enormous intensity of energy, more than that of the surface of the Sun, an impossibility with ordinary light.

In order to appreciate the difference between laser light and ordinary light, one must examine how light of any kind is produced. An ordinary light bulb contains atoms of gas. For the bulb to light up, these atoms must be excited to a state of energy higher then their normal, or ground, state. This is accomplished by sending a current of electricity through the bulb; the current jolts the atoms into the higher-energy state. This excited state is unstable, however, and the atoms will spontaneously return to their ground state by ridding themselves of excess energy.

Scanner device using a laser beam to read shelf labels. (PhotoDisc)

As these atoms emit energy, light is produced. The light emitted by a lamp full of atoms is disorganized and emitted in all directions randomly. This type of light, common to all ordinary sources, from fluorescent lamps to the Sun, is called "incoherent light."

Laser light is different. The excited atoms in a laser emit their excess energy in a unified, controlled manner. The atoms remain in the excited state until there are a great many excited atoms. Then, they are stimulated to emit energy, not independently, but in an organized fashion, with all their light waves traveling in the same direction, crests and troughs perfectly aligned. This type of light is called "coherent light."

Theory to Reality

In 1958, Charles Hard Townes of Columbia University, together with Arthur L. Schawlow, explored the requirements of the laser in a theoretical paper. In the Soviet Union, F. A. Butayeva and V. A. Fabrikant had amplified light in 1957 using mercury; however, their work was not published for two years and was not published in a scientific journal. The work of the Soviet scientists, therefore, re-

MARY SPAETH

Born in 1938, Mary Dietrich Spaeth, inventor of the tunable laser, learned to put things together early. When she was just three years old, her father began giving her tools to play with. She learned to use them well and got interested in science along the way. She studied mathematics and physics at Valparaiso University, graduating in 1960, and earned a master's degree in nuclear physics from Wayne State University in 1962.

The same year she joined Hughes Aircraft Company as a researcher. While waiting for supplies for her regular research in 1966, she examined the lasers in her laboratory. She wondered if, by adding dyes, she could cause the beams to change colors. Cobbling together two lasers—one to boost the power of the test laser—with Duco cement, she added dyes and succeeded at once. She found that she could produce light in a wide range of colors with different dyes. The tunable dye laser afterward was used to separate isotopes in nuclear reactor fuel, to purify plutonium for weapons, and to boost the power of ground-based astronomical telescopes. She also invented a resonant reflector for ruby range finders and performed basic research on passive Q switches used in lasers.

Because Spaeth considered Hughes's promotion policies to discriminate against women scientists, she moved to the Lawrence Livermore National Laboratory in 1974. In 1986 she became the deputy associate director of its Laser Isotope Separation program.

ceived virtually no attention in the Western world.

In 1960, Theodore Harold Maiman constructed the first laser in the United States using a single crystal of synthetic pink ruby, shaped into a cylindrical rod about 4 centimeters long and 0.5 centimeter across. The ends, polished flat and made parallel to within about a millionth of a centimeter, were coated with silver to make them mirrors.

It is a property of stimulated emission that stimulated light waves will be aligned exactly (crest to crest, trough to trough, and with respect to direction) with the radiation that does the stimulating. From the group of excited atoms, one atom returns to its ground state, emitting light. That light hits one of the other exited atoms and stimulates it to fall to its ground state and emit light. The two light waves are exactly in step. The light from these two atoms hits other excited atoms, which respond in the same way, "amplifying" the total sum of light.

If the first atom emits light in a direction parallel to the length of the crystal cylinder, the mirrors at both ends bounce the light waves back and forth, stimulating more light and steadily building up an increasing intensity of light. The mirror at one end of the cylinder is constructed to let through a fraction of the light, enabling the light to emerge as a straight, intense, narrow beam.

CONSEQUENCES

When the laser was introduced, it was an immediate sensation. In the eighteen months following Maiman's announcement that he had succeeded in producing a working laser, about four hundred companies and several government agencies embarked on work involving lasers. Activity centered on improving lasers, as well as on exploring their applications. At the same time, there was equal activity in publicizing the near-miraculous promise of the device, in applications covering the spectrum from "death" rays to sight-saving operations. A popular film in the James Bond series, *Goldfinger* (1964), showed the hero under threat of being sliced in half by a laser beam—an impossibility at the time the film was made because of the low power-output of the early lasers.

In the first decade after Maiman's laser, there was some disappointment. Successful use of lasers was limited to certain areas of medicine, such as repairing detached retinas, and to scientific applications, particularly in connection with standards: The speed of light was measured with great accuracy, as was the distance to the Moon. By 1990, partly because of advances in other fields, essentially all the laser's promise had been fulfilled, including the death ray and James Bond's slicer. Yet the laser continued to find its place in technologies not envisioned at the time of the first laser. For example, lasers are now used in computer printers, in compact disc players, and even in arterial surgery.

See also Atomic clock; Compact disc; Fiber-optics; Holography; Laser-diode recording process; Laser vaporization; Optical disk.

FURTHER READING

Townes, Charles H. *How the Laser Happened: Adventures of a Scientist.* New York: Oxford University Press, 1999.

Weber, Robert L. *Pioneers of Science: Nobel Prize Winners in Physics.* 2d ed. Philadelphia: A. Hilger, 1988.

Yen, W. M., Marc D. Levenson, and Arthur L. Schawlow. *Lasers, Spectroscopy, and New Ideas: A Tribute to Arthur L. Schawlow.* New York: Springer-Verlag, 1987.

LASER-DIODE RECORDING PROCESS

THE INVENTION: Video and audio playback system that uses a low-power laser to decode information digitally stored on reflective disks.

THE ORGANIZATION BEHIND THE INVENTION:
The Philips Corporation, a Dutch electronics firm

THE DEVELOPMENT OF DIGITAL SYSTEMS

Since the advent of the computer age, it has been the goal of many equipment manufacturers to provide reliable digital systems for the storage and retrieval of video and audio programs. A need for such devices was perceived for several reasons. Existing storage media (movie film and 12-inch, vinyl, long-playing records) were relatively large and cumbersome to manipulate and were prone to degradation, breakage, and unwanted noise. Thus, during the late 1960's, two different methods for storing video programs on disc were invented. A mechanical system was demonstrated by the Telefunken Company, while the Radio Corporation of America (RCA) introduced an electrostatic device (a device that used static electricity). The first commercially successful system, however, was developed during the mid-1970's by the Philips Corporation.

Philips devoted considerable resources to creating a digital video system, read by light beams, which could reproduce an entire feature-length film from one 12-inch videodisc. An integral part of this innovation was the fabrication of a device small enough and fast enough to read the vast amounts of greatly compacted data stored on the 12-inch disc without introducing unwanted noise. Although Philips was aware of the other formats, the company opted to use an optical scanner with a small "semiconductor laser diode" to retrieve the digital information. The laser diode is only a fraction of a millimeter in size, operates quite efficiently with high amplitude and relatively low power (0.1 watt), and can be used continuously. Because this configuration operates at a high frequency, its information-carrying capacity is quite large.

Although the digital videodisc system (called "laservision") works well, the low level of noise and the clear images offered by this system were masked by the low quality of the conventional television monitors on which they were viewed. Furthermore, the high price of the playback systems and the discs made them noncompetitive with the videocassette recorders (VCRs) that were then capturing the market for home systems. VCRs had the additional advantage that programs could be recorded or copied easily. The Philips Corporation turned its attention to utilizing this technology in an area where low noise levels and high quality would be more readily apparent—audio disc systems. By 1979, they had perfected the basic compact disc (CD) system, which soon revolutionized the world of stereophonic home systems.

READING DIGITAL DISCS WITH LASER LIGHT

Digital signals (signals composed of numbers) are stored on discs as "pits" impressed into the plastic disc and then coated with a thin reflective layer of aluminum. A laser beam, manipulated by delicate, fast-moving mirrors, tracks and reads the digital information as changes in light intensity. These data are then converted to a varying electrical signal that contains the video or audio information. The data are then recovered by means of a sophisticated pickup that consists of the semiconductor laser diode, a polarizing beam splitter, an objective lens, a collective lens system, and a photodiode receiver. The beam from the laser diode is focused by a collimator lens (a lens that collects and focuses light) and then passes through the polarizing beam splitter (PBS). This device acts like a one-way mirror mounted at 45 degrees to the light path. Light from the laser passes through the PBS as if it were a window, but the light emerges in a polarized state (which means that the vibration of the light takes place in only one plane). For the beam reflected from the CD surface, however, the PBS acts like a mirror, since the reflected beam has an opposite polarization. The light is thus deflected toward the photodiode detector. The objective lens is needed to focus the light onto the disc surface. On the outer surface of the transparent disc, the main spot of light has a diameter of 0.8 millimeter, which narrows to only 0.0017 millimeter at the reflective surface. At the surface, the spot is about three times the size of the microscopic pits (0.0005 millimeter).

The data encoded on the disc determine the relative intensity of the reflected light, on the basis of the presence or absence of pits. When the reflected laser beam enters the photodiode, a modulated light beam is changed into a digital signal that becomes an analog (continuous) audio signal after several stages of signal processing and error correction.

CONSEQUENCES

The development of the semiconductor laser diode and associated circuitry for reading stored information has made CD audio systems practical and affordable. These systems can offer the quality of a live musical performance with a clarity that is undisturbed by noise and distortion. Digital systems also offer several other significant advantages over analog devices. The dynamic range (the difference between the softest and the loudest signals that can be stored and reproduced) is considerably greater in digital systems. In addition, digital systems can be copied precisely; the signal is not degraded by copying, as is the case with analog systems. Finally, error-correcting codes can be used to detect and correct errors in transmitted or reproduced digital signals, allowing greater precision and a higher-quality output sound.

Besides laser video systems, there are many other applications for laser-read CDs. Compact disc read-only memory (CD-ROM) is used to store computer text. One standard CD can store 500 megabytes of information, which is about twenty times the storage of a hard-disk drive on a typical home computer. Compact disc systems can also be integrated with conventional televisions (called CD-V) to present twenty minutes of sound and five minutes of sound with picture. Finally, CD systems connected with a computer (CD-I) mix audio, video, and computer programming. These devices allow the user to stop at any point in the program, request more information, and receive that information as sound with graphics, film clips, or as text on the screen.

See also Compact disc; Laser; Videocassette recorder; Walkman cassette player.

FURTHER READING

Atkinson, Terry. "Picture This: CD's with Video, By Christmas '87." *Los Angeles Times* (February 20, 1987).

Botez, Dan, and Luis Figueroa. *Laser-Diode Technology and Applications II: 16-19 January 1990, Los Angeles, California.* Bellingham, Wash.: SPIE, 1990.

Clemens, Jon K. "Video Disks: Three Choices." *IEEE Spectrum* 19, no. 3 (March, 1982).

"Self-Pulsating Laser for DVD." *Electronics Now* 67, no. 5 (May, 1996).

LASER EYE SURGERY

THE INVENTION: The first significant clinical ophthalmic application of any laser system was the treatment of retinal tears with a pulsed ruby laser.

THE PEOPLE BEHIND THE INVENTION:
Charles J. Campbell (1926-   ), an ophthalmologist
H. Christian Zweng (1925-   ), an ophthalmologist
Milton M. Zaret (1927-   ), an ophthalmologist
Theodore Harold Maiman (1927-   ), the physicist who developed the first laser

MONKEYS AND RABBITS

The term "laser" is an acronym for light amplification by the stimulated emission of radiation. The development of the laser for ophthalmic (eye surgery) surgery arose from the initial concentration of conventional light by magnifying lenses.

Within a laser, atoms are highly energized. When one of these atoms loses its energy in the form of light, it stimulates other atoms to emit light of the same frequency and in the same direction. A cascade of these identical light waves is soon produced, which then oscillate back and forth between the mirrors in the laser cavity. One mirror is only partially reflective, allowing some of the laser light to pass through. This light can be concentrated further into a small burst of high intensity.

On July 7, 1960, Theodore Harold Maiman made public his discovery of the first laser—a ruby laser. Shortly thereafter, ophthalmologists began using ruby lasers for medical purposes.

The first significant medical uses of the ruby laser occurred in 1961, with experiments on animals conducted by Charles J. Campbell in New York, H. Christian Zweng, and Milton M. Zaret. Zaret and his colleagues produced photocoagulation (a thickening or drawing together of substances by use of light) of the eyes of rabbits by flashes from a ruby laser. Sufficient energy was delivered to cause immediate thermal injury to the retina and iris of the rabbit. The beam also was directed to the interior of the rabbit eye, resulting in retinal coagulations. The team examined the retinal lesions and pointed out both the possible advantages of laser as a tool for therapeutic photocoagulation and the potential applications in medical research.

In 1962, Zweng, along with several of his associates, began experimenting with laser photocoagulation on the eyes of monkeys and rabbits in order to establish parameters for the use of lasers on the human eye.

REFLECTED BY BLOOD

The vitreous humor, a transparent jelly that usually fills the vitreous cavity of the eyes of younger individuals, commonly shrinks with age, with myopia, or with certain pathologic conditions. As these conditions occur, the vitreous humor begins to separate from the adjacent retina. In some patients, the separating vitreous humor produces a traction (pulling), causing a retinal tear to form. Through this opening in the retina, liquefied vitreous humor can pass to a site underneath the retina, producing retinal detachment and loss of vision.

A laser can be used to cause photocoagulation of a retinal tear. As a result, an adhesive scar forms between the retina surrounding the tear and the underlying layers so that, despite traction, the retina does not detach. If more than a small area of retina has detached, the laser often is ineffective and major retinal detachment surgery must be performed. Thus, in the experiments of Campbell and Zweng, the ruby laser was used to prevent, rather than treat, retinal detachment.

In subsequent experiments with humans, all patients were treated with the experimental laser photocoagulator without anesthesia. Although usually no attempt was made to seal holes or tears, the diseased portions of the retina were walled off satisfactorily so that no detachments occurred. One problem that arose involved microaneurysms. A "microaneurysm" is a tiny aneurysm, or blood-filled bubble extending from the wall of a blood vessel. When attempts to obliterate microaneurysms were unsuccessful, the researchers postulated that the color of the ruby pulse so resembled the red of blood that the light was reflected rather than absorbed. They believed that another lasing material emitting light in another part of the spectrum might have performed more successfully.

Previously, xenon-arc lamp photocoagulators had been used to treat retinal tears. The long exposure time required of these systems, combined with their broad spectral range emission (versus the single wavelength output of a laser), however, made the retinal spot on which the xenon-arc could be focused too large for many applications. Focused laser spots on the retina could be as small as 50 microns.

CONSEQUENCES

The first laser in ophthalmic use by Campbell, Zweng, and Zaret, among others, was a solid laser—Maiman's ruby laser. While the results they achieved with this laser were more impressive than with the previously used xenon-arc, in the decades following these experiments, argon gas replaced ruby as the most frequently used material in treating retinal tears.

Argon laser energy is delivered to the area around the retinal tear through a slit lamp or by using an intraocular probe introduced directly into the eye. The argon wavelength is transmitted through the clear structures of the eye, such as the cornea, lens, and vitreous. This beam is composed of blue-green light that can be effectively aimed at the desired portion of the eye. Nevertheless, the beam can be absorbed by cataracts and by vitreous or retinal blood, decreasing its effectiveness.

Moreover, while the ruby laser was found to be highly effective in producing an adhesive scar, it was not useful in the treatment of vascular diseases of the eye. A series of laser sources, each with different characteristics, was considered, investigated, and used clinically for various durations during the period that followed Campbell and Zweng's experiments.

Other laser types that are being adapted for use in ophthalmology are carbon dioxide lasers for scleral surgery (surgery on the tough, white, fibrous membrane covering the entire eyeball except the area covered by the cornea) and eye wall resection, dye lasers to kill or slow the growth of tumors, eximer lasers for their ability to break down corneal tissue without heating, and pulsed erbium lasers used to cut intraocular membranes.

See also Contact lenses; Coronary artery bypass surgery; Laser; Laser vaporization.

FURTHER READING

Constable, Ian J., and Arthur Siew Ming Lin. *Laser: Its Clinical Uses in Eye Diseases*. Edinburgh: Churchill Livingstone, 1981.
Guyer, David R. *Retina, Vitreous, Macula*. Philadelphia: Saunders, 1999.
Hecht, Jeff. *Laser Pioneers*. Rev. ed. Boston: Academic Press, 1992.
Smiddy, William E., Lawrence P. Chong, and Donald A. Frambach. *Retinal Surgery and Ocular Trauma*. Philadelphia: Lippincott, 1995.

Laser Vaporization

The invention: Technique using laser light beams to vaporize the plaque that clogs arteries.

The people behind the invention:
Albert Einstein (1879-1955), a theoretical American physicist
Theodore Harold Maiman (1927-   ), inventor of the laser

Light, Lasers, and Coronary Arteries

Visible light, a type of electromagnetic radiation, is actually a form of energy. The fact that the light beams produced by a light bulb can warm an object demonstrates that this is the case. Light beams are radiated in all directions by a light bulb. In contrast, the device called the "laser" produces light that travels in the form of a "coherent" unidirectional beam. Coherent light beams can be focused on very small areas, generating sufficient heat to melt steel.

The term "laser" was invented in 1957 by R. Gordon Gould of Columbia University. It stands for light amplification by stimulated emission of radiation, the means by which laser light beams are made. Many different materials—including solid ruby gemstones, liquid dye solutions, and mixtures of gases—can produce such beams in a process called "lasing." The different types of lasers yield light beams of different colors that have many uses in science, industry, and medicine. For example, ruby lasers, which were developed in 1960, are widely used in eye surgery. In 1983, a group of physicians in Toulouse, France, used a laser for cardiovascular treatment. They used the laser to vaporize the "atheroma" material that clogs the arteries in the condition called "atherosclerosis." The technique that they used is known as "laser vaporization surgery."

Laser Operation, Welding, and Surgery

Lasers are electronic devices that emit intense beams of light when a process called "stimulated emission" occurs. The principles of laser operation, including stimulated emission, were established by Albert Einstein and other scientists in the first third of the twentieth century. In 1960, Theodore H. Maiman of the Hughes Research Center in Malibu, California, built the first laser, using a ruby crystal to produce a laser beam composed of red light.

All lasers are made up of three main components. The first of these, the laser's "active medium," is a solid (like Maiman's ruby crystal), a liquid, or a gas that can be made to lase. The second component is a flash lamp or some other light energy source that puts light into the active medium. The third component is a pair of mirrors that are situated on both sides of the active medium and are designed in such a way that one mirror transmits part of the energy that strikes it, yielding the light beam that leaves the laser.

Lasers can produce energy because light is one of many forms of energy that are called, collectively, electromagnetic radiation (among the other forms of electromagnetic radiation are X rays and radio waves). These forms of electromagnetic radiation have different wavelengths; the smaller the wavelength, the higher the energy level. The energy level is measured in units called "quanta." The emission of light quanta from atoms that are said to be in the "excited state" produces energy, and the absorption of quanta by unexcited atoms—atoms said to be in the "ground state"—excites those atoms.

The familiar light bulb spontaneously and haphazardly emits light of many wavelengths from excited atoms. This emission occurs in all directions and at widely varying times. In contrast, the light reflection between the mirrors at the ends of a laser causes all of the many excited atoms present in the active medium simultaneously to emit light waves of the same wavelength. This process is called "stimulated emission."

Stimulated emission ultimately causes a laser to yield a beam of coherent light, which means that the wavelength, emission time, and direction of all the waves in the laser beam are the same. The use of focusing devices makes it possible to convert an emitted laser beam into a point source that can be as small as a few thousandths of an inch in diameter. Such focused beams are very hot, and they can be used for such diverse functions as cutting or welding metal objects and performing delicate surgery. The nature of the active medium used in a laser determines the wavelength of its emitted light beam; this in turn dictates both the energy of the emitted quanta and the appropriate uses for the laser.

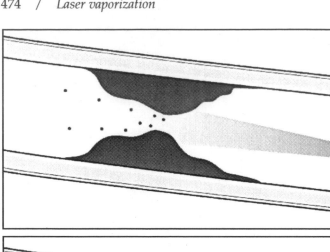

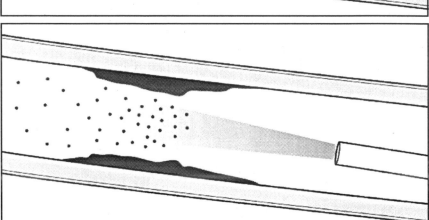

A blocked artery (top) can be threaded with a flexible fiber-optic fiber or bundle of fibers until it reaches the blockage; the fiber then emits laser light, vaporizing the plaque (bottom) and restoring circulation.

Maiman's ruby laser, for example, has been used since the 1960's in eye surgery to reattach detached retinas. This is done by focusing the laser on the tiny retinal tear that causes a retina to become detached. The very hot, high-intensity light beam then "welds" the retina back into place, bloodlessly, by burning it to produce scar tissue. The burning process has no effect on nearby tissues. Other types of lasers have been used in surgeries on the digestive tract and the uterus since the 1970's.

In 1983, a group of physicians began using lasers to treat cardiovascular disease. The original work, which was carried out by a number of physicians in Toulouse, France, involved the vaporization of atheroma deposits (atherosclerotic plaque) in a human artery. This very exciting event added a new method to medical science's arsenal of life-saving techniques.

CONSEQUENCES

Since their discovery, lasers have been used for many purposes in science and industry. Such uses include the study of the laws of chemistry and physics, photography, communications, and surveying. Lasers have been utilized in surgery since the mid-1960's, and their use has had a tremendous impact on medicine. The first type of laser surgery to be conducted was the repair of detached retinas via ruby lasers. This technique has become the method of choice for such eye surgery because it takes only minutes to perform rather than the hours required for conventional surgical methods. It is also beneficial because the lasing of the surgical site cauterizes that site, preventing bleeding.

In the late 1970's, the use of other lasers for abdominal cancer surgery and uterine surgery began and flourished. In these forms of surgery, more powerful lasers are used. In the 1980's, laser vaporization surgery (LVS) began to be used to clear atherosclerotic plaque (atheromas) from clogged arteries. This methodology gives cardiologists a useful new tool. Before LVS was available, surgeons dislodged atheromas by means of "transluminal angioplasty," which involved pushing small, fluoroscope-guided inflatable balloons through clogged arteries.

See also Blood transfusion; CAT scanner; Coronary artery bypass surgery; Electrocardiogram; Laser; Laser eye surgery; Ultrasound.

FURTHER READING

Fackelmann, Kathleen. "Internal Laser Blast Might Ease Heart Pain." *USA Today* (March 8, 1999).
Hecht, Jeff. *Laser Pioneers*. Rev. ed. Boston: Academic Press, 1992.
"Is Cervical Laser Therapy Painful?" *Lancet* no. 8629 (January 14, 1989).

Lothian, Cheri L. "Laser Angioplasty: Vaporizing Coronary Artery Plaque." *Nursing* 22, no. 1 (January, 1992).

"New Cool Laser Procedure Has Promise for Treating Blocked Coronary Arteries." *Wall Street Journal* (May 15, 1989).

Rundle, Rhonda L. "FDA Approves Laser Systems for Angioplasty." *Wall Street Journal* (February 3, 1992).

Sutton, C. J. G., and Michael P. Diamond. *Endoscopic Surgery for Gynecologists*. Philadelphia: W. B. Saunders, 1993.

Long-distance radiotelephony

THE INVENTION: The first radio transmissions from the United States to Europe opened a new era in telecommunications.

THE PEOPLE BEHIND THE INVENTION:

Guglielmo Marconi (1874-1937), Italian inventor of transatlantic telegraphy

Reginald Aubrey Fessenden (1866-1932), an American radio engineer

Lee de Forest (1873-1961), an American inventor

Harold D. Arnold (1883-1933), an American physicist

John J. Carty (1861-1932), an American electrical engineer

An Accidental Broadcast

The idea of commercial transatlantic communication was first conceived by Italian physicist and inventor Guglielmo Marconi, the pioneer of wireless telegraphy. Marconi used a spark transmitter to generate radio waves that were interrupted, or modulated, to form the dots and dashes of Morse code. The rapid generation of sparks created an electromagnetic disturbance that sent radio waves of different frequencies into the air—a broad, noisy transmission that was difficult to tune and detect.

The inventor Reginald Aubrey Fessenden produced an alternative method that became the basis of radio technology in the twentieth century. His continuous radio waves kept to one frequency, making them much easier to detect at long distances. Furthermore, the continuous waves could be modulated by an audio signal, making it possible to transmit the sound of speech.

Fessenden used an alternator to generate electromagnetic waves at the high frequencies required in radio transmission. It was specially constructed at the laboratories of the General Electric Company. The machine was shipped to Brant Rock, Massachusetts, in 1906 for testing. Radio messages were sent to a boat cruising offshore, and the feasibility of radiotelephony was thus demonstrated. Fessenden followed this success with a broadcast of messages and music between Brant Rock and a receiving station constructed at Plymouth, Massachusetts.

The equipment installed at Brant Rock had a range of about 160 kilometers. The transmission distance was determined by the strength of the electric power delivered by the alternator, which was measured in watts. Fessenden's alternator was rated at 500 watts, but it usually delivered much less power.

Yet this was sufficient to send a radio message across the Atlantic. Fessenden had built a receiving station at Machrihanish, Scotland, to test the operation of a large rotary spark transmitter that he had constructed. An operator at this station picked up the voice of an engineer at Brant Rock who was sending instructions to Plymouth. Thus, the first radiotelephone message had been sent across the Atlantic by accident. Fessenden, however, decided not to make this startling development public. The station at Machrihanish was destroyed in a storm, making it impossible to carry out further tests. The successful transmission undoubtedly had been the result of exceptionally clear atmospheric conditions that might never again favor the inventor.

One of the parties following the development of the experiments in radio telephony was the American Telephone and Telegraph (AT&T) Company. Fessenden entered into negotiations to sell his system to the telephone company, but, because of the financial panic of 1907, the sale was never made.

VIRGINIA TO PARIS AND HAWAII

The English physicist John Ambrose Fleming had invented a two-element (diode) vacuum tube in 1904 that could be used to generate and detect radio waves. Two years later, the American inventor Lee de Forest added a third element to the diode to produce his "audion" (triode), which was a more sensitive detector. John J. Carty, head of a research and development effort at AT&T, examined these new devices carefully. He became convinced that an electronic amplifier, incorporating the triode into its design, could be used to increase the strength of telephone signals and to long distances.

On Carty's advice, AT&T purchased the rights to de Forest's audion. A team of about twenty-five researchers, under the leader-

REGINALD AUBREY FESSENDEN

Reginald Aubrey Fessenden was born in Canada in 1866 to a small-town minister and his wife. After graduating from Bishop's College in Lennoxville, Quebec, he took a job as head of Whitney Institute in Bermuda. However, he was brilliant and volatile and had greater ambitions. After two years, he landed a job as a tester for his idol, Thomas Edison. Soon he was working as an engineer and chemist.

Fessenden became a professor of electrical engineering at Purdue University in 1892 and then a year later at the University of Pittsburgh. His ideas were often advanced, so far advanced that some were not developed until much later, and by others. His first patented invention, an electrolyte detector in 1900, was far more sensitive than others in use and made it possible to pick up radio signals carrying complex sound. To transmit such signals, he pioneered the use of carrier waves. During his career he registered more than three hundred patents.

Suspicious and feisty, he also spent a lot of time in disputes, and frequently in court, over his inventions. He sued his backers at the National Electric Signaling Company over rights to operate a connection to Great Britain, and won a $406,000 settlement, which bankrupted the company. He sued Radio Corporation of America (RCA) claiming it prevented him from exploiting his own patents commercially. RCA settled out of court but was enriched by Fessenden's invention.

Having returned to Bermuda, Fessenden died in 1932. He never succeeded in winning the fame and wealth for the radio that he felt was due to him.

ship of physicist Harold D. Arnold, were assigned the job of perfecting the triode and turning it into a reliable amplifier. The improved triode was responsible for the success of transcontinental cable telephone service, which was introduced in January, 1915. The triode was also the basis of AT&T's foray into radio telephony.

Carty's research plan called for a system with three components: an oscillator to generate the radio waves, a modulator to add the audio signals to the waves, and an amplifier to transmit the radio waves. The total power output of the system was 7,500 watts, enough to send the radio waves over thousands of kilometers.

The apparatus was installed in the U.S. Navy's radio tower in Arlington, Virginia, in 1915. Radio messages from Arlington were picked up at a receiving station in California, a distance of 4,000 kilometers, then at a station in Pearl Harbor, Hawaii, which was 7,200 kilometers from Arlington. AT&T's engineers had succeeded in joining the company telephone lines with the radio transmitter at Arlington; therefore, the president of AT&T, Theodore Vail, could pick up his telephone and talk directly with someone in California.

The next experiment was to send a radio message from Arlington to a receiving station set up in the Eiffel Tower in Paris. After several unsuccessful attempts, the telephone engineers in the Eiffel Tower finally heard Arlington's messages on October 21, 1915. The AT&T receiving station in Hawaii also picked up the messages. The two receiving stations had to send their reply by telegraph to the United States because both stations were set up to receive only. Two-way radio communication was still years in the future.

Impact

The announcement that messages had been received in Paris was front-page news and brought about an outburst of national pride in the United States. The demonstration of transatlantic radio telephony was more important as publicity for AT&T than as a scientific advance. All the credit went to AT&T and to Carty's laboratory. Both Fessenden and de Forest attempted to draw attention to their contributions to long-distance radio telephony, but to no avail. The Arlington-to-Paris transmission was a triumph for corporate public relations and corporate research.

The development of the triode had been achieved with large teams of highly trained scientists—in contrast to the small-scale efforts of Fessenden and de Forest, who had little formal scientific training. Carty's laboratory was an example of the new type of industrial research that was to dominate the twentieth century. The golden days of the lone inventor, in the mold of Thomas Edison or Alexander Graham Bell, were gone.

In the years that followed the first transatlantic radio telephone messages, little was done by AT&T to advance the technology or to develop a commercial service. The equipment used in the 1915 demonstration was more a makeshift laboratory apparatus than a prototype for a new radio technology. The messages sent were short and faint. There was a great gulf between hearing "hello" and "goodbye" amid the static. The many predictions of a direct telephone connection between New York and other major cities overseas were premature. It was not until 1927 that a transatlantic radio circuit was opened for public use. By that time, a new technological direction had been taken, and the method used in 1915 had been superseded by shortwave radio communication.

See also Communications satellite; Internet; Long-distance telephone; Radio; Radio crystal sets; Radiotelephony; Television.

FURTHER READING

Marconi, Degna. *My Father: Marconi.* Toronto: Guernica Editions, 1996.
Masini, Giancarlo. *Marconi.* New York: Marsilio, 1995.
Seitz, Frederick. *The Cosmic Inventor: Reginald Aubrey Fessenden.* Philadelphia: American Philosophical Society, 1999.
Streissguth, Thomas. *Communications: Sending the Message.* Minneapolis, Minn.: Oliver Press, 1997.

Long-distance telephone

The invention: System for conveying voice signals via wires over
long distances.

The people behind the invention:
Alexander Graham Bell (1847-1922), a Scottish American
inventor
Thomas A. Watson (1854-1934), an American electrical engineer

The Problem of Distance

The telephone may be the most important invention of the nineteenth century. The device developed by Alexander Graham Bell
and Thomas A. Watson opened a new era in communication and
made it possible for people to converse over long distances for the
first time. During the last two decades of the nineteenth century and
the first decade of the twentieth century, the American Telephone
and Telegraph (AT&T) Company continued to refine and upgrade
telephone facilities, introducing such innovations as automatic dialing and long-distance service.

One of the greatest challenges faced by Bell engineers was to
develop a way of maintaining signal quality over long distances.
Telephone wires were susceptible to interference from electrical
storms and other natural phenomena, and electrical resistance
and radiation caused a fairly rapid drop-off in signal strength,
which made long-distance conversations barely audible or unin-
telligible.

By 1900, Bell engineers had discovered that signal strength could
be improved somewhat by wrapping the main wire conductor with
thinner wires called "loading coils" at prescribed intervals along
the length of the cable. Using this procedure, Bell extended long-
distance service from New York to Denver, Colorado, which was
then considered the farthest point that could be reached with acceptable quality. The result, however, was still unsatisfactory, and
Bell engineers realized that some form of signal amplification would
be necessary to improve the quality of the signal.

A breakthrough came in 1906, when Lee de Forest invented the "audion tube," which could send and amplify radio waves. Bell scientists immediately recognized the potential of the new device for long-distance telephony and began building amplifiers that would be placed strategically along the long-distance wire network.

Work progressed so quickly that by 1909, Bell officials were predicting that the first transcontinental long-distance telephone service, between New York and San Francisco, was imminent. In that year, Bell president Theodore N. Vail went so far as to promise the organizers of the Panama-Pacific Exposition, scheduled to open in San Francisco in 1914, that Bell would offer a demonstration at the exposition. The promise was risky, because certain technical problems associated with sending a telephone signal over a 4,800-kilometer wire had not yet been solved. De Forest's audion tube was a crude device, but progress was being made.

Two more breakthroughs came in 1912, when de Forest improved on his original concept and Bell engineer Harold D. Arnold improved it further. Bell bought the rights to de Forest's vacuum-tube patents in 1913 and completed the construction of the New York-San Francisco circuit. The last connection was made at the Utah-Nevada border on June 17, 1914.

SUCCESS LEADS TO FURTHER IMPROVEMENTS

Bell's long-distance network was tested successfully on June 29, 1914, but the official demonstration was postponed until January 25, 1915, to accommodate the Panama-Pacific Exposition, which had also been postponed. On that date, a connection was established between Jekyll Island, Georgia, where Theodore Vail was recuperating from an illness, and New York City, where Alexander Graham Bell was standing by to talk to his former associate Thomas Watson, who was in San Francisco. When everything was in place, the following conversation took place. Bell: "Hoy! Hoy! Mr. Watson? Are you there? Do you hear me?" Watson: "Yes, Dr. Bell, I hear you perfectly. Do you hear me well?" Bell: "Yes, your voice is perfectly distinct. It is as clear as if you were here in New York."

The first transcontinental telephone conversation transmitted by wire was followed quickly by another that was transmitted via radio. Although the Bell company was slow to recognize the potential of radio wave amplification for the "wireless" transmission of telephone conversations, by 1909 the company had made a significant commitment to conduct research in radio telephony. On April 4, 1915, a wireless signal was transmitted by Bell technicians from Montauk Point on Long Island, New York, to Wilmington, Delaware, a distance of more than 320 kilometers. Shortly thereafter, a similar test was conducted between New York City and Brunswick, Georgia, via a relay station at Montauk Point. The total distance of the transmission was more than 1,600 kilometers. Finally, in September, 1915, Vail placed a successful transcontinental radio-telephone call from his office in New York to Bell engineering chief J. J. Carty in San Francisco.

Only a month later, the first telephone transmission across the Atlantic Ocean was accomplished via radio from Arlington, Virginia, to the Eiffel Tower in Paris, France. The signal was detectable, although its quality was poor. It would be ten years before true transatlantic radio-telephone service would begin.

The Bell company recognized that creating a nationwide long-distance network would increase the volume of telephone calls simply by increasing the number of destinations that could be reached from any single telephone station. As the network expanded, each subscriber would have more reason to use the telephone more often, thereby increasing Bell's revenues. Thus, the company's strategy became one of tying local and regional networks together to create one large system.

IMPACT

Just as the railroads had interconnected centers of commerce, industry, and agriculture all across the continental United States in the nineteenth century, the telephone promised to bring a new kind of interconnection to the country in the twentieth century: instantaneous voice communication. During the first quarter century after the invention of the telephone and during its subsequent commercialization, the emphasis of telephone companies was to set up central office switches that would provide interconnections among subscribers within a fairly limited geographical area. Large cities were wired quickly, and by the beginning of the twentieth century most were served by telephone switches that could accommodate thousands of subscribers.

The development of intercontinental telephone service was a milestone in the history of telephony for two reasons. First, it was a practical demonstration of the almost limitless applications of this innovative technology. Second, for the first time in its brief history, the telephone network took on a national character. It became clear that large central office networks, even in large cities such as New York, Chicago, and Baltimore, were merely small parts of a much larger, universally accessible communication network that spanned a continent. The next step would be to look abroad, to Europe and beyond.

See also Cell phone; Fax machine; Internet; Long-distance radiotelephony; Rotary dial telephone; Telephone switching; Touch-tone telephone.

FURTHER READING

Coe, Lewis. *The Telephone and Its Several Inventors: A History.* Jefferson, N.C.: McFarland, 1995.

Mackay, James A. *Alexander Graham Bell: A Life.* New York: J. Wiley, 1997.

Young, Peter. *Person to Person: The International Impact of the Telephone.* Cambridge: Granta Editions, 1991.

Mammography

THE INVENTION: The first X-ray procedure for detecting and diagnosing breast cancer.

THE PEOPLE BEHIND THE INVENTION:
Albert Salomon, the first researcher to use X-ray technology instead of surgery to identify breast cancer
Jacob Gershon-Cohen (1899-1971), a breast cancer researcher

Studying Breast Cancer

Medical researchers have been studying breast cancer for more than a century. At the end of the nineteenth century, however, no one knew how to detect breast cancer until it was quite advanced. Often, by the time it was detected, it was too late for surgery; many patients who did have surgery died. So after X-ray technology first appeared in 1896, cancer researchers were eager to experiment with it.

The first scientist to use X-ray techniques in breast cancer experiments was Albert Salomon, a German surgeon. Trying to develop a biopsy technique that could tell which tumors were cancerous and thereby avoid unnecessary surgery, he X-rayed more than three thousand breasts that had been removed from patients during breast cancer surgery. In 1913, he published the results of his experiments, showing that X rays could detect breast cancer. Different types of X-ray images, he said, showed different types of cancer.

Though Salomon is recognized as the inventor of breast radiology, he never actually used his technique to diagnose breast cancer. In fact, breast cancer radiology, which came to be known as "mammography," was not taken up quickly by other medical researchers. Those who did try to reproduce his research often found that their results were not conclusive.

During the 1920's, however, more research was conducted in Leipzig, Germany, and in South America. Eventually, the Leipzig researchers, led by Erwin Payr, began to use mammography to diagnose cancer. In the 1930's, a Leipzig researcher named W. Vogel published a paper that accurately described differences between cancerous and noncancerous tumors as they appeared on X-ray photographs. Researchers in the United States paid little attention to mammography until 1926. That year, a physician in Rochester, New York, was using a fluoroscope to examine heart muscle in a patient and discovered that the fluoroscope could be used to make images of breast tissue as well. The physician, Stafford L. Warren, then developed a stereoscopic technique that he used in examinations before surgery. Warren published his findings in 1930; his article also described changes in breast tissue that occurred because of pregnancy, lactation (milk production), menstruation, and breast disease. Yet Stafford's technique was complicated and required equipment that most physicians of the time did not have. Eventually, he lost interest in mammography and went on to other research.

Using the Technique

In the late 1930's, Jacob Gershon-Cohen became the first clinician to advocate regular mammography for all women to detect breast cancer before it became a major problem. Mammography was not very expensive, he pointed out, and it was already quite accurate. A milestone in breast cancer research came in 1956, when Gershon-Cohen and others began a five-year study of more than 1,300 women to test the accuracy of mammography for detecting breast cancer. Each woman studied was screened once every six months. Of the 1,055 women who finished the study, 92 were diagnosed with benign tumors and 23 with malignant tumors. Remarkably, out of all these, only one diagnosis turned out to be wrong.

During the same period, Robert Egan of Houston began tracking breast cancer X rays. Over a span of three years, one thousand X-ray photographs were used to make diagnoses. When these diagnoses were compared to the results of surgical biopsies, it was confirmed that mammography had produced 238 correct diagnoses of cancer, out of 240 cases. Egan therefore joined the crusade for regular breast cancer screening.

Once mammography was finally accepted by doctors in the late 1950's and early 1960's, researchers realized that they needed a way to teach mammography quickly and effectively to those who would use it. A study was done, and it showed that any radiologist could conduct the procedure with only five days of training.

In the early 1970's, the American Cancer Society and the National Cancer Institute joined forces on a nationwide breast cancer screening program called the "Breast Cancer Detection Demonstration Project." Its goal in 1971 was to screen more than 250,000 women over the age of thirty-five.

Since the 1960's, however, some people had argued that mammography was dangerous because it used radiation on patients. In 1976, Ralph Nader, a consumer advocate, stated that women who were to undergo mammography should be given consent forms that would list the dangers of radiation. In the years that followed, mammography was refined to reduced the amount of radiation needed to detect cancer. It became a standard tool for diagnosis, and doctors recommended that women have a mammogram every two or three years after the age of forty.

IMPACT

Radiology is not a science that concerns only breast cancer screening. While it does provide the technical facilities necessary to practice mammography, the photographic images obtained must be interpreted by general practitioners, as well as by specialists. Once

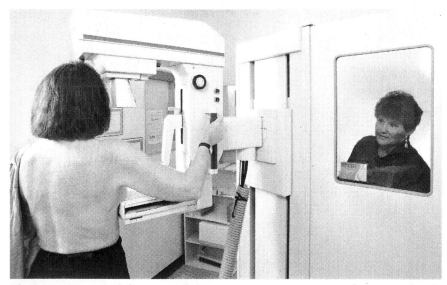

Physicians recommend that women have a mammogram every two or three years after the age of forty. (Digital Stock)

Gershon-Cohen had demonstrated the viability of the technique, a means of training was devised that made it fairly easy for clinicians to learn how to practice mammography successfully. Once all these factors—accuracy, safety, simplicity—were in place, mammography became an important factor in the fight against breast cancer.

The progress made in mammography during the twentieth century was a major improvement in the effort to keep more women from dying of breast cancer. The disease has always been one of the primary contributors to the number of female cancer deaths that occur annually in the United States and around the world. This high figure stems from the fact that women had no way of detecting the disease until tumors were in an advanced state.

Once Salomon's procedure was utilized, physicians had a means by which they could look inside breast tissue without engaging in exploratory surgery, thus giving women a screening technique that was simple and inexpensive. By 1971, a quarter million women over age thirty-five had been screened. Twenty years later, that number was in the millions.

See also Amniocentesis; CAT scanner; Electrocardiogram; Electroencephalogram; Holography; Nuclear magnetic resonance; Pap test; Syphilis test; Ultrasound.

FURTHER READING

"First Digital Mammography System Approved by FDA." *FDA Consumer* 34, no. 3 (May/June, 2000).

Hindle, William H. *Breast Care: A Clinical Guidebook for Women's Primary Health Care Providers.* New York: Springer, 1999.

Okie, Susan. "More Women Are Getting Mammograms: Experts Agree That the Test Has Played Big Role in Reducing Deaths from Breast Cancer." *Washington Post* (January 21, 1997).

Wolbarst, Anthony B. *Looking Within: How X-ray, CT, MRI, Ultrasound, and Other Medical Images Are Created, and How They Help Physicians Save Lives.* Berkeley: University of California Press, 1999.

Mark I calculator

THE INVENTION: Early digital calculator designed to solve differential equations that was a forerunner of modern computers.

THE PEOPLE BEHIND THE INVENTION:

Howard H. Aiken (1900-1973), Harvard University professor and architect of the Mark I

Clair D. Lake (1888-1958), a senior engineer at IBM

Francis E. Hamilton (1898-1972), an IBM engineer

Benjamin M. Durfee (1897-1980), an IBM engineer

The Human Computer

The physical world can be described by means of mathematics. In principle, one can accurately describe nature down to the smallest detail. In practice, however, this is impossible except for the simplest of atoms. Over the years, physicists have had great success in creating simplified models of real physical processes whose behavior can be described by the branch of mathematics called "calculus."

Calculus relates quantities that change over a period of time. The equations that relate such quantities are called "differential equations," and they can be solved precisely in order to yield information about those quantities. Most natural phenomena, however, can be described only by differential equations that can be solved only approximately. These equations are solved by numerical means that involve performing a tremendous number of simple arithmetic operations (repeated additions and multiplications). It has been the dream of many scientists since the late 1700's to find a way to automate the process of solving these equations.

In the early 1900's, people who spent day after day performing the tedious operations that were required to solve differential equations were known as "computers." During the two world wars, these human computers created ballistics tables by solving the differential equations that described the hurling of projectiles and the dropping of bombs from aircraft. The war effort was largely responsible for accelerating the push to automate the solution to these problems.

A Computational Behemoth

The ten-year period from 1935 to 1945 can be considered the prehistory of the development of the digital computer. (In a digital computer, digits represent magnitudes of physical quantities. These digits can have only certain values.) Before this time, all machines for automatic calculation were either analog in nature (in which case, physical quantities such as current or voltage represent the numerical values of the equation and can vary in a continuous fashion) or were simplistic mechanical or electromechanical adding machines.

This was the situation that faced Howard Aiken. At the time, he was a graduate student working on his doctorate in physics. His dislike for the tremendous effort required to solve the differential equations used in his thesis drove him to propose, in the fall of 1937, constructing a machine that would automate the process. He proposed taking existing business machines that were commonly used in accounting firms and combining them into one machine that would be controlled by a series of instructions. One goal was to eliminate all manual intervention in the process in order to maximize the speed of the calculation.

Aiken's proposal came to the attention of Thomas Watson, who was then the president of International Business Machines Corporation (IBM). At that time, IBM was a major supplier of business machines and did not see much of a future in such "specialized" machines. It was the pressure provided by the computational needs of the military in World War II that led IBM to invest in building automated calculators. In 1939, a contract was signed in which IBM agreed to use its resources (personnel, equipment, and finances) to build a machine for Howard Aiken and Harvard University.

IBM brought together a team of seasoned engineers to fashion a working device from Aiken's sketchy ideas. Clair D. Lake, who was selected to manage the project, called on two talented engineers—Francis E. Hamilton and Benjamin M. Durfee—to assist him.

After four years of effort, which was interrupted at times by the demands of the war, a machine was constructed that worked remarkably well. Completed in January, 1943, at Endicott, New York, it was then disassembled and moved to Harvard University in Cambridge, Massachusetts, where it was reassembled. Known as the IBM automatic sequence controlled calculator (ASCC), it began operation in the spring of 1944 and was formally dedicated and revealed to the public on August 7, 1944. Its name indicates the machine's distinguishing feature: the ability to load automatically the instructions that control the sequence of the calculation. This capability was provided by punching holes, representing the instructions, in a long, ribbonlike paper tape that could be read by the machine.

Computers of that era were big, and the ASCC I was particularly impressive. It was 51 feet long by 8 feet tall, and it weighed 5 tons. It contained more than 750,000 parts, and when it was running, it sounded like a room filled with sewing machines. The ASCC later became known as the Harvard Mark I.

IMPACT

Although this machine represented a significant technological achievement at the time and contributed ideas that would be used in subsequent machines, it was almost obsolete from the start. It was electromechanical, since it relied on relays, but it was built at the dawn of the electronic age. Fully electronic computers offered better reliability and faster speeds. Howard Aiken continued, without the help of IBM, to develop successors to the Mark I. Because he resisted using electronics, however, his machines did not significantly affect the direction of computer development.

For all its complexity, the Mark I operated reasonably well, first solving problems related to the war effort and then turning its attention to the more mundane tasks of producing specialized mathematical tables. It remained in operation at the Harvard Computational Laboratory until 1959, when it was retired and disassembled. Parts of this landmark computational tool are now kept at the Smithsonian Institute.

See also BASIC programming language; Differential analyzer; Personal computer; Pocket calculator; UNIVAC computer.

FURTHER READING

Cohen, I. Bernard. *Howard Aiken: Portrait of a Computer Pioneer.* Cambridge, Mass.: MIT Press, 1999.

Ritchie, David. *The Computer Pioneers: The Making of the Modern Computer.* New York: Simon and Schuster, 1986.

Slater, Robert. *Portraits in Silicon.* Cambridge, Mass.: MIT Press, 1987.

Mass spectrograph

The invention: The first device used to measure the mass of atoms, which was found to be the result of the combination of isotopes.

The people behind the invention:
Francis William Aston (1877-1945), an English physicist who was awarded the 1922 Nobel Prize in Chemistry
Sir Joseph John Thomson (1856-1940), an English physicist
William Prout (1785-1850), an English biochemist
Ernest Rutherford (1871-1937), an English physicist

Same Element, Different Weights

Isotopes are different forms of a chemical element that act similarly in chemical or physical reactions. Isotopes differ in two ways: They possess different atomic weights and different radioactive transformations. In 1803, John Dalton proposed a new atomic theory of chemistry that claimed that chemical elements in a compound combine by weight in whole number proportions to one another. By 1815, William Prout had taken Dalton's hypothesis one step further and claimed that the atomic weights of elements were integral (the integers are the positive and negative whole numbers and zero) multiples of the hydrogen atom. For example, if the weight of hydrogen was 1, then the weight of carbon was 12, and that of oxygen 16. Over the next decade, several carefully controlled experiments were conducted to determine the atomic weights of a number of elements. Unfortunately, the results of these experiments did not support Prout's hypothesis. For example, the atomic weight of chlorine was found to be 35.5. It took a theory of isotopes, developed in the early part of the twentieth century, to verify Prout's original theory.

After his discovery of the electron, Sir Joseph John Thomson, the leading physicist at the Cavendish Laboratory in Cambridge, England, devoted much of his remaining research years to determining the nature of "positive electricity." (Since electrons are negatively charged, most electricity is negative.) While developing an instrument sensitive enough to analyze the positive electron, Thomson invited Francis William Aston to work with him at the Cavendish Laboratory. Recommended by J. H. Poynting, who had taught Aston physics at Mason College, Aston began a lifelong association at Cavendish, and Trinity College became his home.

When electrons are stripped from an atom, the atom becomes positively charged. Through the use of magnetic and electrical fields, it is possible to channel the resulting positive rays into parabolic tracks. By examining photographic plates of these tracks, Thomson was able to identify the atoms of different elements. Aston's first contribution at Cavendish was to improve the instrument used to photograph the parabolic tracks. He developed a more efficient pump to create the required vacuum and devised a camera that would provide sharper photographs. By 1912, the improved apparatus had provided proof that the individual molecules of a substance have the same mass. While working on the element neon, however, Thomson obtained two parabolas, one with a mass of 20 and the other with a mass of 22, which seemed to contradict the previous findings that molecules of any substance have the same mass. Aston was given the task of resolving this mystery.

Treating Particles Like Light

In 1919, Aston began to build a device called a "mass spectrograph." The idea was to treat ionized or positive atoms like light. He reasoned that, because light can be dispersed into a rainbowlike spectrum and analyzed by means of its different colors, the same procedure could be used with atoms of an element such as neon. By creating a device that used magnetic fields to focus the stream of particles emitted by neon, he was able to create a mass spectrum and record it on a photographic plate. The heavier mass of neon (the first neon isotope) was collected on one part of a spectrum and the lighter neon (the second neon isotope) showed up on another. This mass spectrograph was a magnificent apparatus: The masses could be analyzed without reference to the velocity of the particles, which was a problem with the parabola method devised by Thomson. Neon possessed two isotopes: one with a mass of 20 and the other with a mass of 22, in a ratio of 10:1. When combined, this gave the atomic weight 20.20, which was the accepted weight of neon.

Francis William Aston

Francis W. Aston was born near Birmingham, England, in 1877 to William Aston, a farmer and metals dealer, and Fanny Charlotte Hollis, a gunmaker's daughter. As a boy he loved to perform experiments by himself in his own small laboratory at home. His diligence helped him earn top marks in school, and he attended Mason College (later the University of Birmingham). However, he failed to win a scholarship to continue his studies after graduation in 1901.

He did not give up on experiments, however, even while holding a job as the chemist for a local brewery. He built his own equipment and investigated the nature of electricity. This work attracted the attention of the most famous researchers of the day. He finally got a scholarship in 1903 to the University of Birmingham and then joined the staff of Joseph John Thomson at the Royal Institution in London and Cambridge University, which remained his home until his death in 1945.

Aston liked to work alone as much as possible. Given his unflagging attention to the details of measurement and his inventiveness with experimental equipment, his colleagues respected his lone-dog approach. Their trust was rewarded. After refining the mass spectrograph, Aston was able to explain a thorny problem in chemistry by showing that elements are composed of differing percentages of isotopes and that atomic weight varied slightly depending on the density of their atoms' nuclei. The research earned him the Nobel Prize in Chemistry in 1922.

Aston's solitude extended into his private life. He never married, lavishing his affection instead on animals, outdoor sports, photography, travel, and music.

Aston's accomplishment in developing the mass spectrograph was recognized immediately by the scientific community. His was a simple device that was capable of accomplishing a large amount of research quickly. The field of isotope research, which had been opened up by Aston's research, ultimately played an important part in other areas of physics.

IMPACT

The years following 1919 were highly charged with excitement, since month after month new isotopes were announced. Chlorine had two; bromine had isotopes of 79 and 81, which gave an almost exact atomic weight of 80; krypton had six isotopes; and xenon had even more. In addition to the discovery of nonradioactive isotopes, the "whole-number rule" for chemistry was verified: Protons were the basic building blocks for different atoms, and they occurred exclusively in whole numbers.

Aston's original mass spectrograph had an accuracy of 1 in 1,000. In 1927, he built an even more accurate instrument, which was ten times more accurate. The new apparatus was sensitive enough to measure Albert Einstein's law of mass energy conversion during a nuclear reaction. Between 1927 and 1935, Aston reviewed all the elements that he had worked on earlier and published updated results. He also began to build a still more accurate instrument, which proved to be of great value to nuclear chemistry.

The discovery of isotopes opened the way to further research in nuclear physics and completed the speculations begun by Prout during the previous century. Although radioactivity was discovered separately, isotopes played a central role in the field of nuclear physics and chain reactions.

See also Cyclotron; Electron microscope; Neutrino detector; Scanning tunneling microscope; Synchrocyclotron; Tevatron accelerator; Ultramicroscope.

FURTHER READING

Aston, Francis William. "Mass Spectra and Isotopes" [Nobel lecture]. In *Chemistry, 1922-1941*. River Edge, N.J.: World Scientific, 1999.

Squires, Gordon. "Francis Aston and the Mass Spectrograph." *Journal of the Chemical Society. Dalton Transactions* no. 23 (1998).

Thackray, Arnold. *Atoms and Powers: An Essay on Newtonian Matter-Theory and the Development of Chemistry*. Cambridge, Mass.: Harvard University Press, 1970.

Memory metal

THE INVENTION: Known as nitinol, a metal alloy that returns to its
original shape, after being deformed, when it is heated to the
proper temperature.

THE PERSON BEHIND THE INVENTION:
  William Buehler (1923-   ), an American metallurgist

The Alloy with a Memory

In 1960, William Buehler developed an alloy that consisted of 53
to 57 percent nickel (by weight) and the balance titanium. This alloy,
which is called nitinol, turned out to have remarkable properties.
Nitinol is a "memory metal," which means that, given the proper
conditions, objects made of nitinol can be restored to their original
shapes even after they have been radically deformed. The return to
the original shape is triggered by heating the alloy to a moderate
temperature. As the metal "snaps back" to its original shape, considerable force is exerted and mechanical work can be done.

Alloys made of nickel and titanium have great potential in a
wide variety of industrial and government applications. These include: for the computer market, a series of high-performance electronic connectors; for the medical market, intravenous fluid devices
that feature precise fluid control; for the consumer market, eyeglass
frame components; and, for the industrial market, power cable couplings that provide durability at welded joints.

The Uncoiling Spring

At one time, the "uncoiling spring experiment" was used to
amuse audiences, and a number of scientists have had fun with
nitinol in front of unsuspecting viewers. It is now generally recognized that the shape memory effect involves a thermoelastic transformation at the atomic level. This process is unique in that the
transformation back to the original shape occurs as a result of stored
elastic energy that assists the chemical driving force that is un-
leashed by heating the metal.

The mechanism, simply stated, is that shape memory alloys are rather easily deformed below their "critical temperature." Provided that the extent of the deformation is not too great, the original, undeformed state can be recovered by heating the alloy to a temperature just below the critical temperature. It is also significant that substantial stresses are generated when a deformed specimen "springs back" to its original shape. This phenomenon is very peculiar compared to the ordinary behavior of most materials.

Researchers at the Naval Ordnance Laboratory discovered nitinol by accident in the process of trying to learn how to make titanium less brittle. They tried adding nickel, and when they were showing a wire of the alloy to some administrators, someone smoking a cigar held his match too close to the sample, causing the nitinol to spring back into shape. One of the first applications of the discovery was a new way to link hydraulic lines on the Navy's F-14 fighter jets. The nitinol "sleeve" was cooled with liquid nitrogen, which enlarged the sample. Then it was slipped into place between two pipes. When the sleeve was warmed up, it contracted, clamping the pipes together and keeping them clamped with a force of nearly 50,000 pounds per square inch.

Nitinol is not an easy alloy with which to work. When it is drilled or passed through a lathe, it becomes hardened and resists change. Welding nitinol and electroplating it have become manufacturing nightmares. It also resists taking on a desired shape. The frictional forces of many processes heat the nitinol, which activates its memory. Its fantastic elasticity also causes difficulties. If it is placed in a press with too little force, the spring comes out of the die unchanged. With too much force, the metal breaks into fragments. Using oil as a cooling lubricant and taking a step-wise approach to altering the alloy, however, allows it to be fashioned into particular shapes.

One unique use of nitinol occurs in cardiac surgery. Surgical tools made of nitinol can be bent up to 90 degrees, allowing them to be passed into narrow vessels and then retrieved. The tools are then straightened out in an autoclave so that they can be reused.

CONSEQUENCES

Many of the technical problems of working with nitinol have been solved, and manufacturers of the alloy are selling more than twenty different nitinol products to countless companies in the fields of medicine, transportation, consumer products, and toys.

Nitinol toys include blinking movie posters, butterflies with flapping wings, and dinosaurs whose tails move; all these applications are driven by a contracting bit of wire that is connected to a watch battery. The "Thermobile" and the "Icemobile" are toys whose wheels are set in motion by hot water or by ice cubes.

Orthodontists sometimes use nitinol wires and springs in braces because the alloy pulls with a force that is more gentle and even than that of stainless steel, thus causing less pain. Nitinol does not react with organic materials, and it is also useful as a new type of blood-clot filter. Best of all, however, is the use of nitinol for eyeglass frames. If the wearer deforms the frames by sitting on them (and people do so frequently), the optometrist simply dips the crumpled frames in hot water and the frames regain their original shape.

From its beginnings as an "accidental" discovery, nitinol has gone on to affect various fields of science and technology, from the "Cryofit" couplings used in the hydraulic tubing of aircraft to the pin-and-socket contacts used in electrical circuits. Nitinol has also found its way into integrated circuit packages.

In an age of energy conservation, the unique phase transformation of nickel-titanium alloys allows them to be used in low-temperature heat engines. The world has abundant resources of low-grade thermal energy, and the recovery of this energy can be accomplished by the use of materials such as nitinol. Despite the limitations imposed on heat engines working at low temperatures across a small temperature change, sources of low-grade heat are so widespread that the economical conversion of a fractional percentage of that energy could have a significant impact on the world's energy supply.

Nitinol has also become useful as a material capable of absorbing internal vibrations in structural materials, and it has been used as "Harrington rods" to treat scoliosis (curvature of the spine).

See also Disposable razor; Neoprene; Plastic; Steelmaking process; Teflon; Tungsten filament.

FURTHER READING

Gisser, Kathleen R. C., et al. "Nickel-Titanium Memory Metal." *Journal of Chemical Education* 71, no. 4 (April, 1994).

Iovine, John. "The World's 'Smartest' Metal." *Poptronics* 1, no. 12 (December, 2000).

Jackson, Curtis M., H. J. Wagner, and Roman Jerzy Wasilewski. *55-Nitinol: The Alloy with a Memory: Its Physical Metallurgy, Properties, and Applications.* Washington: Technology Utilization Office, 1972.

Walker, Jearl. "The Amateur Scientist." *Scientific American* 254, no. 5 (May, 1986).

Microwave cooking

THE INVENTION: System of high-speed cooking that uses microwave radition to agitate liquid molecules to raise temperatures by friction.

THE PEOPLE BEHIND THE INVENTION:
Percy L. Spencer (1894-1970), an American engineer
Heinrich Hertz (1857-1894), a German physicist
James Clerk Maxwell (1831-1879), a Scottish physicist

The Nature of Microwaves

Microwaves are electromagnetic waves, as are radio waves, X rays, and visible light. Water waves and sound waves are wave-shaped disturbances of particles in the media—water in the case of water waves and air or water in the case of sound waves—through which they travel. Electromagnetic waves, however, are wavelike variations of intensity in electric and magnetic fields.

Electromagnetic waves were first studied in 1864 by James Clerk Maxwell, who explained mathematically their behavior and velocity. Electromagnetic waves are described in terms of their "wavelength" and "frequency." The wavelength is the length of one cycle, which is the distance from the highest point of one wave to the highest point of the next wave, and the frequency is the number of cycles that occur in one second. Frequency is measured in units called "hertz," named for the German physicist Heinrich Hertz. The frequencies of microwaves run from 300 to 3,000 megahertz (1 megahertz equals 1 million hertz, or 1 million cycles per second), corresponding to wavelengths of 100 to 10 centimeters.

Microwaves travel in the same way that light waves do; they are reflected by metallic objects, absorbed by some materials, and transmitted by other materials. When food is subjected to microwaves, it heats up because the microwaves make the water molecules in foods (water is the most common compound in foods) vibrate. Water is a "dipole molecule," which means that it contains both positive and negative charges. When the food is subjected to microwaves, the dipole water molecules try to align themselves with the alternating electromagnetic field of the microwaves. This causes the water molecules to collide with one another and with other molecules in the food. Consequently, heat is produced as a result of friction.

Development of the Microwave Oven

Percy L. Spencer apparently discovered the principle of microwave cooking while he was experimenting with a radar device at the Raytheon Company. A candy bar in his pocket melted after being exposed to microwaves. After realizing what had happened, Spencer made the first microwave oven from a milk can and applied for two patents, "Method of Treating Foodstuffs" and "Means for Treating Foodstuffs," on October 8, 1945, giving birth to microwave-oven technology.

Spencer wrote that his invention "relates to the treatment of foodstuffs and, more particularly, to the cooking thereof through the use of electromagnetic energy." Though the use of electromagnetic energy for heating was recognized at that time, the frequencies that were used were lower than 50 megahertz. Spencer discovered that heating at such low frequencies takes a long time. He eliminated the time disadvantage by using shorter wavelengths in the microwave region. Wavelengths of 10 centimeters or shorter were comparable to the average dimensions of foods. When these wavelengths were used, the heat that was generated became intense, the energy that was required was minimal, and the process became efficient enough to be exploited commercially.

Although Spencer's patents refer to the cooking of foods with microwave energy, neither deals directly with a microwave oven. The actual basis for a microwave oven may be patents filed by other researchers at Raytheon. A patent by Karl Stiefel in 1949 may be the forerunner of the microwave oven, and in 1950, Fritz Gross received a patent entitled "Cooking Apparatus," which specifically describes an oven that is very similar to modern microwave ovens.

Perhaps the first mention of a commercial microwave oven was made in the November, 1946, issue of *Electronics* magazine. This article described the newly developed Radarange as a device that could bake biscuits in 29 seconds, cook hamburgers in 35 seconds,

Percy L. Spencer

Percy L. Spencer (1894-1970) had an unpromising background for the inventor of the twentieth century's principal innovation in the technology of cooking. He was orphaned while still a young boy and never completed grade school. However, he possessed a keen curiosity and the imaginative intelligence to educate himself and recognize how to make things better.

In 1941 the magnetron, which produces microwaves, was so complex and difficult to make that fewer than two dozen were produced in a day. This pace delayed the campaign to improve radar, which used magnetrons, so Spencer, while working for Raytheon Corporation, set out to speed things along. He simplified the design and made it more efficient at the same time. Production of magnetrons soon increased more than a thousandfold. In 1945 he discovered by accident that microwaves could heat chocolate past the melting point. He immediately tried an experiment by training microwaves on popcorn kernels and was delighted to see them puff up straight away.

The first microwave oven based on his discovery stood five feet, six inches tall and weighed 750 pounds, suitable only for restaurants. However, it soon got smaller, thanks to researchers at Raytheon. And after some initial hostility from cooks, it became popular. Raytheon bought Amana Refrigeration in 1965 to manufacture the home models and marketed them worldwide. Meanwhile, Spencer had become a senior vice president at the company and a member of its board of directors. Raytheon named one of its buildings after him, the U.S. Navy presented him with the Distinguished Service Medal for his contributions, and in 1999 he entered the Inventors Hall of Fame.

and grill a hot dog in 8 to 10 seconds. Another article that appeared a month later mentioned a unit that had been developed specifically for airline use. The frequency used in this oven was 3,000 megahertz. Within a year, a practical model 13 inches wide, 14 inches deep, and 15 inches high appeared, and several new models were operating in and around Boston. In June, 1947, *Electronics* magazine reported the installation of a Radarange in a restaurant, signaling the commercial use of microwave cooking. It was reported that this method more than tripled the speed of service. The Radarange became an important addition to a number of restaurants, and in 1948, Bernard Proctor and Samuel Goldblith used it for the first time to conduct research into microwave cooking.

In the United States, the radio frequencies that can be used for heating are allocated by the Federal Communications Commission (FCC). The two most popular frequencies for microwave cooking are 915 and 2,450 megahertz, and the 2,450 frequency is used in home microwave ovens. It is interesting that patents filed by Spencer in 1947 mention a frequency on the order of 2,450 megahertz. This fact is another example of Spencer's vision in the development of microwave cooking principles. The Raytheon Company concentrated on using 2,450 megahertz, and in 1955, the first domestic microwave oven was introduced. It was not until the late 1960's, however, that the price of the microwave oven decreased sufficiently for the device to become popular. The first patent describing a microwave heating system being used in conjunction with a conveyor was issued to Spencer in 1952. Later, based on this development, continuous industrial applications of microwaves were developed.

Impact

Initially, microwaves were viewed as simply an efficient means of rapidly converting electric energy to heat. Since that time, however, they have become an integral part of many applications. Because of the pioneering efforts of Percy L. Spencer, microwave applications in the food industry for cooking and for other processing operations have flourished. In the early 1970's, there were eleven microwave oven companies worldwide, two of which specialized in food processing operations, but the growth of the microwave oven industry has paralleled the growth in the radio and television industries. In 1984, microwave ovens accounted for more shipments than had ever been achieved by any appliance—9.1 million units.

By 1989, more than 75 percent of the homes in the United States had microwave ovens, and in the 1990's, microwavable foods were among the fastest-growing products in the food industry. Microwave energy facilitates reductions in operating costs and required energy, higher-quality and more reliable products, and positive environmental effects. To some degree, the use of industrial micro-wave energy remains in its infancy. New and improved applications of microwaves will continue to appear.

See also Electric refrigerator; Fluorescent lighting; Food freez-ing; Robot (household); Television; Tupperware; Vacuum cleaner; Washing machine.

FURTHER READING

Baird, Davis, R. I. G. Hughes, and Alfred Nordmann. *Heinrich Hertz: Classical Physicist, Modern Philosopher.* Boston: Kluwer Academic, 1998.
Roman, Mark. "That Marvelous Machine in Your Kitchen." *Reader's Digest* (February, 1990).
Scott, Otto. *The Creative Ordeal: The Story of Raytheon.* New York: Atheneum, 1974.
Simpson, Thomas K. *Maxwell on the Electromagnetic Field: A Guided Study.* New Brunswick, N.J.: Rutgers University Press, 1997.
Tolstoy, Ivan. *James Clerk Maxwell: A Biography.* Chicago: University of Chicago Press, 1982.

Neoprene

THE INVENTION: The first commercially practical synthetic rubber, Neoprene gave a boost to polymer chemistry and the search for new materials.

THE PEOPLE BEHIND THE INVENTION:

Wallace Hume Carothers (1896-1937), an American chemist

Arnold Miller Collins (1899-   ), an American chemist

Elmer Keiser Bolton (1886-1968), an American chemist

Julius Arthur Nieuwland (1879-1936), a Belgian American priest, botanist, and chemist

Synthetic Rubber: A Mirage?

The growing dependence of the industrialized nations upon elastomers (elastic substances) and the shortcomings of natural rubber motivated the twentieth century quest for rubber substitutes. By 1914, rubber had become nearly as indispensable as coal or iron. The rise of the automobile industry, in particular, had created a strong demand for rubber. Unfortunately, the availability of rubber was limited by periodic shortages and spiraling prices. Furthermore, the particular properties of natural rubber, such as its lack of resistance to oxygen, oils, and extreme temperatures, restrict its usefulness in certain applications. These limitations stimulated a search for special-purpose rubber substitutes.

Interest in synthetic rubber dates back to the 1860 discovery by the English chemist Greville Williams that the main constituent of rubber is isoprene, a liquid hydrocarbon. Nineteenth century chemists attempted unsuccessfully to transform isoprene into rubber. The first large-scale production of a rubber substitute occurred during World War I. A British blockade forced Germany to begin to manufacture methyl rubber in 1916, but methyl rubber turned out to be a poor substitute for natural rubber. When the war ended in 1918, a practical synthetic rubber was still only a mirage. Nevertheless, a breakthrough was on the horizon.

MIRAGE BECOMES REALITY

In 1930, chemists at E. I. Du Pont de Nemours discovered the elastomer known as neoprene. Of the more than twenty chemists who helped to make this discovery possible, four stand out: Elmer Bolton, Julius Nieuwland, Wallace Carothers, and Arnold Collins.

Bolton directed Du Pont's drystuffs department in the mid-1920's. Largely because of the rapidly increasing price of rubber, he initiated a project to synthesize an elastomer from acetylene, a gaseous hydrocarbon. In December, 1925, Bolton attended the American Chemical Society's convention in Rochester, New York, and heard a presentation dealing with acetylene reactions. The presenter was Julius Nieuwland, the foremost authority on the chemistry of acetylene.

Nieuwland was a professor of organic chemistry at the University of Notre Dame. (One of his students was the legendary football coach Knute Rockne.) The priest-scientist had been investigating acetylene reactions for more than twenty years. Using a copper chloride catalyst he had discovered, he isolated a new compound, divinylacetylene (DVA). He later treated DVA with a vulcanizing (hardening) agent and succeeded in producing a rubberlike substance, but the substance proved to be too soft for practical use.

Bolton immediately recognized the importance of Nieuwland's discoveries and discussed with him the possibility of using DVA as a raw material for a synthetic rubber. Seven months later, an alliance was formed that permitted Du Pont researchers to use Nieuwland's copper catalyst. Bolton hoped that the catalyst would be the key to making an elastomer from acetylene. As it turned out, Nieuwland's catalyst was indispensable for manufacturing neoprene.

Over the next several years, Du Pont scientists tried unsuccessfully to produce rubberlike materials. Using Nieuwland's catalyst, they managed to prepare DVA and also to isolate monovinylacetylene (MVA), a new compound that eventually proved to be the vital intermediate chemical in the making of neoprene. Reactions of MVA and DVA, however, produced only hard, brittle materials.

In 1928, Du Pont hired a thirty-one-year-old Harvard instructor, Wallace Carothers, to direct the organic chemicals group. He began a systematic exploration of polymers (complex molecules). In early

1930, he accepted an assignment to investigate the chemistry of DVA. He appointed one of his assistants, Arnold Collins, to conduct the laboratory experiments. Carothers suggested that Collins should explore the reaction between MVA and hydrogen chloride. His suggestion would lead to the discovery of neoprene.

One of Collins's experiments yielded a new liquid, and on April 17, 1930, he recorded in his laboratory notebook that the liquid had solidified into a rubbery substance. When he dropped it on a bench, it bounced. This was the first batch of neoprene. Carothers named Collins's liquid "chloroprene." Chloroprene is analogous structurally to isoprene, but it polymerizes much more rapidly. Carothers conducted extensive investigations of the chemistry of chloroprene and related compounds. His studies were the foundation for Du Pont's development of an elastomer that was superior to all previously known synthetic rubbers.

Du Pont chemists, including Carothers and Collins, formally introduced neoprene—originally called "DuPrene"—on November 3, 1931, at the meeting of the American Chemical Society in Akron, Ohio. Nine months later, the new elastomer began to be sold.

Impact

The introduction of neoprene was a milestone in humankind's development of new materials. It was the first synthetic rubber worthy of the name. Neoprene possessed higher tensile strength than rubber and much better resistance to abrasion, oxygen, heat, oils, and chemicals. Its main applications included jacketing for electric wires and cables, work-shoe soles, gasoline hoses, and conveyor and power-transmission belting. By 1939, when Adolf Hitler's troops invaded Poland, nearly every major industry in America was using neoprene. After the Japanese bombing of Pearl Harbor, in 1941, the elastomer became even more valuable to the United States. It helped the United States and its allies survive the critical shortage of natural rubber that resulted when Japan seized Malayan rubber plantations.

A scientifically and technologically significant side effect of the introduction of neoprene was the stimulus that the breakthrough gave to polymer research. Chemists had long debated whether polymers were mysterious aggregates of smaller units or were genuine molecules. Carothers ended the debate by demonstrating in a series of now-classic papers that polymers were indeed ordinary—but very large—molecules. In the 1930's, he put polymer studies on a firm footing. The advance of polymer science led, in turn, to the development of additional elastomers and synthetic fibers, including nylon, which was invented by Carothers himself in 1935.

See also Buna rubber; Nylon; Orlon; Plastic; Polyester; Polyethylene; Polystyrene; Silicones; Teflon.

FURTHER READING

Furukawa, Yasu. *Inventing Polymer Science: Staudinger, Carothers, and the Emergence of Macromolecular Chemistry*. Philadelphia: University of Pennsylvania Press, 1998.

Hermes, Matthew E. *Enough for One Lifetime: Wallace Carothers, Inventor of Rayon*. Washington, D.C.: American Chemical Society and the Chemical Heritage Foundation, 1996.

Taylor, Graham D., and Patricia E. Sudnik. *Du Pont and the International Chemical Industry*. Boston, Mass.: Twayne, 1984.

Neutrino detector

THE INVENTION: A device that provided the first direct evidence that the Sun runs on thermonuclear power and challenged existing models of the Sun.

THE PEOPLE BEHIND THE INVENTION:
Raymond Davis, Jr. (1914-   ), an American chemist
John Norris Bahcall (1934-   ), an American astrophysicist

MISSING ENERGY

In 1871, Hermann von Helmholtz, the German physicist, anatomist, and physiologist, suggested that no ordinary chemical reaction could be responsible for the enormous energy output of the Sun. By the 1920's, astrophysicists had realized that the energy radiated by the Sun must come from nuclear fusion, in which protons or nuclei combine to form larger nuclei and release energy. These reactions were assumed to be taking place deep in the interior of the Sun, in an immense thermonuclear furnace, where the pressures and temperatures were high enough to allow fusion to proceed.

Conventional astronomical observations could record only the particles of light emitted by the much cooler outer layers of the Sun and could not provide evidence for the existence of a thermonuclear furnace in the interior. Then scientists realized that the neutrino might be used to prove that this huge furnace existed. Of all the particles released in the fusion process, only one type—the neutrino—interacts so infrequently with matter that it can pass through the Sun and reach the earth. These neutrinos provide a way to verify directly the hypothesis of thermonuclear energy generated in stars.

The neutrino was "invented" in 1930 by the American physicist Wolfgang Pauli to account for the apparent missing energy in the beta decay, or emission of an electron, from radioactive nuclei. He proposed that an unseen nuclear particle, which he called a neutrino, was also emitted in beta decay, and that it carried off the "missing" energy. To balance the energy but not be observed in the decay process, Pauli's hypothetical particle had to have no electrical charge, have little or no mass, and interact only very weakly with ordinary matter. Typical neutrinos would have to be able to pass through millions of miles of ordinary matter in order to reach the earth. Scientists' detectors, and even the whole earth or Sun, were essentially transparent as far as Pauli's neutrinos were concerned.

Because the neutrino is so difficult to detect, it took more than twenty-five years to confirm its existence. In 1956, Clyde Cowan and Frederick Reines, both physicists at the Los Alamos National Laboratory, built the world's largest scintillation counter, a device to detect the small flash of light given off when the neutrino strikes ("interacts" with) a certain substance in the apparatus. They placed this scintillation counter near the Savannah River Nuclear Reactor, which was producing about 1 trillion neutrinos every second. Although only one neutrino interaction was observed in their detector every twenty minutes, Cowan and Reines were able to confirm the existence of Pauli's elusive particle.

The task of detecting the solar neutrinos was even more formidable. If an apparatus similar to the Cowan and Reines detector were employed to search for the neutrinos from the Sun, only one interaction could be expected every few thousand years.

Missing Neutrinos

At about the same time that Cowan and Reines performed their experiment, another type of neutrino detector was under development by Raymond Davis, Jr., a chemist at the Brookhaven National Laboratory. Davis employed an idea, originally suggested in 1948 by the nuclear physicist Bruno Pontecorvo, that when a neutrino interacts with a chlorine-37 nucleus, it produces a nucleus of argon 37. Any argon so produced could then be extracted from large volumes of chlorine-rich liquid by passing helium gas through the liquid. Since argon 37 is radioactive, it is relatively easy to detect.

Davis tested a version of this neutrino detector, containing about 3,785 liters of carbon tetrachloride liquid, near a nuclear reactor at the Brookhaven National Laboratory from 1954 to 1956. In the scientific paper describing his results, Davis suggested that this type of neutrino detector could be made large enough to permit detection of solar neutrinos.

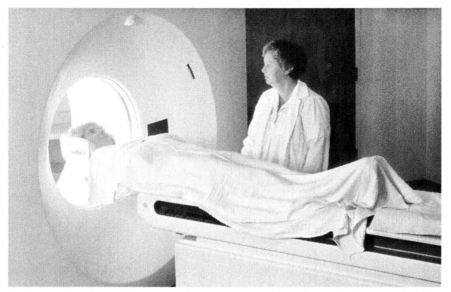

Patients undergoing nuclear magnetic resonance image (MRI) examinations are placed inside cylindrical chambers in which their bodies are held rigidly in place. (Digital Stock)

Although Davis's first attempt to detect solar neutrinos from a limestone mine at Barberton, Ohio, failed, he continued his search with a much larger detector 1,478 meters underground in the Homestake Gold Mine in Lead, South Dakota. The cylindrical tank (6.1 meters in diameter, 16 meters long, and containing 378,540 liters of perchloroethylene) was surrounded by water to shield the detector from neutrons emitted by trace quantities of uranium and thorium in the walls of the mine. The experiment was conducted underground to shield it from cosmic radiation.

To describe his results, Davis coined a new unit, the "solar neutrino unit" (SNU), with 1 SNU indicating the production of one atom of argon 37 every six days. Astrophysicist John Norris Bahcall, using the best available astronomical models of the nuclear reactions going on in the sun's interior, as well as the physical properties of the neutrinos, had predicted a capture rate of 50 SNUs in 1963. The 1967 results from Davis's detector, however, had an upper limit of only 3 SNUs.

CONSEQUENCES

The main significance of the detection of solar neutrinos by Davis was the direct confirmation that thermonuclear fusion must be occurring at the center of the Sun. The low number of solar neutrinos Davis detected, however, has called into question some of the fundamental beliefs of astrophysics. As Bahcall explained: "We know more about the Sun than about any other star. . . . The Sun is also in what is believed to be the best-understood stage of stellar evolution. . . . If we are to have confidence in the many astronomical and cosmological applications of the theory of stellar evolution, it ought at least to give the right answers about the Sun."

Many solutions to the problem of the "missing" solar neutrinos have been proposed. Most of these solutions can be divided into two broad classes: those that challenge the model of the sun's interior and those that challenge the understanding of the behavior of the neutrino. Since the number of neutrinos produced is very sensitive to the temperature of the sun's interior, some astrophysicists have suggested that the true solar temperature may be lower than expected. Others suggest that the sun's outer layer may absorb more neutrinos than expected. Some physicists, however, believe neutrinos may occur in several different forms, only one of which can be detected by the chlorine detectors.

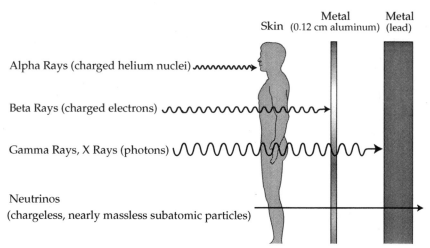

Neutrinos can pass through most forms of matter without interacting with other nuclear particles.

Davis's discovery of the low number of neutrinos reaching Earth has focused years of attention on a better understanding of how the Sun generates its energy and how the neutrino behaves. New and more elaborate solar neutrino detectors have been built with the aim of understanding stars, including the Sun, as well as the physics and behavior of the elusive neutrino.

See also Radio interferometer; Weather satellite.

FURTHER READING

Bartusiak, Marcia. "Underground Astronomer." *Astronomy* 28, no. 1 (January, 2000).
"Neutrino Test to Probe Sun." *New Scientist* 140, no. 1898 (November 6, 1993).
"Pioneering Neutrino Astronomers to Share 2000 Wolf Prize in Physics." *Physics Today* 53, no. 3 (March, 2000).
Schwarzschild, Bertram. "Can Helium Mixing Explain the Solar Neutrino Shortages?" *Physics Today* 50, no. 3 (March, 1997).
Zimmerman, Robert. "The Shadow Boxer." *The Sciences* 36, no. 1 (January/February, 1996).

NUCLEAR MAGNETIC RESONANCE

THE INVENTION: Procedure that uses hydrogen atoms in the human body, strong electromagnets, radio waves, and detection equipment to produce images of sections of the brain.

THE PEOPLE BEHIND THE INVENTION:

Raymond Damadian (1936-   ), an American physicist and inventor

Paul C. Lauterbur (1929-   ), an American chemist

Peter Mansfield (1933-   ), a scientist at the University of Nottingham, England

PEERING INTO THE BRAIN

Doctors have always wanted the ability to look into the skull and see the human brain without harming the patient who is being examined. Over the years, various attempts were made to achieve this ability. At one time, the use of X rays, which were first used by Wilhelm Conrad Röntgen in 1895, seemed to be an option, but it was found that X rays are absorbed by bone, so the skull made it impossible to use X-ray technology to view the brain. The relatively recent use of computed tomography (CT) scanning, a computer-assisted imaging technology, made it possible to view sections of the head and other areas of the body, but the technique requires that the part of the body being "imaged," or viewed, be subjected to a small amount of radiation, thereby putting the patient at risk. Positron emission tomography (PET) could also be used, but it requires that small amounts of radiation be injected into the patient, which also puts the patient at risk. Since the early 1940's, however, a new technology had been developing.

This technology, which appears to pose no risk to patients, is called "nuclear magnetic resonance spectroscopy." It was first used to study the molecular structures of pure samples of chemicals. This method developed until it could be used to follow one chemical as it changed into another, and then another, in a living cell. By 1971, Raymond Damadian had proposed that body images that were more vivid and more useful than X rays could be produced by means of nuclear magnetic resonance spectroscopy. In 1978, he founded his own company, FONAR, which manufactured the scanners that are necessary for the technique.

MAGNETIC RESONANCE IMAGES

The first nuclear magnetic resonance images (MRIs) were published by Paul Lauterbur in 1973. Although there seemed to be no possibility that MRI could be harmful to patients, everyone involved in MRI research was very cautious. In 1976, Peter Mansfield, at the University of Nottingham, England, obtained an MRI of his partner's finger. The next year, Paul Bottomley, a member of Waldo Hinshaw's research group at the same university, put his left wrist into an experimental machine that the group had developed. A vivid cross section that showed layers of skin, muscle, bone, muscle, and skin, in that order, appeared on the machine's monitor. Studies with animals showed no apparent memory or other brain problems. In 1978, Electrical and Musical Industries (EMI), a British corporate pioneer in electronics that merged with Thorn in 1980, obtained the first MRI of the human head. It took six minutes.

An MRI of the brain, or any other part of the body, is made possible by the water content of the body. The gray matter of the brain contains more water than the white matter does. The blood vessels and the blood itself also have water contents that are different from those of other parts of the brain. Therefore, the different structures and areas of the brain can be seen clearly in an MRI. Bone contains very little water, so it does not appear on the monitor. This is why the skull and the backbone cause no interference when the brain or the spinal cord is viewed.

Every water molecule contains two hydrogen atoms and one oxygen atom. A strong electromagnetic field causes the hydrogen molecules to line up like marchers in a parade. Radio waves can be used to change the position of these parallel hydrogen molecules. When the radio waves are discontinued, a small radio signal is produced as the molecules return to their marching position. This distinct radio signal is the basis for the production of the image on a computer screen.

Hydrogen was selected for use in MRI work because it is very abundant in the human body, it is part of the water molecule, and it has the proper magnetic qualities. The nucleus of the hydrogen atom consists of a single proton, a particle with a positive charge. The signal from the hydrogen's proton is comparatively strong.

There are several methods by which the radio signal from the hydrogen atom can be converted into an image. Each method uses a computer to create first a two-dimensional, then a three-dimensional, image.

Peter Mansfield's team at the University of Nottingham holds the patent for the slice-selection technique that makes it possible to excite and image selectively a specific cross section of the brain or any other part of the body. This is the key patent in MRI technology. Damadian was granted a patent that described the use of two coils, one to drive and one to pick up signals across selected portions of the human body. EMI, the company that introduced the X-ray scanner for CT images, developed a commercial prototype for the MRI. The British Technology Group, a state-owned company that helps to bring innovations to the marketplace, has sixteen separate MRI-related patents. Ten years after EMI produced the first image of the human brain, patents and royalties were still being sorted out.

CONSEQUENCES

MRI technology has revolutionized medical diagnosis, especially in regard to the brain and the spinal cord. For example, in multiple sclerosis, the loss of the covering on nerve cells can be detected. Tumors can be identified accurately. The painless and noninvasive use of MRI has almost completely replaced the myelogram, which involves using a needle to inject dye into the spine.

Although there is every indication that the use of MRI is very safe, there are some people who cannot benefit from this valuable tool. Those whose bodies contain metal cannot be placed into the MRI machine. No one instrument can meet everyone's needs.

The development of MRI stands as an example of the interaction of achievements in various fields of science. Fundamental physics, biochemistry, physiology, electronic image reconstruction, advances in superconducting wires, the development of computers, and advancements in anatomy all contributed to the development of MRI. Its development is also the result of international efforts. Scientists and laboratories in England and the United States pioneered the technology, but contributions were also made by scientists in France, Switzerland, and Scotland. This kind of interaction and cooperation can only lead to greater understanding of the human brain.

See also Amniocentesis; CAT scanner; Electrocardiogram; Electroencephalogram; Mammography; Ultrasound; X-ray image intensifier.

FURTHER READING

Elster, Allen D., and Jonathan H. Burdette. *Questions and Answers in Magnetic Resonance Imaging.* 2d ed. St. Louis, Mo.: Mosby, 2001.

Mackay, R. Stuart. *Medical Images and Displays: Comparisons of Nuclear Magnetic Resonance, Ultrasound, X-rays, and Other Modalities.* New York: Wiley, 1984.

Mattson, James, and Merrill Simon. *The Story of MRI: The Pioneers of NMR and Magnetic Resonance in Medicine.* Jericho, N.Y.: Dean Books, 1996.

Wakefield, Julie. "The 'Indomitable' MRI." *Smithsonian* 31, no. 3 (June, 2000).

Wolbarst, Anthony B. *Looking Within: How X-ray, CT, MRI, Ultrasound, and Other Medical Images Are Created, and How they Help Physicians Save Lives.* Berkeley: University of California Press, 1999.

Nuclear power plant

The invention: The first full-scale commercial nuclear power plant, which gave birth to the nuclear power industry.

The people behind the invention:
Enrico Fermi (1901-1954), an Italian American physicist who won the 1938 Nobel Prize in Physics
Otto Hahn (1879-1968), a German physical chemist who won the 1944 Nobel Prize in Chemistry
Lise Meitner (1878-1968), an Austrian Swedish physicist
Hyman G. Rickover (1898-1986), a Polish American naval officer

Discovering Fission

Nuclear fission involves the splitting of an atomic nucleus, leading to the release of large amounts of energy. Nuclear fission was discovered in Germany in 1938 by Otto Hahn after he had bombarded uranium with neutrons and observed traces of radioactive barium. When Hahn's former associate, Lise Meitner, heard of this, she realized that the neutrons may have split the uranium nuclei (each of which holds 92 protons) into two smaller nuclei to produce barium (56 protons) and krypton (36 protons). Meitner and her nephew, Otto Robert Frisch, were able to calculate the enormous energy that would be released in this type of reaction. They published their results early in 1939.

Nuclear fission was quickly verified in several laboratories, and the Danish physicist Niels Bohr soon demonstrated that the rare uranium 235 (U-235) isotope is much more likely to fission than the common uranium 238 (U-238) isotope, which makes up 99.3 percent of natural uranium. It was also recognized that fission would produce additional neutrons that could cause new fissions, producing even more neutrons and thus creating a self-sustaining chain reaction. In this process, the fissioning of one gram of U-235 would release about as much energy as the burning of three million tons of coal.

The first controlled chain reaction was demonstrated on December 2, 1942, in a nuclear reactor at the University of Chicago, under the leadership of Enrico Fermi. He used a graphite moderator to slow the neutrons by collisions with carbon atoms. "Critical mass" was achieved when the mass of graphite and uranium assembled was large enough that the number of neutrons not escaping from the pile would be sufficient to sustain a U-235 chain reaction. Cadmium control rods could be inserted to absorb neutrons and slow the reaction.

It was also recognized that the U-238 in the reactor would absorb accelerated neutrons to produce the new element plutonium, which is also fissionable. During World War II (1939-1945), large reactors were built to "breed" plutonium, which was easier to separate than U-235. An experimental breeder reactor at Arco, Idaho, was the first to use the energy of nuclear fission to produce a small amount of electricity (about 100 watts) on December 20, 1951.

NUCLEAR ELECTRICITY

Power reactors designed to produce substantial amounts of electricity use the heat generated by fission to produce steam or hot gas to drive a turbine connected to an ordinary electric genera tor. The first power reactor design to be developed in the United States was the pressurized water reactor (PWR). In the PWR, water under high pressure is used both as the moderator and as the coolant. After circulating through the reactor core, the hot pressurized water flows through a heat exchanger to produce steam. Reactors moderated by "heavy water" (in which the hydrogen in the water is replaced with deuterium, which contains an extra neutron) can operate with natural uranium.

The pressurized water system was used in the first reactor to produce substantial amounts of power, the experimental Mark I reactor. It was started up on May 31, 1953, at the Idaho National Engineering Laboratory. The Mark I became the prototype for the reactor used in the first nuclear-powered submarine. Under the leadership of Hyman G. Rickover, who was head of the Division of Naval Reactors of the Atomic Energy Commission (AEC), Westinghouse Electric Corporation was engaged to build a PWR system to power the submarine USS *Nautilus*. It began sea trials in January of 1955 and ran for two years before refueling.

Cooling towers of a nuclear power plant. (PhotoDisc)

In the meantime, the first experimental nuclear power plant for generating electricity was completed in the Soviet Union in June of 1954, under the direction of the Soviet physicist Igor Kurchatov. It produced 5 megawatts of electric power. The first full-scale nuclear power plant was built in England under the direction of the British nuclear engineer Sir Christopher Hinton. It began producing about 90 megawatts of electric power in October, 1956.

On December 2, 1957, on the fifteenth anniversary of the first controlled nuclear chain reaction, the Shippingport Atomic Power Station in Shippingport, Pennsylvania, became the first full-scale commercial nuclear power plant in the United States. It produced about 60 megawatts of electric power for the Duquesne Light Company until 1964, when its reactor core was replaced, increasing its power to 100 megawatts with a maximum capacity of 150 megawatts.

CONSEQUENCES

The opening of the Shippingport Atomic Power Station marked the beginning of the nuclear power industry in the United States, with all of its glowing promise and eventual problems. It was predicted that electrical energy would become too cheap to meter. The AEC hoped to encourage the participation of industry, with government support limited to research and development. They encouraged a variety of reactor types in the hope of extending technical knowledge.

The Dresden Nuclear Power Station, completed by Commonwealth Edison in September, 1959, at Morris, Illinois, near Chicago, was the first full-scale privately financed nuclear power station in the United States. By 1973, forty-two plants were in operation producing 26,000 megawatts, fifty more were under construction, and about one hundred were on order. Industry officials predicted that 50 percent of the nation's electric power would be nuclear by the end of the twentieth century.

The promise of nuclear energy has not been completely fulfilled. Growing concerns about safety and waste disposal have led to increased efforts to delay or block the construction of new plants. The cost of nuclear plants rose as legal delays and inflation pushed costs higher, so that many in the planning stages could no longer be competitive. The 1979 Three Mile Island accident in Pennsylvania and the much more serious 1986 Chernobyl accident in the Soviet Union increased concerns about the safety of nuclear power. Nevertheless, by 1986, more than one hundred nuclear power plants were operating in the United States, producing about 60,000 megawatts of power. More than three hundred reactors in twenty-five countries provide about 200,000 megawatts of electric power worldwide.

Many believe that, properly controlled, nuclear energy offers a clean-energy solution to the problem of environmental pollution.

See also Breeder reactor; Compressed-air-accumulating power plant; Fuel cell; Geothermal power; Nuclear reactor; Solar thermal engine; Tidal power plant.

FURTHER READING

Henderson, Harry. *Nuclear Power: A Reference Handbook*. Santa Barbara, Calif.: ABC-CLIO, 2000.
Rockwell, Theodore. *The Rickover Effect: The Inside Story of How Admiral Hyman Rickover Built the Nuclear Navy*. New York: J. Wiley, 1995.
Shea, William R. *Otto Hahn and the Rise of Nuclear Physics*. Boston: D. Reidel, 1983.
Sime, Ruth Lewin. *Lise Meitner: A Life in Physics*. Berkeley: University of California Press, 1996.

NUCLEAR REACTOR

THE INVENTION: The first nuclear reactor to produce substantial quantities of plutonium, making it practical to produce usable amounts of energy from a chain reaction.

THE PEOPLE BEHIND THE INVENTION:

Enrico Fermi (1901-1954), an American physicist

Martin D. Whitaker (1902-1960), the first director of Oak Ridge National Laboratory

Eugene Paul Wigner (1902-1995), the director of research and development at Oak Ridge

THE TECHNOLOGY TO END A WAR

The construction of the nuclear reactor at Oak Ridge National Laboratory in 1943 was a vital part of the Manhattan Project, the effort by the United States during World War II (1939-1945) to develop an atomic bomb. The successful operation of that reactor was a major achievement not only for the project itself but also for the general development and application of nuclear technology. The first director of the Oak Ridge National Laboratory was Martin D. Whitaker; the director of research and development was Eugene Paul Wigner.

The nucleus of an atom is made up of protons and neutrons. "Fission" is the process by which the nucleus of certain elements is split in two by a neutron from some material that emits an occasional neutron naturally. When an atom splits, two things happen: A tremendous amount of thermal energy is released, and two or three neutrons, on the average, escape from the nucleus. If all the atoms in a kilogram of "uranium 235" were to fission, they would produce as much heat energy as the burning of 3 million kilograms of coal. The neutrons that are released are important, because if at least one of them hits another atom and causes it to fission (and thus to release more energy and more neutrons), the process will continue. It will become a self-sustaining chain reaction that will produce a continuing supply of heat.

Inside a reactor, a nuclear chain reaction is controlled so that it proceeds relatively slowly. The most familiar use for the heat thus released is to boil water and make steam to turn the turbine generators that produce electricity to serve industrial, commercial, and residential needs. The fissioning process in a weapon, however, proceeds very rapidly, so that all the energy in the atoms is produced and released virtually at once. The first application of nuclear technology, which used a rapid chain reaction, was to produce the two atomic bombs that ended World War II.

BREEDING BOMB FUEL

The work that began at Oak Ridge in 1943 was made possible by a major event that took place in 1942. At the University of Chicago, Enrico Fermi had demonstrated for the first time that it was possible to achieve a self-sustaining atomic chain reaction. More important, the reaction could be controlled: It could be started up, it could generate heat and sufficient neutrons to keep itself going, and it could be turned off. That first chain reaction was very slow, and it generated very little heat; but it demonstrated that controlled fission was possible.

Any heat-producing nuclear reaction is an energy conversion process that requires fuel. There is only one readily fissionable element that occurs naturally and can be used as fuel. It is a form of uranium called uranium 235. It makes up less than 1 percent of all naturally occurring uranium. The remainder is uranium 238, which does not fission readily. Even uranium 235, however, must be enriched before it can be used as fuel.

The process of enrichment increases the concentration of uranium 235 sufficiently for a chain reaction to occur. Enriched uranium is used to fuel the reactors used by electric utilities. Also, the much more plentiful uranium 238 can be converted into plutonium 239, a form of the human-made element plutonium, which does fission readily. That conversion process is the way fuel is produced for a nuclear weapon. Therefore, the major objective of the Oak Ridge effort was to develop a pilot operation for separating plutonium from the uranium in which it was produced. Large-scale plutonium production, which had never been attempted before, eventually would be done at the Hanford Engineer Works in Washington. First, however, plutonium had to be pro-

Part of the Oak Ridge National Laboratory, where plutonium was separated to create the first atomic bomb. (Martin Marietta)

duced successfully on a small scale at Oak Ridge.

The reactor was started up on November 4, 1943. By March 1, 1944, the Oak Ridge laboratory had produced several grams of plutonium. The material was sent to the Los Alamos laboratory in New Mexico for testing. By July, 1944, the reactor operated at four times its original power level. By the end of that year, however, plutonium production at Oak Ridge had ceased, and the reactor thereafter was used principally to produce radioisotopes for physical and biological research and for medical treatment. Ultimately, the Hanford Engineer Works' reactors produced the plutonium for the bomb that was dropped on Nagasaki, Japan, on August 9, 1945.

The original objectives for which Oak Ridge had been built had been achieved, and subsequent activity at the facility was directed toward peacetime missions that included basic studies of the structure of matter.

IMPACT

The most immediate impact of the work done at Oak Ridge was its contribution to ending World War II. When the atomic bombs were dropped, the war ended, and the United States emerged intact. The immediate and long-range devastation to the people of Japan, however, opened the public's eyes to the almost unimaginable death and destruction that could be caused by a nuclear war. Fears of such a war remain to this day, especially as more and more nations develop the technology to build nuclear weapons.

On the other hand, great contributions to human civilization have resulted from the development of nuclear energy. Electric power generation, nuclear medicine, spacecraft power, and ship propulsion have all profited from the pioneering efforts at the Oak Ridge National Laboratory. Currently, the primary use of nuclear energy is to produce electric power. Handled properly, nuclear energy may help to solve the pollution problems caused by the burning of fossil fuels.

See also Breeder reactor; Compressed-air-accumulating power plant; Fuel cell; Geothermal power; Heat pump; Nuclear power plant; Solar thermal engine; Tidal power plant.

FURTHER READING

Epstein, Sam, Beryl Epstein, and Raymond Burns. *Enrico Fermi: Father of Atomic Power.* Champaign, Ill.: Garrard, 1970.

Johnson, Leland, and Daniel Schaffer. *Oak Ridge National Laboratory: The First Fifty Years.* Knoxville: University of Tennessee Press, 1994.

Morgan, K. Z., and Ken M. Peterson. *The Angry Genie: One Man's Walk Through the Nuclear Age.* Norman: University of Oklahoma Press, 1999.

Wagner, Francis S. *Eugene P. Wigner, An Architect of the Atomic Age.* Toronto: Rákóczi Foundation, 1981.

Nylon

THE INVENTION: A resilient, high-strength polymer with applications ranging from women's hose to safety nets used in space flights.

THE PEOPLE BEHIND THE INVENTION:

Wallace Hume Carothers (1896-1937), an American organic chemist

Charles M. A. Stine (1882-1954), an American chemist and director of chemical research at Du Pont

Elmer Keiser Bolton (1886-1968), an American industrial chemist

Pure Research

In the twentieth century, American corporations created industrial research laboratories. Their directors became the organizers of inventions, and their scientists served as the sources of creativity. The research program of E. I. Du Pont de Nemours and Company (Du Pont), through its most famous invention—nylon—became the model for scientifically based industrial research in the chemical industry.

During World War I (1914-1918), Du Pont tried to diversify, concerned that after the war it would not be able to expand with only explosives as a product. Charles M. A. Stine, Du Pont's director of chemical research, proposed that Du Pont should move into fundamental research by hiring first-rate academic scientists and giving them freedom to work on important problems in organic chemistry. He convinced company executives that a program to explore the fundamental science underlying Du Pont's technology would ultimately result in discoveries of value to the company. In 1927, Du Pont gave him a new laboratory for research. Stine visited universities in search of brilliant, but not-yet-established, young scientists. He hired Wallace Hume Carothers. Stine suggested that Carothers do fundamental research in polymer chemistry.

Before the 1920's, polymers were a mystery to chemists. Polymeric materials were the result of ingenious laboratory practice, and this practice ran far ahead of theory and understanding. German chemists debated whether polymers were aggregates of smaller units held together by some unknown special force or genuine molecules held together by ordinary chemical bonds.

German chemist Hermann Staudinger asserted that they were large molecules with endlessly repeating units. Carothers shared this view, and he devised a scheme to prove it by synthesizing very large molecules by simple reactions in such a way as to leave no doubt about their structure. Carothers's synthesis of polymers revealed that they were ordinary molecules but giant in size.

The Longest Molecule

In April, 1930, Carothers's research group produced two major innovations: neoprene synthetic rubber and the first laboratory-synthesized fiber. Neither result was the goal of their research. Neoprene was an incidental discovery during a project to study short polymers of acetylene. During experimentation, an unexpected substance appeared that polymerized spontaneously. Carothers studied its chemistry and developed the process into the first successful synthetic rubber made in the United States.

The other discovery was an unexpected outcome of the group's project to synthesize polyesters by the reaction of acids and alcohols. Their goal was to create a polyester that could react indefinitely to form a substance with high molecular weight. The scientists encountered a molecular weight limit of about 5,000 units to the size of the polyesters, until Carothers realized that the reaction also produced water, which was decomposing polyesters back into acid and alcohol. Carothers and his associate Julian Hill devised an apparatus to remove the water as it formed. The result was a polyester with a molecular weight of more than 12,000, far higher than any previous polymer.

Hill, while removing a sample from the apparatus, found that he could draw it out into filaments that on cooling could be stretched to form very strong fibers. This procedure, called "cold-drawing," oriented the molecules from a random arrangement into a long, linear one of great strength. The polyester fiber, however, was unsuitable for textiles because of its low melting point.

In June, 1930, Du Pont promoted Stine; his replacement as research director was Elmer Keiser Bolton. Bolton wanted to control fundamental research more closely, relating it to projects that would pay off and not allowing the research group freedom to pursue purely theoretical questions.

Despite their differences, Carothers and Bolton shared an interest in fiber research. On May 24, 1934, Bolton's assistant Donald Coffman "drew" a strong fiber from a new polyamide. This was the first nylon fiber, although not the one commercialized by Du Pont. The nylon fiber was high-melting and tough, and it seemed that a practical synthetic fiber might be feasible.

By summer of 1934, the fiber project was the heart of the research group's activity. The one that had the best fiber properties was nylon 5-10, the number referring to the number of carbon atoms in the amine and acid chains. Yet the nylon 6-6 prepared on February 28, 1935, became Du Pont's nylon. Nylon 5-10 had some advantages, but Bolton realized that its components would be unsuitable for commercial production, whereas those of nylon 6-6 could be obtained from chemicals in coal.

A determined Bolton pursued nylon's practical development, a process that required nearly four years. Finally, in April, 1937, Du Pont filed a patent for synthetic fibers, which included a statement by Carothers that there was no previous work on polyamides; this was a major breakthrough. After Carothers's death on April 29, 1937, the patent was issued posthumously and assigned to Du Pont. Du Pont made the first public announcement of nylon on October 27, 1938.

IMPACT

Nylon was a generic term for polyamides, and several types of nylon became commercially important in addition to nylon 6-6. These nylons found widespread use as both a fiber and a moldable plastic. Since it resisted abrasion and crushing, was nonabsorbent, was stronger than steel on a weight-for-weight basis, and was almost nonflammable, it embraced an astonishing range of uses: in laces, screens, surgical sutures, paint, toothbrushes, violin strings, coatings for electrical wires, lingerie, evening gowns, leotards, athletic equipment, outdoor furniture, shower curtains, handbags, sails, luggage, fish nets, carpets, slip covers, bus seats, and even safety nets on the space shuttle.

The invention of nylon stimulated notable advances in the chemistry and technology of polymers. Some historians of technology have even dubbed the postwar period as the "age of plastics," the age of synthetic products based on the chemistry of giant molecules made by ingenious chemists and engineers.

The success of nylon and other synthetics, however, has come at a cost. Several environmental problems have surfaced, such as those created by the nondegradable feature of some plastics, and there is the problem of the increasing utilization of valuable, vanishing resources, such as petroleum, which contains the essential chemicals needed to make polymers. The challenge to reuse and recycle these polymers is being addressed by both scientists and policymakers.

See also Buna rubber; Neoprene; Orlon; Plastic; Polyester; Polyethylene; Polystyrene.

FURTHER READING

Furukawa, Yasu. *Inventing Polymer Science: Staudinger, Carothers, and the Emergence of Macromolecular Chemistry*. Philadelphia: University of Pennsylvania Press, 1998.

Handley, Susannah. *Nylon: The Story of a Fashion Revolution: A Celebration of Design from Art Silk to Nylon and Thinking Fibres*. Baltimore: Johns Hopkins University Press, 1999.

Hermes, Matthew E. *Enough for One Lifetime: Wallace Carothers, Inventor of Rayon*. Washington, D.C.: American Chemical Society and the Chemical Heritage Foundation, 1996.

Joyce, Robert M. *Elmer Keiser Bolton: June 23, 1886-July 30, 1968*. Washington, D.C.: National Academy Press, 1983.

Oil-well drill bit

The invention: A rotary cone drill bit that enabled oil-well drillers to penetrate hard rock formations.

The people behind the invention:
Howard R. Hughes (1869-1924), an American lawyer, drilling engineer, and inventor
Walter B. Sharp (1860-1912), an American drilling engineer, inventor, and partner to Hughes

Digging for Oil

A rotary drill rig of the 1990's is basically unchanged in its essential components from its earlier versions of the 1900's. A drill bit is attached to a line of hollow drill pipe. The latter passes through a hole on a rotary table, which acts essentially as a horizontal gear wheel and is driven by an engine. As the rotary table turns, so do the pipe and drill bit.

During drilling operations, mud-laden water is pumped under high pressure down the sides of the drill pipe and jets out with great force through the small holes in the rotary drill bit against the bottom of the borehole. This fluid then returns outside the drill pipe to the surface, carrying with it rock material cuttings from below. Circulated rock cuttings and fluids are regularly examined at the surface to determine the precise type and age of rock formation and for signs of oil and gas.

A key part of the total rotary drilling system is the drill bit, which has sharp cutting edges that make direct contact with the geologic formations to be drilled. The first bits used in rotary drilling were paddlelike "fishtail" bits, fairly successful for softer formations, and tubular coring bits for harder surfaces. In 1893, M. C. Baker and C. E. Baker brought a rotary water-well drill rig to Corsicana, Texas, for modification to deeper oil drilling. This rig led to the discovery of the large Corsicana-Powell oil field in Navarro County, Texas. This success also motivated its operators, the American Well and Prospecting Company, to begin the first large-scale manufacture of rotary drilling rigs for commercial sale.

In the earliest rotary drilling for oil, short fishtail bits were the tool of choice, insofar as they were at that time the best at being able to bore through a wide range of geologic strata without needing frequent replacement. Even so, in the course of any given oil well, many bits were required typically in coastal drilling in the Gulf of Mexico. Especially when encountering locally harder rock units such as limestone, dolomite, or gravel beds, fishtail bits would typically either curl backward or break off in the hole, requiring the time-consuming work of pulling out all drill pipe and "fishing" to retrieve fragments and clear the hole.

Because of the frequent bit wear and damage, numerous small blacksmith shops established themselves near drill rigs, dressing or sharpening bits with a hand forge and hammer. Each bit-forging shop had its own particular way of shaping bits, producing a wide variety of designs. Nonstandard bit designs were frequently modified further as experiments to meet the specific requests of local drillers encountering specific drilling difficulties in given rock layers.

SPEEDING THE PROCESS

In 1907 and 1908, patents were obtained in New Jersey and Texas for steel, cone-shaped drill bits incorporating a roller-type coring device with many serrated teeth. Later in 1908, both patents were bought by lawyer Howard R. Hughes.

Although comparatively weak rocks such as sands, clays, and soft shales could be drilled rapidly (at rates exceeding 30 meters per hour), in harder shales, lime-dolostones, and gravels, drill rates of 1 meter per hour or less were not uncommon. Conventional drill bits of the time had average operating lives of three to twelve hours. Economic drilling mandated increases in both bit life and drilling rate. Directly motivated by his petroleum prospecting interests, Hughes and his partner, Walter B. Sharp, undertook what were probably the first recorded systematic studies of drill bit performance while matched against specific rock layers.

Although many improvements in detail and materials have been made to the Hughes cone bit since its inception in 1908, its basic design is still used in rotary drilling. One of Hughes's major innovations was the much larger size of the cutters, symmetrically distrib-

HOWARD R. HUGHES

Howard Hughes (1905-1976) is famous for having been one of the most dashing, innovative, quirky tycoons of the twentieth century. It all started with his father, Howard R. Hughes. In fact it was the father's enterprise, Hughes Tool Company, that the son took over at age eighteen and built into an immense financial empire based on high-tech products.

The senior Hughes was born in Lancaster, Missouri, in 1869. He spent his boyhood in Keokuk, Iowa, where his own father practiced law. He himself studied law at Harvard University and the University of Iowa and then joined his father's practice, but not for long. In 1901 news came of a big oil strike near Beaumont, Texas. Like hundreds of other ambitious men, Hughes headed there. By 1906 he had immersed himself in the technical problems of drilling and began experimenting to improve drill bits. He produced a wooden model of the roller-type drill two years later while in Oil City, Louisiana. With business associate Walter Sharp he successfully tested a prototype in an oil well in the Goose Creek field near Houston. It drilled faster and more efficiently than those then in use.

Hughes and Sharp opened the Sharp-Hughes Tool Company to manufacture the drills and related equipment, and their products quickly became the industry standard. A shrewd business strategist, Hughes leased, rather than sold, his drill bits for $30,000 per well, retaining his patents to preserve his monopoly over the rotary drill technology. After Sharp died in 1912, Hughes changed the company to the Hughes Tool Company. When Hughes himself died in 1924, he left his son, then a student at Rice Institute (later Rice University), the company and a million-dollar fortune, which Hughes junior would eventually multiply hundreds of times over.

uted as a large number of small individual teeth on the outer face of two or more cantilevered bearing pins. In addition, "hard facing" was employed to drill bit teeth to increase usable life. Hard facing is a metallurgical process basically consisting of wedding a thin layer of a hard metal or alloy of special composition to a metal surface to increase its resistance to abrasion and heat. A less noticeable but equally essential innovation, not included in other drill bit patents, was an ingeniously designed gauge surface that provided strong uniform support for all the drill teeth. The force-fed oil lubrication was another new feature included in Hughes's patent and prototypes, reducing the power necessary to rotate the bit by 50 percent over that of prior mud or water lubricant designs.

IMPACT

In 1925, the first superhard facing was used on cone drill bits. In addition, the first so-called self-cleaning rock bits appeared from Hughes, with significant advances in roller bearings and bit tooth shape translating into increased drilling efficiency. The much larger teeth were more adaptable to drilling in a wider variety of geological formations than earlier models. In 1928, tungsten carbide was introduced as an additional bit facing hardener by Hughes metallurgists. This, together with other improvements, resulted in the Hughes ACME tooth form, which has been in almost continuous use since 1926.

Many other drilling support technologies, such as drilling mud, mud circulation pumps, blowout detectors and preventers, and pipe properties and connectors have enabled rotary drilling rigs to reach new depths (exceeding 5 kilometers in 1990). The successful experiments by Hughes in 1908 were critical initiators of these developments.

See also Geothermal power; Steelmaking process; Thermal cracking process.

FURTHER READING

Brantly, John Edward. *History of Oil Well Drilling*. Houston: Gulf Publishing, 1971.
Charlez, Philippe A. *Rock Mechanics*. Vol. 2: *Petroleum Applications*. Paris: Editions Technip, 1997.
Rao, Karanam Umamaheshwar, and Misra Banabihari. *Principles of Rock Drilling*. Brookfield, Vt.: Balkema, 1998.

OPTICAL DISK

THE INVENTION: A nonmagnetic storage medium for computers that can hold much greater quantities of data than similar size magnetic media, such as hard and floppy disks.

THE PEOPLE BEHIND THE INVENTION:

Klaas Compaan, a Dutch physicist

Piet Kramer, head of Philips' optical research laboratory

Lou F. Ottens, director of product development for Philips' musical equipment division

George T. de Kruiff, manager of Philips' audio-product development department

Joop Sinjou, a Philips project leader

HOLOGRAMS CAN BE COPIED INEXPENSIVELY

Holography is a lensless photographic method that uses laser light to produce three-dimensional images. This is done by splitting a laser beam into two beams. One of the beams is aimed at the object whose image is being reproduced so that the laser light will reflect from the object and strike a photographic plate or film. The second beam of light is reflected from a mirror near the object and also strikes the photographic plate or film. The "interference pattern," which is simply the pattern created by the differences between the two reflected beams of light, is recorded on the photographic surface. The recording that is made in this way is called a "hologram." When laser light or white light strikes the hologram, an image is created that appears to be a three-dimensional object.

Early in 1969, Radio Corporation of America (RCA) engineers found a way to copy holograms inexpensively by impressing interference patterns on a nickel sheet that then became a mold from which copies could be made. Klaas Compaan, a Dutch physicist, learned of this method and had the idea that images could be recorded in a similar way and reproduced on a disk the size of a phonograph record. Once the images were on the disk, they could be projected onto a screen in any sequence. Compaan saw the possibilities of such a technology in the fields of training and education.

COMPUTER DATA STORAGE BREAKTHROUGH

In 1969, Compaan shared his idea with Piet Kramer, who was the head of Philips' optical research laboratory. The idea intrigued Kramer. Between 1969 and 1971, Compaan spent much of his time working on the development of a prototype.

By September, 1971, Compaan and Kramer, together with a handful of others, had assembled a prototype that could read a black-and-white video signal from a spinning glass disk. Three months later, they demonstrated it for senior managers at Philips. In July, 1972, a color prototype was demonstrated publicly. After the demonstration, Philips began to consider putting sound, rather than images, on the disks. The main attraction of that idea was that the 12-inch (305-millimeter) disks would hold up to forty-eight hours of music. Very quickly, however, Lou F. Ottens, director of product development for Philips' musical equipment division, put an end to any talk of a long-playing audio disk.

Ottens had developed the cassette-tape cartridge in the 1960's. He had plenty of experience with the recording industry, and he had no illusions that the industry would embrace that new medium. He was convinced that the recording companies would consider forty-eight hours of music unmarketable. He also knew that any new medium would have to offer a dramatic improvement over existing vinyl records.

In 1974, only three years after the first microprocessor (the basic element of computers) was invented, designing a digital consumer product—rather than an analog product such as those that were already commonly accepted—was risky. (Digital technology uses numbers to represent information, whereas analog technology represents information by mechanical or physical means.) When George T. de Kruiff became Ottens's manager of audio-product development in June, 1974, he was amazed that there were no digital circuit specialists in the audio department. De Kruiff recruited new digital engineers, bought computer-aided design tools, and decided that the project should go digital.

Within a few months, Ottens's engineers had rigged up a digital system. They used an audio signal that was representative of an acoustical wave, sampled it to change it to digital form, and encoded it as a series of pulses. On the disk itself, they varied the length of the "dimples" that were used to represent the sound so that the rising and falling edges of the series of pulses corresponded to the dimples' walls. A helium-neon laser was reflected from the dimples to photodetectors that were connected to a digital-to-analog converter.

In 1978, Philips demonstrated a prototype for Polygram (a West German company) and persuaded Polygram to develop an inexpensive disk material with the appropriate optical qualities. Most important was that the material could not warp. Polygram spent about $150,000 and three months to develop the disk. In addition, it was determined that the gallium-arsenide (GaAs) laser would be used in the project. Sharp Corporation agreed to manufacture a long-life GaAs diode laser to Philips' specifications.

The optical-system designers wanted to reduce the number of parts in order to decrease manufacturing costs and improve reliability. Therefore, the lenses were simplified and considerable work was devoted to developing an error-correction code. Philips and Sony engineers also worked together to create a standard format. In 1983, Philips made almost 100,000 units of optical disks.

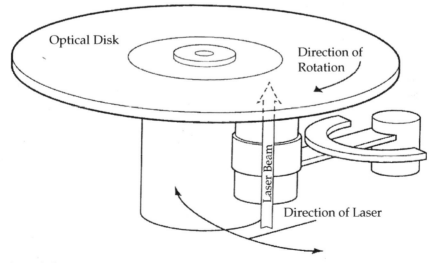

An optical memory.

CONSEQUENCES

In 1983, one of the most successful consumer products of all time was introduced: the optical-disk system. The overwhelming success of optical-disk reproduction led to the growth of a multibillion-dollar industry around optical information and laid the groundwork for a whole crop of technologies that promise to revolutionize computer data storage. Common optical-disk products are the compact disc (CD), the compact disc read-only memory (CD-ROM), the write-once, read-many (WORM) erasable disk, and CD-I (interactive CD).

The CD-ROM, the WORM, and the erasable optical disk, all of which are used in computer applications, can hold more than 550 megabytes, from 200 to 800 megabytes, and 650 megabytes of data, respectively.

The CD-ROM is a nonerasable disc that is used to store computer data. After the write-once operation is performed, a WORM becomes a read-only optical disk. An erasable optical disk can be erased and rewritten easily. CD-ROMs, coupled with expert-system technology, are expected to make data retrieval easier. The CD-ROM, the WORM, and the erasable optical disk may replace magnetic hard and floppy disks as computer data storage devices.

See also Bubble memory; Compact disc; Computer chips; Floppy disk; Hard disk; Holography.

FURTHER READING

Fox, Barry. "Head to Head in the Recording Wars." *New Scientist* 136, no. 1843 (October 17, 1992).

Goff, Leslie. "Philips' Eye on the Future." *Computerworld* 33, no. 32 (August 9, 1999).

Kolodziej, Stan. "Optical Discs: The Dawn of a New Era in Mass Storage." *Canadian Datasystems* 14, no. 9 (September, 1982). 36-39.

Savage, Maria. "Beyond Film." *Bulletin of the American Society for Information Science* 7, no. 1 (October, 1980).

Orlon

THE INVENTION: A synthetic fiber made from polyacrylonitrile that has become widely used in textiles and in the preparation of high-strength carbon fibers.

THE PEOPLE BEHIND THE INVENTION:
Herbert Rein (1899-1955), a German chemist
Ray C. Houtz (1907-   ), an American chemist

A DIFFICULT PLASTIC

"Polymers" are large molecules that are made up of chains of many smaller molecules, called "monomers." Materials that are made of polymers are also called polymers, and some polymers, such as proteins, cellulose, and starch, occur in nature. Most polymers, however, are synthetic materials, which means that they were created by scientists.

The twenty-year period beginning in 1930 was the age of great discoveries in polymers by both chemists and engineers. During this time, many of the synthetic polymers, which are also known as plastics, were first made and their uses found. Among these polymers were nylon, polyester, and polyacrylonitrile. The last of these materials, polyacrylonitrile (PAN), was first synthesized by German chemists in the late 1920's. They linked more than one thousand of the small, organic molecules of acrylonitrile to make a polymer. The polymer chains of this material had the properties that were needed to form strong fibers, but there was one problem. Instead of melting when heated to a high temperature, PAN simply decomposed. This made it impossible, with the technology that existed then, to make fibers.

The best method available to industry at that time was the process of melt spinning, in which fibers were made by forcing molten polymer through small holes and allowing it to cool. Researchers realized that, if PAN could be put into a solution, the same apparatus could be used to spin PAN fibers. Scientists in Germany and the United States tried to find a solvent or liquid that would dissolve PAN, but they were unsuccessful until World War II began.

Fibers for War

In 1938, the German chemist Walter Reppe developed a new class of organic solvents called "amides." These new liquids were able to dissolve many materials, including some of the recently discovered polymers. When World War II began in 1940, both the Germans and the Allies needed to develop new materials for the war effort. Materials such as rubber and fibers were in short supply. Thus, there was increased governmental support for chemical and industrial research on both sides of the war. This support was to result in two independent solutions to the PAN problem.

In 1942, Herbert Rein, while working for I. G. Farben in Germany, discovered that PAN fibers could be produced from a solution of polyacrylonitrile dissolved in the newly synthesized solvent dimethylformamide. At the same time Ray C. Houtz, who was working for E. I. Du Pont de Nemours in Wilmington, Delaware, found that the related solvent dimethylacetamide would also form excellent PAN fibers. His work was patented, and some fibers were produced for use by the military during the war. In 1950, Du Pont began commercial production of a form of polyacrylonitrile fibers called Orlon. The Monsanto Company followed with a fiber called Acrilon in 1952, and other companies began to make similar products in 1958.

There are two ways to produce PAN fibers. In both methods, polyacrylonitrile is first dissolved in a suitable solvent. The solution is next forced through small holes in a device called a "spinneret." The solution emerges from the spinneret as thin streams of a thick, gooey liquid. In the "wet spinning method," the streams then enter another liquid (usually water or alcohol), which extracts the solvent from the solution, leaving behind the pure PAN fiber. After air drying, the fiber can be treated like any other fiber. The "dry spinning method" uses no liquid. Instead, the solvent is evaporated from the emerging streams by means of hot air, and again the PAN fiber is left behind.

In 1944, another discovery was made that is an important part of the polyacrylonitrile fiber story. W. P. Coxe of Du Pont and L. L. Winter at Union Carbide Corporation found that, when PAN fibers are heated under certain conditions, the polymer decomposes and changes into graphite (one of the elemental forms of carbon) but still keeps its fiber form. In contrast to most forms of graphite, these fibers were exceptionally strong. These were the first carbon fibers ever made. Originally known as "black Orlon," they were first produced commercially by the Japanese in 1964, but they were too weak to find many uses. After new methods of graphitization were developed jointly by labs in Japan, Great Britain, and the United States, the strength of the carbon fibers was increased, and the fibers began to be used in many fields.

IMPACT

As had been predicted earlier, PAN fibers were found to have some very useful properties. Their discovery and commercialization helped pave the way for the acceptance and wide use of polymers. The fibers derive their properties from the stiff, rodlike structure of polyacrylonitrile. Known as acrylics, these fibers are more durable than cotton, and they are the best alternative to wool for sweaters. Acrylics are resistant to heat and chemicals, can be dyed easily, resist fading or wrinkling, and are mildew-resistant. Thus, after their introduction, PAN fibers were very quickly made into yarns, blankets, draperies, carpets, rugs, sportswear, and various items of clothing. Often, the fibers contain small amounts of other polymers that give them additional useful properties.

A significant amount of PAN fiber is used in making carbon fibers. These lightweight fibers are stronger for their weight than any known material, and they are used to make high-strength composites for applications in aerospace, the military, and sports. A "fiber composite" is a material made from two parts: a fiber, such as carbon or glass, and something to hold the fibers together, which is usually a plastic called an "epoxy." Fiber composites are used in products that require great strength and light weight. Their applications can be as ordinary as a tennis racket or fishing pole or as exotic as an airplane tail or the body of a spacecraft.

See also Buna rubber; Neoprene; Nylon; Plastic; Polyester; Polyethylene; Polystyrene.

FURTHER READING

Handley, Susannah. *Nylon: The Story of a Fashion Revolution: A Celebration of Design from Art Silk to Nylon and Thinking Fibres.* Baltimore: Johns Hopkins University Press, 1999.
Hunter, David. "Du Pont Bids Adieu to Acrylic Fibers." *Chemical Week* 146, no. 24 (June 20, 1990).
Kornheiser, Tony. "So Long, Orlon." *Washington Post* (June 13, 1990).
Seymour, Raymond Benedict, and Roger Stephen Porter. *Manmade Fibers: Their Origin and Development.* New York: Elsevier Applied Science, 1993.

Pacemaker

THE INVENTION: A small device using transistor circuitry that regulates the heartbeat of the patient in whom it is surgically emplaced.

THE PEOPLE BEHIND THE INVENTION:
Ake Senning (1915-   ), a Swedish physician
Rune Elmquist, co-inventor of the first pacemaker
Paul Maurice Zoll (1911-   ), an American cardiologist

Cardiac Pacing

The fundamentals of cardiac electrophysiology (the electrical activity of the heart) were determined during the eighteenth century; the first successful cardiac resuscitation by electrical stimulation occurred in 1774. The use of artificial pacemakers for resuscitation was demonstrated in 1929 by Mark Lidwell. Lidwell and his coworkers developed a portable apparatus that could be connected to a power source. The pacemaker was used successfully on several stillborn infants after other methods of resuscitation failed. Nevertheless, these early machines were unreliable.

Ake Senning's first experience with the effect of electrical stimulation on cardiac physiology was memorable; grasping a radio ground wire, Senning felt a brief episode of ventricular arrhythmia (irregular heartbeat). Later, he was able to apply a similar electrical stimulation to control a heartbeat during surgery.

The principle of electrical regulation of the heart was valid. It was shown that pacemakers introduced intravenously into the sinus node area of a dog's heart could be used to control the heartbeat rate. Although Paul Maurice Zoll utilized a similar apparatus in several patients with cardiac arrhythmia, it was not appropriate for extensive clinical use; it was large and often caused unpleasant sensations or burns. In 1957, however, Ake Senning observed that attaching stainless steel electrodes to a child's heart made it possible to regulate the heart's rate of contraction. Senning considered this to represent the beginning of the era of clinical pacing.

DEVELOPMENT OF CARDIAC PACEMAKERS

Senning's observations of the successful use of the cardiac pacemaker had allowed him to identify the problems inherent in the device. He realized that the attachment of the device to the lower, ventricular region of the heart made possible more reliable control, but other problems remained unsolved. It was inconvenient, for example, to carry the machine externally; a cord was wrapped around the patient that allowed the pacemaker to be recharged, which had to be done frequently. Also, for unknown reasons, heart resistance would increase with use of the pacemaker, which meant that increasingly large voltages had to be used to stimulate the heart. Levels as high as 20 volts could cause quite a "start" in the patient. Furthermore, there was a continuous threat of infection.

In 1957, Senning and his colleague Rune Elmquist developed a pacemaker that was powered by rechargeable nickel-cadmium batteries, which had to be recharged once a month. Although Senning and Elmquist did not yet consider the pacemaker ready for human testing, fate intervened. A forty-three-year-old man was admitted to the hospital suffering from an atrioventricular block, an inability of the electrical stimulus to travel along the conductive fibers of the "bundle of His" (a band of cardiac muscle fibers). As a result of this condition, the patient required repeated cardiac resuscitation. Similar types of heart block were associated with a mortality rate higher than 50 percent per year and nearly 95 percent over five years.

Senning implanted two pacemakers (one failed) into the myocardium of the patient's heart, one of which provided a regulatory rate of 64 beats per minute. Although the pacemakers required periodic replacement, the patient remained alive and active for twenty years. (He later became president of the Swedish Association for Heart and Lung Disease.)

During the next five years, the development of more reliable and more complex pacemakers continued, and implanting the pacemaker through the vein rather than through the thorax made it simpler to use the procedure. The first pacemakers were of the "asynchronous" type, which generated a regular charge that overrode the natural pacemaker in the heart. The rate could be set by the physician but could not be altered if the need arose. In 1963, an atrial-triggered synchronous pacemaker was installed by a Swedish team. The advantage of this apparatus lay in its ability to trigger a heart contraction only when the normal heart rhythm was interrupted. Most of these pacemakers contained a sensing device that detected the atrial impulse and generated an electrical discharge only when the heart rate fell below 68 to 72 beats per minute.

The biggest problems during this period lay in the size of the pacemaker and the short life of the battery. The expiration of the electrical impulse sometimes caused the death of the patient. In addition, the most reliable method of checking the energy level of the battery was to watch for a decreased pulse rate. As improvements were made in electronics, the pacemaker became smaller, and in 1972, the more reliable lithium-iodine batteries were introduced. These batteries made it possible to store more energy and to monitor the energy level more effectively. The use of this type of power source essentially eliminated the battery as the limiting factor in the longevity of the pacemaker. The period of time that a pacemaker could operate continuously in the body increased from a period of days in 1958 to five to ten years by the 1970's.

CONSEQUENCES

The development of electronic heart pacemakers revolutionized cardiology. Although the initial machines were used primarily to control cardiac bradycardia, the often life-threatening slowing of the heartbeat, a wide variety of arrhythmias and problems with cardiac output can now be controlled through the use of these devices. The success associated with the surgical implantation of pacemakers is attested by the frequency of its use. Prior to 1960, only three pacemakers had been implanted. During the 1990's, however, some 300,000 were implanted each year throughout the world. In the United States, the prevalence of implants is on the order of 1 per 1,000 persons in the population.

Pacemaker technology continues to improve. Newer models can sense pH and oxygen levels in the blood, as well as respiratory rate. They have become further sensitized to minor electrical disturbances and can adjust accordingly. The use of easily sterilized circuitry has eliminated the danger of infection. Once the pacemaker has been installed in the patient, the basic electronics require no additional attention. With the use of modern pacemakers, many forms of electrical arrhythmias need no longer be life-threatening.

See also Artificial heart; Contact lenses; Coronary artery bypass surgery; Electrocardiogram; Hearing aid; Heart-lung machine.

FURTHER READING

Bigelow, W. G. *Cold Hearts: The Story of Hypothermia and the Pacemaker in Heart Surgery.* Toronto: McClelland and Stewart, 1984.
Greatbatch, Wilson. *The Making of the Pacemaker: Celebrating a Lifesaving Invention.* Amherst, N.Y.: Prometheus Books, 2000.
"The Pacemaker." *Newsweek* 130, no. 24A (Winter, 1997/1998).
Thalen, H. J. *The Artificial Cardiac Pacemaker: Its History, Development and Clinical Application.* London: Heinemann Medical, 1969.

Pap test

THE INVENTION: A cytologic technique the diagnosing uterine cancer, the second most common fatal cancer in American women.

THE PEOPLE BEHIND THE INVENTION:

George N. Papanicolaou (1883-1962), a Greek-born American physician and anatomist

Charles Stockard (1879-1939), an American anatomist

Herbert Traut (1894-1972), an American gynecologist

Cancer in History

Cancer, first named by the ancient Greek physician Hippocrates of Cos, is one of the most painful and dreaded forms of human disease. It occurs when body cells run wild and interfere with the normal activities of the body. The early diagnosis of cancer is extremely important because early detection often makes it possible to effect successful cures. The modern detection of cancer is usually done by the microscopic examination of the cancer cells, using the techniques of the area of biology called "cytology, " or cell biology.

Development of cancer cytology began in 1867, after L. S. Beale reported tumor cells in the saliva from a patient who was afflicted with cancer of the pharynx. Beale recommended the use in cancer detection of microscopic examination of cells shed or removed (exfoliated) from organs including the digestive, the urinary, and the reproductive tracts. Soon, other scientists identified numerous striking differences, including cell size and shape, the size of cell nuclei, and the complexity of cell nuclei.

Modern cytologic detection of cancer evolved from the work of George N. Papanicolaou, a Greek physician who trained at the University of Athens Medical School. In 1913, he emigrated to the United States.

In 1917, he began studying sex determination of guinea pigs with Charles Stockard at New York's Cornell Medical College. Papanicolaou's efforts required him to obtain ova (egg cells) at a precise period in their maturation cycle, a process that required an indicator of the time at which the animals ovulated. In search of this indicator, Papanicolaou designed a method that involved microscopic examination of the vaginal discharges from female guinea pigs.

Initially, Papanicolaou sought traces of blood, such as those seen in the menstrual discharges from both primates and humans. Papanicolaou found no blood in the guinea pig vaginal discharges. Instead, he noticed changes in the size and the shape of the uterine cells shed in these discharges. These changes recurred in a fifteen-to-sixteen-day cycle that correlated well with the guinea pig menstrual cycle.

"New Cancer Detection Method"

Papanicolaou next extended his efforts to the study of humans. This endeavor was designed originally to identify whether comparable changes in the exfoliated cells of the human vagina occurred in women. Its goal was to gain an understanding of the human menstrual cycle. In the course of this work, Papanicolaou observed distinctive abnormal cells in the vaginal fluid from a woman afflicted with cancer of the cervix. This led him to begin to attempt to develop a cytologic method for the detection of uterine cancer, the second most common type of fatal cancer in American women of the time.

In 1928, Papanicolaou published his cytologic method of cancer detection in the *Proceedings of the Third Race Betterment Conference*, held in Battle Creek, Michigan. The work was received well by the news media (for example, the January 5, 1928, *New York World* credited him with a "new cancer detection method"). Nevertheless, the publication—and others he produced over the next ten years—was not very interesting to gynecologists of the time. Rather, they preferred use of the standard methodology of uterine cancer diagnosis (cervical biopsy and curettage).

Consequently, in 1932, Papanicolaou turned his energy toward studying human reproductive endocrinology problems related to the effects of hormones on cells of the reproductive system. One example of this work was published in a 1933 issue of *The American Journal of Anatomy*, where he described "the sexual cycle in the human female." Other such efforts resulted in better understanding of reproductive problems that include amenorrhea and menopause.

It was not until Papanicolaou's collaboration with gynecologist Herbert Traut (beginning in 1939), which led to the publication of *Diagnosis of Uterine Cancer by the Vaginal Smear* (1943), that clinical acceptance of the method began to develop. Their monograph documented an impressive, irrefutable group of studies of both normal and disease states that included nearly two hundred cases of cancer of the uterus.

Soon, many other researchers began to confirm these findings; by 1948, the newly named American Cancer Society noted that the "Pap" smear seemed to be a very valuable tool for detecting vaginal cancer. Wide acceptance of the Pap test followed, and, beginning in 1947, hundreds of physicians from all over the world flocked to Papanicolaou's course on the subject. They learned his smear/diagnosis techniques and disseminated them around the world.

IMPACT

The Pap test has been cited by many physicians as being the most significant and useful modern discovery in the field of cancer research. One way of measuring its impact is the realization that the test allows the identification of uterine cancer in the earliest stages, long before other detection methods can be used. Moreover, because of resultant early diagnosis, the disease can be cured in more than 80 percent of all cases identified by the test. In addition, Pap testing allows the identification of cancer of the uterine cervix so early that its cure rate can be nearly 100 percent.

Papanicolaou extended the use of the smear technique from examination of vaginal discharges to diagnosis of cancer in many other organs from which scrapings, washings, and discharges can be obtained. These tissues include the colon, the kidney, the bladder, the prostate, the lung, the breast, and the sinuses. In most cases, such examination of these tissues has made it possible to diagnose cancer much sooner than is possible by using other existing methods. As a result, the smear method has become a basis of cancer control in national health programs throughout the world.

See also Amniocentesis; Birth control pill; Mammography; Syphilis test; Ultrasound.

FURTHER READING

Apgar, Barbara, Lawrence L. Gabel, and Robert T. Brown. *Oncology*. Philadelphia: W. B. Saunders, 1998.
Entman, Stephen S., and Charles B. Rush. *Office Gynecology*. Philadelphia: Saunders, 1995.
Glass, Robert H., Michèle G. Curtis, and Michael P. Hopkins. *Glass's Office Gynecology*. 5th ed. Baltimore: Williams & Wilkins, 1999.
Rushing, Lynda, and Nancy Joste. *Abnormal Pap Smears: What Every Woman Needs to Know*. Amherst, N.Y.: Prometheus Books, 2001.

Penicillin

THE INVENTION: The first successful and widely used antibiotic drug, penicillin has been called the twentieth century's greatest "wonder drug."

THE PEOPLE BEHIND THE INVENTION:
Sir Alexander Fleming (1881-1955), a Scottish bacteriologist, cowinner of the 1945 Nobel Prize in Physiology or Medicine
Baron Florey (1898-1968), an Australian pathologist, cowinner of the 1945 Nobel Prize in Physiology or Medicine
Ernst Boris Chain (1906-1979), an émigré German biochemist, cowinner of the 1945 Nobel Prize in Physiology or Medicine

THE SEARCH FOR THE PERFECT ANTIBIOTIC

During the early twentieth century, scientists were aware of antibacterial substances but did not know how to make full use of them in the treatment of diseases. Sir Alexander Fleming discovered penicillin in 1928, but he was unable to duplicate his laboratory results of its antibiotic properties in clinical tests; as a result, he did not recognize the medical potential of penicillin. Between 1935 and 1940, penicillin was purified, concentrated, and clinically tested by pathologist Baron Florey, biochemist Ernst Boris Chain, and members of their Oxford research group. Their achievement has since been regarded as one of the greatest medical discoveries of the twentieth century.

Florey was a professor at Oxford University in charge of the Sir William Dunn School of Pathology. Chain had worked for two years at Cambridge University in the laboratory of Frederick Gowland Hopkins, an eminent chemist and discoverer of vitamins. Hopkins recommended Chain to Florey, who was searching for a candidate to lead a new biochemical unit in the Dunn School of Pathology.

In 1938, Florey and Chain formed a research group to investigate the phenomenon of antibiosis, or the antagonistic association between different forms of life. The union of Florey's medical knowledge and Chain's biochemical expertise proved to be an ideal combination for exploring the antibiosis potential of penicillin. Florey and Chain began their investigation with a literature search in which Chain came across Fleming's work and added penicillin to their list of potential antibiotics.

Their first task was to isolate pure penicillin from a crude liquid extract. A culture of Fleming's original *Penicillium notatum* was maintained at Oxford and was used by the Oxford group for penicillin production. Extracting large quantities of penicillin from the medium was a painstaking task, as the solution contained only one part of the antibiotic in ten million. When enough of the raw juice was collected, the Oxford group focused on eliminating impurities and concentrating the penicillin. The concentrated liquid was then freeze-dried, leaving a soluble brown powder.

SPECTACULAR RESULTS

In May, 1940, Florey's clinical tests of the crude penicillin proved its value as an antibiotic. Following extensive controlled experiments with mice, the Oxford group concluded that they had discovered an antibiotic that was nontoxic and far more effective against pathogenic bacteria than any of the known sulfa drugs. Furthermore, penicillin was not inactivated after injection into the bloodstream but was excreted unchanged in the urine. Continued tests showed that penicillin did not interfere with white blood cells and had no adverse effect on living cells. Bacteria susceptible to the antibiotic included those responsible for gas gangrene, pneumonia, meningitis, diphtheria, and gonorrhea. American researchers later proved that penicillin was also effective against syphilis.

In January, 1941, Florey injected a volunteer with penicillin and found that there were no side effects to treatment with the antibiotic. In February, the group began treatment of Albert Alexander, a forty-three-year-old policeman with a serious staphylococci and streptococci infection that was resisting massive doses of sulfa drugs. Alexander had been hospitalized for two months after an infection in the corner of his mouth had spread to his face, shoulder, and lungs. After receiving an injection of 200 milligrams of penicillin, Alexander showed remarkable progress, and for the next ten days his condition improved. Unfortunately, the Oxford

SIR ALEXANDER FLEMING

In 1900 Alexander Fleming (1881-1955) enlisted in the London Scottish Regiment, hoping to see action in the South African (Boer) War then underway between Great Britain and South Africa's independent Afrikaner republics. However, the war ended too soon for him. So, having come into a small inheritance, he decided to become a physician instead. Accumulating honors and prizes along the way, he succeeded and became a fellow of the Royal College of Surgeons of England in 1909.

His mentor was Sir Almroth Wright. Fleming assisted him at St. Mary's Hospital in Paddington, and they were at the forefront of the burgeoning field of bacteriology. They were, for example, among the first to treat syphilis with the newly discovered Salvarsan, and they championed immunization through vaccination. With the outbreak of World War I, Fleming followed Wright into the Royal Army Medical Corps, conducting research on battlefield wounds at a laboratory near Boulogne. The infections Fleming inspected horrified him. After the war, again at St. Mary's Hospital, he dedicated himself to finding anti-bacterial agents.

He succeed twice: "lysozyme" in 1921 and penicillin in 1928. To his great disappointment, he was unable to produce pure, potent concentrations of the drug. That had to await the work of Ernst Chain and Howard Florey in 1940. Meanwhile, Fleming studied the antibacterial properties of sulfa drugs. He was overjoyed that Chain and Florey succeeded where he had failed and that penicillin saved lives during World War II and afterward, but he was taken aback when with them he began to receive a stream of tributes, awards, decorations, honorary degrees, and fellowships, including the Nobel Prize in Physiology or Medicine in 1945. He was by nature a reserved man.

However, he adjusted to his role as one of the most lionized medical researchers of his generation and continued his work, both as a professor of medicine at the University of London from 1928 until 1948 and as director of the same St. Mary's Hospital laboratory where he had started his career (renamed the Wright-Fleming Institute in 1948). He died soon after he retired in 1955.

production facility was unable to generate enough penicillin to overcome Alexander's advanced infection completely, and he died on March 15. A later case involving a fourteen-year-old boy with staphylococcal septicemia and osteomyelitis had a more spectacular result: The patient made a complete recovery in two months. In all the early clinical treatments, patients showed vast improvement, and most recovered completely from infections that resisted all other treatment.

IMPACT

Penicillin is among the greatest medical discoveries of the twentieth century. Florey and Chain's chemical and clinical research brought about a revolution in the treatment of infectious disease. Almost every organ in the body is vulnerable to bacteria. Before penicillin, the only antimicrobial drugs available were quinine, arsenic, and sulfa drugs. Of these, only the sulfa drugs were useful for treatment of bacterial infection, but their high toxicity often limited their use. With this small arsenal, doctors were helpless to treat thousands of patients with bacterial infections.

The work of Florey and Chain achieved particular attention because of World War II and the need for treatments of such scourges as gas gangrene, which had infected the wounds of numerous World War I soldiers. With the help of Florey and Chain's Oxford group, scientists at the U.S. Department of Agriculture's Northern Regional Research Laboratory developed a highly efficient method for producing penicillin using fermentation. After an extended search, scientists were also able to isolate a more productive penicillin strain, *Penicillium chrysogenum*. By 1945, a strain was developed that produced five hundred times more penicillin than Fleming's original mold had.

Penicillin, the first of the "wonder drugs," remains one of the most powerful antibiotic in existence. Diseases such as pneumonia, meningitis, and syphilis are still treated with penicillin. Penicillin and other antibiotics also had a broad impact on other fields of medicine, as major operations such as heart surgery, organ transplants, and management of severe burns became possible once the threat of bacterial infection was minimized.

Florey and Chain received numerous awards for their achievement, the greatest of which was the 1945 Nobel Prize in Physiology or Medicine, which they shared with Fleming for his original discovery. Florey was among the most effective medical scientists of his generation, and Chain earned similar accolades in the science of biochemistry. This combination of outstanding medical and chemical expertise made possible one of the greatest discoveries in human history.

See also Antibacterial drugs; Artificial hormone; Genetically engineered insulin; Polio vaccine (Sabin); Polio vaccine (Salk); Reserpine; Salvarsan; Tuberculosis vaccine; Typhus vaccine; Yellow fever vaccine.

FURTHER READING

Bickel, Lennard. *Florey, The Man Who Made Penicillin.* Carlton South, Victoria, Australia: Melbourne University Press, 1995.

Clark, Ronald William. *The Life of Ernst Chain: Penicillin and Beyond.* New York: St. Martin's Press, 1985.

Hughes, William Howard. *Alexander Fleming and Penicillin.* Hove: Wayland, 1979.

Mateles, Richard I. *Penicillin: A Paradigm for Biotechnology.* Chicago: Canadida Corporation, 1998.

Personal computer

The invention: Originally a tradename of the IBM Corporation, "personal computer" has become a generic term for increasingly powerful desktop computing systems using microprocessors.

The people behind the invention:
Tom J. Watson, (1874-1956), the founder of IBM, who set corporate philosophy and marketing principles
Frank Cary (1920-   ), the chief executive officer of IBM at the time of the decision to market a personal computer
John Opel (1925-   ), a member of the Corporate Management Committee
George Belzel, a member of the Corporate Management Committee
Paul Rizzo, a member of the Corporate Management Committee
Dean McKay (1921-   ), a member of the Corporate Management Committee
William L. Sydnes, the leader of the original twelve-member design team

Shaking up the System

For many years, the International Business Machines (IBM) Corporation had been set in its ways, sticking to traditions established by its founder, Tom Watson, Sr. If it hoped to enter the new microcomputer market, however, it was clear that only nontraditional methods would be useful. Apple Computer was already beginning to make inroads into large IBM accounts, and IBM stock was starting to stagnate on Wall Street. A 1979 *Business Week* article asked: "Is IBM just another stodgy, mature company?" The microcomputer market was expected to grow more than 40 percent in the early 1980's, but IBM would have to make some changes in order to bring a competitive personal computer (PC) to the market.

The decision to build and market the PC was made by the company's Corporate Management Committee (CMC). CMC members included chief executive officer Frank Cary, John Opel, George

Belzel, Paul Rizzo, Dean McKay, and three senior vice presidents. In July of 1980, Cary gave the order to proceed. He wanted the PC to be designed and built within a year. The CMC approved the initial design of the PC one month later. Twelve engineers, with William L. Sydnes as their leader, were appointed as the design team. At the end of 1980, the team had grown to 150.

Most parts of the PC had to be produced outside IBM. Microsoft Corporation won the contract to produce the PC's disk operating system (DOS) and the BASIC (*B*eginner's *A*ll-purpose *S*ymbolic *I*nstruction Code) language that is built into the PC's read-only memory (ROM). Intel Corporation was chosen to make the PC's central processing unit (CPU) chip, the "brains" of the machine. Outside programmers wrote software for the PC. Ten years earlier, this strategy would have been unheard of within IBM since all aspects of manufacturing, service, and repair were traditionally taken care of in-house.

Marketing the System

IBM hired a New York firm to design a media campaign for the new PC. Readers of magazines and newspapers saw the character of Charlie Chaplin advertising the new PC. The machine was delivered on schedule on August 12, 1981. The price of the basic "system unit" was $1,565. A system with 64 kilobytes of random access memory (RAM), a 13-centimeter single-sided disk drive holding 160 kilobytes, and a monitor was priced at about $3,000. A system with color graphics, a second disk drive, and a dot matrix printer cost about $4,500.

Many useful computer programs had been adapted to the PC and were available when it was introduced. VisiCalc from Personal Software—the program that is credited with "making" the microcomputer revolution—was one of the first available. Other packages included a comprehensive accounting system by Peachtree Software and a word processing package called Easywriter by Information Unlimited Software.

As the selection of software grew, so did sales. In the first year after its introduction, the IBM PC went from a zero market share to 28 percent of the market. Yet the credit for the success of the PC does not go to IBM alone. Many hundreds of companies were able to produce software and hardware for the PC. Within two years, powerful products such as Lotus Corporation's 1-2-3 business spreadsheet had come to the market. Many believed that Lotus 1-2-3 was the program that caused the PC to become so phenomenally successful. Other companies produced hardware features (expansion boards) that increased the PC's memory storage or enabled the machine to "drive" audiovisual presentations such as slide shows. Business especially found the PC to be a powerful tool. The PC has survived because of its expansion capability.

IBM has continued to upgrade the PC. In 1983, the PC/XT was introduced. It had more expansion slots and a fixed disk offering 10 million bytes of storage for programs and data. Many of the companies that made expansion boards found themselves able to make whole PCs. An entire range of PC-compatible systems was introduced to the market, many offering features that IBM did not include in the original PC. The original PC has become a whole family of computers, sold by both IBM and other companies. The hardware and software continue to evolve; each generation offers more computing power and storage with a lower price tag.

Consequences

IBM's entry into the microcomputer market gave microcomputers credibility. Apple Computer's earlier introduction of its computer did not win wide acceptance with the corporate world. Apple did, however, thrive within the educational marketplace. IBM's name already carried with it much clout, because IBM was a successful company. Apple Computer represented all that was great about the "new" microcomputer, but the IBM PC benefited from IBM's image of stability and success.

IBM coined the term *personal computer* and its acronym *PC*. The acronym PC is now used almost universally to refer to the microcomputer. It also had great significance with users who had previously used a large mainframe computer that had to be shared with the whole company. This was their personal computer. That was important to many PC buyers, since the company mainframe was perceived as being complicated and slow. The PC owner now had complete control.

See also Apple II computer; BINAC computer; Colossus computer; ENIAC computer; Floppy disk; Hard disk; IBM Model 1401 computer; Internet; Supercomputer; UNIVAC computer.

FURTHER READING

Cerruzi, Paul E. *A History of Modern Computing.* Cambridge, Mass.: MIT Press, 2000.

Chposky, James, and Ted Leonsis. *Blue Magic: The People, Power, and Politics Behind the IBM Personal Computer.* New York: Facts on File, 1988.

Freiberger, Paul, and Michael Swaine. *Fire in the Valley: The Making of the Personal Computer.* New York: McGraw-Hill, 2000.

Grossman. Wendy. *Remembering the Future: Interviews from Personal Computer World.* New York: Springer, 1997.

Photoelectric cell

The invention: The first devices to make practical use of the photo-electric effect, photoelectric cells were of decisive importance in the electron theory of metals.

The people behind the invention:
Julius Elster (1854-1920), a German experimental physicist
Hans Friedrich Geitel (1855-1923), a German physicist
Wilhelm Hallwachs (1859-1922), a German physicist

Early Photoelectric Cells

The photoelectric effect was known to science in the early nineteenth century when the French physicist Alexandre-Edmond Becquerel wrote of it in connection with his work on glass-enclosed primary batteries. He discovered that the voltage of his batteries increased with intensified illumination and that green light produced the highest voltage. Since Becquerel researched batteries exclusively, however, the liquid-type photocell was not discovered until 1929, when the Wein and Arcturus cells were introduced commercially. These cells were miniature voltaic cells arranged so that light falling on one side of the front plate generated a considerable amount of electrical energy. The cells had short lives, unfortunately; when subjected to cold, the electrolyte froze, and when subjected to heat, the gas generated would expand and explode the cells.

What came to be known as the photoelectric cell, a device connecting light and electricity, had its beginnings in the 1880's. At that time, scientists noticed that a negatively charged metal plate lost its charge much more quickly in the light (especially ultraviolet light) than in the dark. Several years later, researchers demonstrated that this phenomenon was not an "ionization" effect because of the air's increased conductivity, since the phenomenon took place in a vacuum but did not take place if the plate were positively charged. Instead, the phenomenon had to be attributed to the light that excited the electrons of the metal and caused them to fly off: A neutral plate even acquired a slight positive charge under the influence of strong light. Study of this effect not only contributed evidence to an electronic theory of matter—and, as a result of some brilliant mathematical work by the physicist Albert Einstein, later increased knowledge of the nature of radiant energy—but also further linked the studies of light and electricity. It even explained certain chemical phenomena, such as the process of photography. It is important to note that all the experimental work on photoelectricity accomplished prior to the work of Julius Elster and Hans Friedrich Geitel was carried out before the existence of the electron was known.

EXPLAINING PHOTOELECTRIC EMISSION

After the English physicist Sir Joseph John Thomson's discovery of the electron in 1897, investigators soon realized that the photoelectric effect was caused by the emission of electrons under the influence of radiation. The fundamental theory of photoelectric emission was put forward by Einstein in 1905 on the basis of the German physicist Max Planck's quantum theory (1900). Thus, it was not surprising that light was found to have an electronic effect. Since it was known that the longer radio waves could shake electrons into resonant oscillations and the shorter X rays could detach electrons from the atoms of gases, the intermediate waves of visual light would have been expected to have some effect upon electrons—such as detaching them from metal plates and therefore setting up a difference of potential. The photoelectric cell, developed by Elster and Geitel in 1904, was a practical device that made use of this effect.

In 1888, Wilhelm Hallwachs observed that an electrically charged zinc electrode loses its charge when exposed to ultraviolet radiation if the charge is negative, but is able to retain a positive charge under the same conditions. The following year, Elster and Geitel discovered a photoelectric effect caused by visible light; however, they used the alkali metals potassium and sodium for their experiments instead of zinc.

The Elster-Geitel photocell (a vacuum emission cell, as opposed to a gas-filled cell) consisted of an evacuated glass bulb containing two electrodes. The cathode consisted of a thin film of a rare, chemically active metal (such as potassium) that lost its electrons fairly readily;

JULIUS ELSTER AND HANS GEITEL

Nicknamed the Castor and Pollux of physics after the twins of Greek mythology, Johann Philipp Ludwig Julius Elster and Hans Friedrich Geitel were among the most productive teams in the history of science. Elster, born in 1854, and Geitel, born in 1855, met in 1875 while attending university in Heidelberg, Germany. Graduate studies took them to separate cities, but then in 1881 they were together again as mathematics and physics teachers at Herzoglich Gymnasium in Wolfenbüttel. In 1884 they began their scientific collaboration, which lasted more than thirty years and produced more than 150 reports.

Essentially experimentalists, they investigated phenomena that were among the greatest mysteries of the times. Their first works concerned the electrification of flames and the electrical properties of thunderstorms. They went on to study the photoelectric effect, thermal electron emission, practical uses for photocells, and Becquerel rays in the earth and air. They developed a method for measuring electrical phenomena in gases that remained the standard for the following forty years.

Their greatest achievements, however, lay with radioactivity and radiation. Their demonstration that incandescent filaments emitted "negative electricity" proved beyond doubt that electrons, which J. J. Thomson had recently claimed to have detected, did in fact exist. They also proved that radioactivity, such as that from uranium, came wholly from within the atom, not from environmental influences. Ernest Rutherford, the great English physicist, said in 1913 that Elster and Geitel had contributed more to the understanding of terrestrial and atmospheric radioactivity than anyone else.

The pair were practically inseparable until Elster died in 1920. Geitel died three years later.

the anode was simply a wire sealed in to complete the circuit. This anode was maintained at a positive potential in order to collect the negative charges released by light from the cathode. The Elster-Geitel photocell resembled two other types of vacuum tubes in existence at the time: the cathode-ray tube, in which the cathode emitted electrons under the influence of a high potential, and the thermionic valve (a valve that permits the passage of current in one direction only), in which it emitted electrons under the influence of heat. Like both of these vacuum tubes, the photoelectric cell could be classified as an "electronic" device.

The new cell, then, emitted electrons when stimulated by light, and at a rate proportional to the intensity of the light. Hence, a current could be obtained from the cell. Yet Elster and Geitel found that their photoelectric currents fell off gradually; they therefore spoke of "fatigue" (instability). It was discovered later that most of this change was not a direct effect of a photoelectric current's passage; it was not even an indirect effect but was caused by oxidation of the cathode by the air. Since all modern cathodes are enclosed in sealed vessels, that source of change has been completely abolished. Nevertheless, the changes that persist in modern cathodes often are indirect effects of light that can be produced independently of any photoelectric current.

IMPACT

The Elster-Geitel photocell was, for some twenty years, used in all emission cells adapted for the visible spectrum, and throughout the twentieth century, the photoelectric cell has had a wide variety of applications in numerous fields. For example, if products leaving a factory on a conveyor belt were passed between a light and a cell, they could be counted as they interrupted the beam. Persons entering a building could be counted also, and if invisible ultraviolet rays were used, those persons could be detected without their knowledge. Simple relay circuits could be arranged that would automatically switch on street lamps when it grew dark. The sensitivity of the cell with an amplifying circuit enabled it to "see" objects too faint for the human eye, such as minor stars or certain lines in the spectra of elements excited by a flame or discharge. The fact that the current depended on the intensity of the light made it possible to construct photoelectric meters that could judge the strength of illumination without risking human error—for example, to determine the right exposure for a photograph.

A further use for the cell was to make talking films possible. The early "talkies" had depended on gramophone records, but it was very difficult to keep the records in time with the film. Now, the waves of speech and music could be recorded in a "sound track" by turning the sound first into current through a microphone and then into light with a neon tube or magnetic shutter; next, the variations in the intensity of this light on the side of the film were photographed. By reversing the process and running the film between a light and a photoelectric cell, the visual signals could be converted back to sound.

See also Alkaline storage battery; Photovoltaic cell; Solar thermal engine.

FURTHER READING

Hoberman, Stuart. *Solar Cell and Photocell Experimenters Guide*. Indianapolis, Ind.: H. W. Sams, 1965.
Perlin, John. *From Space to Earth: The Story of Solar Electricity*. Ann Arbor, Mich.: Aatec Publications, 1999.
Walker, R. C., and T. M. C. Lance. *Photoelectric Cell Applications: A Practical Book Describing the Uses of Photoelectric Cells in Television, Talking Pictures, Electrical Alarms, Counting Devices, Etc*. 3d ed. London: Sir I. Pitman & Sons, 1938.

Photovoltaic cell

THE INVENTION: Drawing their energy directly from the Sun, the first photovoltaic cells powered instruments on early space vehicles and held out hope for future uses of solar energy.

THE PEOPLE BEHIND THE INVENTION:
Daryl M. Chapin (1906-1995), an American physicist
Calvin S. Fuller (1902-1994), an American chemist
Gerald L. Pearson (1905-   ), an American physicist

UNLIMITED ENERGY SOURCE

All the energy that the world has at its disposal ultimately comes from the Sun. Some of this solar energy was trapped millions of years ago in the form of vegetable and animal matter that became the coal, oil, and natural gas that the world relies upon for energy. Some of this fuel is used directly to heat homes and to power factories and gasoline vehicles. Much of this fossil fuel, however, is burned to produce the electricity on which modern society depends.

The amount of energy available from the Sun is difficult to imagine, but some comparisons may be helpful. During each forty-hour period, the Sun provides the earth with as much energy as the earth's total reserves of coal, oil, and natural gas. It has been estimated that the amount of energy provided by the sun's radiation matches the earth's reserves of nuclear fuel every forty days. The annual solar radiation that falls on about twelve hundred square miles of land in Arizona matched the world's estimated total annual energy requirement for 1960. Scientists have been searching for many decades for inexpensive, efficient means of converting this vast supply of solar radiation directly into electricity.

THE BELL SOLAR CELL

Throughout its history, Bell Systems has needed to be able to transmit, modulate, and amplify electrical signals. Until the 1930's, these tasks were accomplished by using insulators and metallic conductors. At that time, semiconductors, which have electrical properties that are between those of insulators and those of conductors, were developed. One of the most important semiconductor materials is silicon, which is one of the most common elements on the earth. Unfortunately, silicon is usually found in the form of compounds such as sand or quartz, and it must be refined and purified before it can be used in electrical circuits. This process required much initial research, and very pure silicon was not available until the early 1950's.

Electric conduction in silicon is the result of the movement of negative charges (electrons) or positive charges (holes). One way of accomplishing this is by deliberately adding to the silicon phosphorus or arsenic atoms, which have five outer electrons. This addition creates a type of semiconductor that has excess negative charges (an n-type semiconductor). Adding boron atoms, which have three outer electrons, creates a semiconductor that has excess positive charges (a p-type semiconductor). Calvin Fuller made an important study of the formation of p-n junctions, which are the points at which p-type and n-type semiconductors meet, by using the process of diffusing impurity atoms—that is, adding atoms of materials that would increase the level of positive or negative charges, as described above. Fuller's work stimulated interested in using the process of impurity diffusion to create cells that would turn solar energy into electricity. Fuller and Gerald Pearson made the first large-area p-n junction by using the diffusion process. Daryl Chapin, Fuller, and Pearson made a similar p-n junction very close to the surface of a silicon crystal, which was then exposed to sunlight.

The cell was constructed by first making an ingot of arsenic-doped silicon that was then cut into very thin slices. Then a very thin layer of p-type silicon was formed over the surface of the n-type wafer, providing a p-n junction close to the surface of the cell. Once the cell cooled, the p-type layer was removed from the back of the cell and lead wires were attached to the two surfaces. When light was absorbed at the p-n junction, electron-hole pairs were produced, and the electric field that was present at the junction forced the electrons to the n side and the holes to the p side.

The recombination of the electrons and holes takes place after the electrons have traveled through the external wires, where they do

Parabolic mirrors at a solar power plant. (PhotoDisc)

useful work. Chapin, Fuller, and Pearson announced in 1954 that the resulting photovoltaic cell was the most efficient (6 percent) means then available for converting sunlight into electricity.

The first experimental use of the silicon solar battery was in amplifiers for electrical telephone signals in rural areas. An array of 432 silicon cells, capable of supplying 9 watts of power in bright sunlight, was used to charge a nickel-cadmium storage battery. This, in turn, powered the amplifier for the telephone signal. The electrical energy derived from sunlight during the day was sufficient to keep the storage battery charged for continuous operation. The system was successfully tested for six months of continuous use in Americus, Georgia, in 1956. Although it was a technical success, the silicon solar cell was not ready to compete economically with conventional means of producing electrical power.

CONSEQUENCES

One of the immediate applications of the solar cell was to supply electrical energy for Telstar satellites. These cells are used extensively on all satellites to generate power. The success of the U.S. satellite program prompted serious suggestions in 1965 for the use of an orbiting power satellite. A large satellite could be placed into a synchronous orbit of the earth. It would collect sunlight, convert it to microwave radiation, and beam the energy to an Earth-based receiving station. Many technical problems must be solved, however, before this dream can become a reality.

Solar cells are used in small-scale applications such as power sources for calculators. Large-scale applications are still not economically competitive with more traditional means of generating electric power. The development of the Third World countries, however, may provide the incentive to search for less-expensive solar cells that can be used, for example, to provide energy in remote villages. As the standards of living in such areas improve, the need for electric power will grow. Solar cells may be able to provide the necessary energy while safeguarding the environment for future generations.

See also Alkaline storage battery; Fluorescent lighting; Fuel cell; Photoelectric cell; Solar thermal engine.

FURTHER READING

Green, Martin A. *Power to the People: Sunlight to Electricity Using Solar Cells*. Sydney, Australia: University of South Wales Press, 2000.
_____. "Photovoltaics: Technology Overview." *Energy Policy* 28, no. 14 (November, 2000).
Perlin, John. *From Space to Earth: The Story of Solar Electricity*. Ann Arbor, Mich.: Aatec Publications, 1999.

PLASTIC

THE INVENTION: The first totally synthetic thermosetting plastic, which paved the way for modern materials science.

THE PEOPLE BEHIND THE INVENTION:
John Wesley Hyatt (1837-1920), an American inventor
Leo Hendrik Baekeland (1863-1944), a Belgian-born chemist, consultant, and inventor
Christian Friedrich Schönbein (1799-1868), a German chemist who produced guncotton, the first artificial polymer
Adolf von Baeyer (1835-1917), a German chemist

EXPLODING BILLIARD BALLS

In the 1860's, the firm of Phelan and Collender offered a prize of ten thousand dollars to anyone producing a substance that could serve as an inexpensive substitute for ivory, which was somewhat difficult to obtain in large quantities at reasonable prices. Earlier, Christian Friedrich Schönbein had laid the groundwork for a breakthrough in the quest for a new material in 1846 by the serendipitous discovery of nitrocellulose, more commonly known as "guncotton," which was produced by the reaction of nitric acid with cotton.

An American inventor, John Wesley Hyatt, while looking for a substitute for ivory as a material for making billiard balls, discovered that the addition of camphor to nitrocellulose under certain conditions led to the formation of a white material that could be molded and machined. He dubbed this substance "celluloid," and this product is now acknowledged as the first synthetic plastic. Celluloid won the prize for Hyatt, and he promptly set out to exploit his product. Celluloid was used to make baby rattles, collars, dentures, and other manufactured goods.

As a billiard ball substitute, however, it was not really adequate, for various reasons. First, it is thermoplastic—in other words, a material that softens when heated and can then be easily deformed or molded. It was thus too soft for billiard ball use. Second, it was highly flammable, hardly a desirable characteristic. A widely circulated, perhaps apocryphal, story claimed that celluloid billiard balls detonated when they collided.

Truly Artificial

Since celluloid can be viewed as a derivative of a natural product, it is not a completely synthetic substance. Leo Hendrik Baekeland has the distinction of being the first to produce a completely artificial plastic. Born in Ghent, Belgium, Baekeland emigrated to the United States in 1889 to pursue applied research, a pursuit not encouraged in Europe at the time. One area in which Baekeland hoped to make an inroad was in the development of an artificial shellac. Shellac at the time was a natural and therefore expensive product, and there would be a wide market for any reasonably priced substitute. Baekeland's research scheme, begun in 1905, focused on finding a solvent that could dissolve the resinous products from a certain class of organic chemical reaction.

The particular resins he used had been reported in the mid-1800's by the German chemist Adolf von Baeyer. These resins were produced by the condensation reaction of formaldehyde with a class of chemicals called "phenols." Baeyer found that frequently the major product of such a reaction was a gummy residue that was virtually impossible to remove from glassware. Baekeland focused on finding a material that could dissolve these resinous products. Such a substance would prove to be the shellac substitute he sought.

These efforts proved frustrating, as an adequate solvent for these resins could not be found. After repeated attempts to dissolve these residues, Baekeland shifted the orientation of his work. Abandoning the quest to dissolve the resin, he set about trying to develop a resin that would be impervious to any solvent, reasoning that such a material would have useful applications.

Baekeland's experiments involved the manipulation of phenol-formaldehyde reactions through precise control of the temperature and pressure at which the reactions were performed. Many of these experiments were performed in a 1.5-meter-tall reactor vessel, which he called a "Bakelizer." In 1907, these meticulous experiments paid off when Baekeland opened the reactor to reveal a clear solid that was heat resistant, nonconducting, and machinable. Experimentation proved that the material could be dyed practically any color in the manufacturing process, with no effect on the physical properties of the solid.

Baekeland filed a patent for this new material in 1907. (This patent was filed one day before that filed by James Swinburne, a British

JOHN WESLEY HYATT

John Wesley Hyatt's parents wanted him to be a minister, a step up in status from his father's job as a blacksmith. Born in 1837 in Starkey, New York, Hyatt received the standard primary education and then obediently went to a seminary as a teenager. However, his mind was on making things rather than spirituality; he was especially ingenious with machinery. The seminary held him only a year. He became a printer's apprentice at sixteen and later set up shop in Albany.

His mind ranged beyond printing too. He invented a method to make emery wheels for sharpening cutlery, which brought him his first patent at twenty-four. In an attempt to win the Phelan and Collender Company contest for artificial billiard balls, he developed several moldable compounds from wood pulp. He started the Embossing Company in Albany to make chess and checker pieces from the compounds and put his youngest brother in charge. With another brother he experimented with guncotton until he invented celluloid. In 1872, he and his brothers started the Celluloid Manufacturing Company. They designed new milling machinery for the new substance and turned out billiard balls, bowling balls, golf club heads and other sporting goods but then branched out into domestic items, such as boxes, handles, combs, and even collars. Celluloid became the basic material of photographic film and, later, motion picture film.

Meanwhile, Hyatt continued to invent—machinery for cutting and molding plastic and rolling steel, a water purification system, a method for squeezing juice from sugar cane, an industrial sewing machine, roller bearings for heavy machinery—registering more than 250 patents, which is impressive for a person with no formal scientific or technical training. The Society of Chemical Industry awarded Hyatt its prestigious Perkin Medal in 1914. Hyatt died in 1920.

electrical engineer who had developed a similar material in his quest to produce an insulating material.) Baekeland dubbed his new creation "Bakelite" and announced its existence to the scientific community on February 15, 1909, at the annual meeting of the American Chemical Society. Among its first uses was in the manufacture of ignition parts for the rapidly growing automobile industry.

IMPACT

Bakelite proved to be the first of a class of compounds called "synthetic polymers." Polymers are long chains of molecules chemically linked together. There are many natural polymers, such as cotton. The discovery of synthetic polymers led to vigorous research into the field and attempts to produce other useful artificial materials. These efforts met with a fair amount of success; by 1940, a multitude of new products unlike anything found in nature had been discovered. These included such items as polystyrene and low-density polyethylene. In addition, artificial substitutes for natural polymers, such as rubber, were a goal of polymer chemists. One of the results of this research was the development of neoprene.

Industries also were interested in developing synthetic polymers to produce materials that could be used in place of natural fibers such as cotton. The most dramatic success in this area was achieved by Du Pont chemist Wallace Carothers, who had also developed neoprene. Carothers focused his energies on forming a synthetic fiber similar to silk, resulting in the synthesis of nylon.

Synthetic polymers constitute one branch of a broad area known as "materials science." Novel, useful materials produced synthetically from a variety of natural materials have allowed for tremendous progress in many areas. Examples of these new materials include high-temperature superconductors, composites, ceramics, and plastics. These materials are used to make the structural components of aircraft, artificial limbs and implants, tennis rackets, garbage bags, and many other common objects.

See also Buna rubber; Contact lenses; Laminated glass; Neoprene; Nylon; Orlon; Polyester; Polyethylene; Polystyrene; Pyrex glass; Silicones; Teflon; Velcro.

FURTHER READING

Amato, Ivan. "Chemist: Leo Baekeland." *Time* 153, no. 12 (March 29, 1999).

Clark, Tessa. *Bakelite Style*. Edison, N.J.: Chartwell Books, 1997.

Fenichell, Stephen. *Plastic: The Making of a Synthetic Century*. New York: HarperBusiness, 1997.

Sparke, Penny. *The Plastics Age: From Bakelite to Beanbags and Beyond*. Woodstock, N.Y.: Overlook Press, 1990.

Pocket calculator

THE INVENTION: The first portable and reliable hand-held calculator capable of performing a wide range of mathematical computations.

THE PEOPLE BEHIND THE INVENTION:
Jack St. Clair Kilby (1923-), the inventor of the semiconductor microchip
Jerry D. Merryman (1932-), the first project manager of the team that invented the first portable calculator
James Van Tassel (1929-), an inventor and expert on semiconductor components

AN ANCIENT DREAM

In the earliest accounts of civilizations that developed number systems to perform mathematical calculations, evidence has been found of efforts to fashion a device that would permit people to perform these calculations with reduced effort and increased accuracy. The ancient Babylonians are regarded as the inventors of the first abacus (or counting board, from the Greek *abakos*, meaning "board" or "tablet"). It was originally little more than a row of shallow grooves with pebbles or bone fragments as counters.

The next step in mechanical calculation did not occur until the early seventeenth century. John Napier, a Scottish baron and mathematician, originated the concept of "logarithms" as a mathematical device to make calculating easier. This concept led to the first slide rule, created by the English mathematician William Oughtred of Cambridge. Oughtred's invention consisted of two identical, circular logarithmic scales held together and adjusted by hand. The slide rule made it possible to perform rough but rapid multiplication and division. Oughtred's invention in 1623 was paralleled by the work of a German professor, Wilhelm Schickard, who built a "calculating clock" the same year. Because the record of Schickard's work was lost until 1935, however, the French mathematician Blaise Pascal was generally thought to have built the first mechanical calculator, the "Pascaline," in 1645.

Other versions of mechanical calculators were built in later centuries, but none was rapid or compact enough to be useful beyond specific laboratory or mercantile situations. Meanwhile, the dream of such a machine continued to fascinate scientists and mathematicians.

The development that made a fast, small calculator possible did not occur until the middle of the twentieth century, when Jack St. Clair Kilby of Texas Instruments invented the silicon microchip (or integrated circuit) in 1958. An integrated circuit is a tiny complex of electronic components and their connections that is produced in or on a small slice of semiconductor material such as silicon. Patrick Haggerty, then president of Texas Instruments, wrote in 1964 that "integrated electronics" would "remove limitations" that determined the size of instruments, and he recognized that Kilby's invention of the microchip made possible the creation of a portable, hand-held calculator. He challenged Kilby to put together a team to design a calculator that would be as powerful as the large, electromechanical models in use at the time but small enough to fit into a coat pocket. Working with Jerry D. Merryman and James Van Tassel, Kilby began to work on the project in October, 1965.

AN AMAZING REALITY

At the outset, there were basically five elements that had to be designed. These were the logic designs that enabled the machine to perform the actual calculations, the keyboard or keypad, the power supply, the readout display, and the outer case. Kilby recalls that once a particular size for the unit had been determined (something that could be easily held in the hand), project manager Merryman was able to develop the initial logic designs in three days. Van Tassel contributed his experience with semiconductor components to solve the problems of packaging the integrated circuit. The display required a thermal printer that would work on a low power source. The machine also had to include a microencapsulated ink source so that the paper readouts could be imprinted clearly. Then the paper had to be advanced for the next calculation. Kilby, Merryman, and Van Tassel filed for a patent on their work in 1967.

Although this relatively small, working prototype of the minicalculator made obsolete the transistor-operated design of the much

JERRY D. MERRYMAN

In 1965 Texas Instruments assigned two engineers to join Jack St. Clair Kilby, inventor of the integrated circuit, in an effort to produce a pocket-sized calculator: James H. Van Tassel, a specialist in semiconductor components, and Jerry D. Merryman, a versatile engineer who became the project manager. It took Merryman only seventy-two hours to work out the logic design for the calculator, and the team set about designing, fabricating, and testing its components. After two years, it had a prototype, the first pocket calculator. However, it required a large, strong pocket. It measured 4.25 inches by 6.12 inches by 1.76 inches and weighed 2.8 pounds. Kilby, Van Tassel, and Merry filed for a patent and received it in 1975. In 1989 the team was jointly presented the Holley Medical for the achievement by the American Society of Mechanical Engineers. By then Merryman held sixty other patents, foreign and domestic.

Born in 1932, Merryman grew up in Hearne, Texas, and after high school went to Texas A&M University. He never graduated, but he did become extraordinarily adept at electrical engineering, teaching himself what he needed to know while doing small jobs on his own. He was said to have almost an intuitive sense for circuitry. After he joined Texas Instruments in 1963 he quickly earned a reputation for solving complex problems, one of the reasons he was made part of the hand calculator team. He became a Texas Instruments Fellow in 1975 and helped design semiconductor manufacturing equipment, particularly by adapting high-speed lasers for use in extremely fine optical lithography. He also invented thermal data systems.

Along with Kilby and Van Tassel, Merryman received the George R. Stibitz Computer Pioneer Award in 1997.

larger desk calculators, the cost of setting up new production lines and the need to develop a market made it impractical to begin production immediately. Instead, Texas Instruments and Canon of Tokyo formed a joint venture, which led to the introduction of the Canon Pocketronic Printing Calculator in Japan in April, 1970, and in the United States that fall. Built entirely of Texas Instruments parts, this four-function machine with three metal oxide semicon-

True pocket calculators fit as easily in shirt pockets as pencils and pens. (PhotoDisc)

ductor (MOS) circuits was similar to the prototype designed in 1967. The calculator was priced at $400, weighed 740 grams, and measured 101 millimeters wide by 208 millimeters long by 49 millimeters high. It could perform twelve-digit calculations and worked up to four decimal places.

In September, 1972, Texas Instruments put the Datamath, its first commercial hand-held calculator using a single MOS chip, on the retail market. It weighed 340 grams and measured 75 millimeters wide by 137 millimeters long by 42 millimeters high. The Datamath was priced at $120 and included a full-floating decimal point that could appear anywhere among the numbers on its eight-digit, light-emitting diode (LED) display. It came with a rechargeable battery that could also be connected to a standard alternating current (AC) outlet. The Datamath also had the ability to conserve power while awaiting the next keyboard entry. Finally, the machine had a built-in limited amount of memory storage.

CONSEQUENCES

Prior to 1970, most calculating machines were of such dimensions that professional mathematicians and engineers were either tied to their desks or else carried slide rules whenever they had to be away from their offices. By 1975, Keuffel & Esser, the largest slide rule manufacturer in the world, was producing its last model, and mechanical engineers found that problems that had previously taken a week could now be solved in an hour using the new machines.

That year, the Smithsonian Institution accepted the world's first miniature electronic calculator for its permanent collection, noting that it was the forerunner of more than one hundred million pocket calculators then in use. By the 1990's, more than fifty million portable units were being sold each year in the United States. In general, the electronic pocket calculator revolutionized the way in which people related to the world of numbers.

Moreover, the portability of the hand-held calculator made it ideal for use in remote locations, such as those a petroleum engineer might have to explore. Its rapidity and reliability made it an indispensable instrument for construction engineers, architects, and real estate agents, who could figure the volume of a room and other building dimensions almost instantly and then produce cost estimates almost on the spot.

See also Cell phone; Differential analyzer; Mark I calculator; Personal computer; Transistor radio; Walkman cassette player.

FURTHER READING

Ball, Guy. *Collector's Guide to Pocket Calculators.* Tustin, Calif.: Wilson/Barnett Publishing, 1996.
Clayton, Mark. "Calculators in Class: Freedom from Scratch Paper or 'Crutch'?" *Christian Science Monitor* (May 23, 2000).
Lederer, Victor. "Calculators: The Applications Are Unlimited. *Administrative Management* 38 (July, 1977).
Lee, Jennifer. "Throw Teachers a New Curve." *New York Times* (September 2, 1999).
"The Semiconductor Becomes a New Marketing Force." *Business Week* (August 24, 1974).

Polio vaccine (Sabin)

THE INVENTION: Albert Bruce Sabin's vaccine was the first to stimulate long-lasting immunity against polio without the risk of causing paralytic disease.

THE PEOPLE BEHIND THE INVENTION:
Albert Bruce Sabin (1906-1993), a Russian-born American virologist
Jonas Edward Salk (1914-1995), an American physician, immunologist, and virologist
Renato Dulbecco (1914-   ), an Italian-born American virologist who shared the 1975 Nobel Prize in Physiology or Medicine

THE SEARCH FOR A LIVING VACCINE

Almost a century ago, the first major poliomyelitis (polio) epidemic was recorded. Thereafter, epidemics of increasing frequency and severity struck the industrialized world. By the 1950's, as many as sixteen thousand individuals, most of them children, were being paralyzed by the disease each year.

Poliovirus enters the body through ingestion by the mouth. It replicates in the throat and the intestines and establishes an infection that normally is harmless. From there, the virus can enter the bloodstream. In some individuals it makes its way to the nervous system, where it attacks and destroys nerve cells crucial for muscle movement. The presence of antibodies in the bloodstream will prevent the virus from reaching the nervous system and causing paralysis. Thus, the goal of vaccination is to administer poliovirus that has been altered so that it cannot cause disease but nevertheless will stimulate the production of antibodies to fight the disease.

Albert Bruce Sabin received his medical degree from New York University College of Medicine in 1931. Polio was epidemic in 1931, and for Sabin polio research became a lifelong interest. In 1936, while working at the Rockefeller Institute, Sabin and Peter Olinsky successfully grew poliovirus using tissues cultured in vitro. Tissue culture proved to be an excellent source of virus. Jonas Edward Salk soon developed an inactive polio vaccine consisting of virus grown from tissue culture that had been inactivated (killed) by chemical treatment. This vaccine became available for general use in 1955, almost fifty years after poliovirus had first been identified.

Sabin, however, was not convinced that an inactivated virus vaccine was adequate. He believed that it would provide only temporary protection and that individuals would have to be vaccinated repeatedly in order to maintain protective levels of antibodies. Knowing that natural infection with poliovirus induced lifelong immunity, Sabin believed that a vaccine consisting of a living virus was necessary to produce long-lasting immunity. Also, unlike the inactive vaccine, which is injected, a living virus (weakened so that it would not cause disease) could be taken orally and would invade the body and replicate of its own accord.

Sabin was not alone in his beliefs. Hilary Koprowski and Harold Cox also favored a living virus vaccine and had, in fact, begun searching for weakened strains of poliovirus as early as 1946 by repeatedly growing the virus in rodents. When Sabin began his search for weakened virus strains in 1953, a fiercely competitive contest ensued to achieve an acceptable live virus vaccine.

Rare, Mutant Polioviruses

Sabin's approach was based on the principle that, as viruses acquire the ability to replicate in a foreign species or tissue (for example, in mice), they become less able to replicate in humans and thus less able to cause disease. Sabin used tissue culture techniques to isolate those polioviruses that grew most rapidly in monkey kidney cells. He then employed a technique developed by Renato Dulbecco that allowed him to recover individual virus particles. The recovered viruses were injected directly into the brains or spinal cords of monkeys in order to identify those viruses that did not damage the nervous system. These meticulously performed experiments, which involved approximately nine thousand monkeys and more than one hundred chimpanzees, finally enabled Sabin to isolate rare mutant polioviruses that would replicate in the intestinal tract but not in the nervous systems of chimpanzees or, it was hoped, of humans. In addition, the weakened virus strains were shown to stimulate antibodies when they were fed to chimpanzees; this was a critical attribute for a vaccine strain.

By 1957, Sabin had identified three strains of attenuated viruses that were ready for small experimental trials in humans. A small group of volunteers, including Sabin's own wife and children, were fed the vaccine with promising results. Sabin then gave his vaccine to virologists in the Soviet Union, Eastern Europe, Mexico, and Holland for further testing. Combined with smaller studies in the United States, these trials established the effectiveness and safety of his oral vaccine.

During this period, the strains developed by Cox and by Koprowski were being tested also in millions of persons in field trials around the world. In 1958, two laboratories independently compared the vaccine strains and concluded that the Sabin strains were superior. In 1962, after four years of deliberation by the U.S. Public Health Service, all three of Sabin's vaccine strains were licensed for general use.

ALBERT SABIN

Born in Bialystok, Poland, in 1906, Albert Bruce Sabin emigrated with his family to the United States in 1921. Like Jonas Salk—the other great inventor of a polio vaccine—Sabin earned his medical degree at New York University (1931), where he began his research on polio.

While in the U.S. Army Medical Corps during World War II, he helped produce vaccines for dengue fever and Japanese encephalitis. After the war he returned to his professorship at the University of Cincinnati College of Medicine and Children's Hospital Research Foundation. The polio vaccine he developed there saved millions of children worldwide from paralytic polio. Many of these lives were doubtless saved because of his refusal to patent the vaccine, thereby making it simpler to produce and distribute and less expensive to administer

Sabin's work brought him more than forty honorary degrees from American and foreign universities and medals from the governments of the United States and Soviet Union. He was president of the Weizmann Institute of Science after 1970 and later became a professor of biomedicine at the Medical University of South Carolina. He died in 1993.

CONSEQUENCES

The development of polio vaccines ranks as one of the triumphs of modern medicine. In the early 1950's, paralytic polio struck 13,500 out of every 100 million Americans. The use of the Salk vaccine greatly reduced the incidence of polio, but outbreaks of paralytic disease continued to occur: Fifty-seven hundred cases were reported in 1959 and twenty-five hundred cases in 1960. In 1962, the oral Sabin vaccine became the vaccine of choice in the United States. Since its widespread use, the number of paralytic cases in the United States has dropped precipitously, eventually averaging fewer than ten per year. Worldwide, the oral vaccine prevented an estimated 5 million cases of paralytic poliomyelitis between 1970 and 1990.

The oral vaccine is not without problems. Occasionally, the living virus mutates to a disease-causing (virulent) form as it multiplies in the vaccinated person. When this occurs, the person may develop paralytic poliomyelitis. The inactive vaccine, in contrast, cannot mutate to a virulent form. Ironically, nearly every incidence of polio in the United States is caused by the vaccine itself.

In the developing countries of the world, the issue of vaccination is more pressing. Millions receive neither form of polio vaccine; as a result, at least 250,000 individuals are paralyzed or die each year. The World Health Organization and other health providers continue to work toward the very practical goal of completely eradicating this disease.

See also Antibacterial drugs; Birth control pill; Iron lung; Penicillin; Polio vaccine (Salk); Reserpine; Salvarsan; Tuberculosis vaccine; Typhus vaccine; Yellow fever vaccine.

FURTHER READING

DeJauregui, Ruth. *100 Medical Milestones That Shaped World History.* San Mateo, Calif.: Bluewood Books, 1998.
Grady, Denise. "As Polio Fades, Dr. Salk's Vaccine Re-emerges." *New York Times* (December 14, 1999).
Plotkin, Stanley A., and Edward A. Mortimer. *Vaccines.* 2d ed. Philadelphia: W. B. Saunders, 1994.
Seavey, Nina Gilden, Jane S. Smith, and Paul Wagner. *A Paralyzing Fear: The Triumph over Polio in America.* New York: TV Books, 1998.

Polio vaccine (Salk)

THE INVENTION: Jonas Salk's vaccine was the first that prevented polio, resulting in the virtual eradication of crippling polio epidemics.

THE PEOPLE BEHIND THE INVENTION:
Jonas Edward Salk (1914-1995), an American physician, immunologist, and virologist
Thomas Francis, Jr. (1900-1969), an American microbiologist

Cause for Celebration

Poliomyelitis (polio) is an infectious disease that can adversely affect the central nervous system, causing paralysis and great muscle wasting due to the destruction of motor neurons (nerve cells) in the spinal cord. Epidemiologists believe that polio has existed since ancient times, and evidence of its presence in Egypt, circa 1400 B.C.E., has been presented. Fortunately, the Salk vaccine and the later vaccine developed by the American virologist Albert Bruce Sabin can prevent the disease. Consequently, except in underdeveloped nations, polio is rare. Moreover, although once a person develops polio, there is still no cure for it, a large number of polio cases end without paralysis or any observable effect.

Polio is often called "infantile paralysis." This results from the fact that it is seen most often in children. It is caused by a virus and begins with body aches, a stiff neck, and other symptoms that are very similar to those of a severe case of influenza. In some cases, within two weeks after its onset, the course of polio begins to lead to muscle wasting and paralysis.

On April 12, 1955, the world was thrilled with the announcement that Jonas Edward Salk's poliomyelitis vaccine could prevent the disease. It was reported that schools were closed in celebration of this event. Salk, the son of a New York City garment worker, has since become one of the most well-known and publicly venerated medical scientists in the world.

Vaccination is a method of disease prevention by immunization, whereby a small amount of virus is injected into the body to prevent

Jonas Salk

The son of a garment industry worker, Jonas Edward Salk was born in New York City in 1914. He worked his way through school, graduating from New York University School of Medicine in 1938. Afterward he joined microbiologist Thomas Francis, Jr., in developing a vaccine for influenza.

In 1942, Salk began a research fellowship at the University of Michigan and subsequently joined the epidemiology faculty. He moved to the University of Pittsburgh in 1947, directing its Viral Research Lab, and while there developed his vaccine for poliomyelitis. The discovery catapulted Salk into worldwide fame, but he was a controversial figure among scientists.

Although Salk received the Presidential Medal of Freedom, a Congressional gold medal, and the Nehru Award for International Understanding, he was turned down for membership in the National Academy of Sciences. In 1963 he opened the Salk Institute for Biological Sciences in La Jolla, California. Well aware of his reputation among medical researchers, he once joked, "I couldn't possibly have become a member of this institute if I hadn't founded it myself." He died in 1995.

a viral disease. The process depends on the production of antibodies (body proteins that are specifically coded to prevent the disease spread by the virus) in response to the vaccination. Vaccines are made of weakened or killed virus preparations.

Electrifying Results

The Salk vaccine was produced in two steps. First, polio viruses were grown in monkey kidney tissue cultures. These polio viruses were then killed by treatment with the right amount of formaldehyde to produce an effective vaccine. The killed-virus polio vaccine was found to be safe and to cause the production of antibodies against the disease, a sign that it should prevent polio.

In early 1952, Salk tested a prototype vaccine against Type I polio virus on children who were afflicted with the disease and were thus deemed safe from reinfection. This test showed that the vaccination greatly elevated the concentration of polio antibodies in these children. On July 2, 1952, encouraged by these results, Salk vaccinated forty-three children who had never had polio with vaccines against each of the three virus types (Type I, Type II, and Type III). All inoculated children produced high levels of polio antibodies, and none of them developed the disease. Consequently, the vaccine appeared to be both safe in humans and likely to become an effective public health tool.

In 1953, Salk reported these findings in the *Journal of the American Medical Association*. In April, 1954, nationwide testing of the Salk vaccine began, via the mass vaccination of American schoolchildren. The results of the trial were electrifying. The vaccine was safe, and it greatly reduced the incidence of the disease. In fact, it was estimated that Salk's vaccine gave schoolchildren 60 to 90 percent protection against polio.

Salk was instantly praised. Then, however, several cases of polio occurred as a consequence of the vaccine. Its use was immediately suspended by the U.S. surgeon general, pending a complete examination. Soon, it was evident that all the cases of vaccine-derived polio were attributable to faulty batches of vaccine made by one pharmaceutical company. Salk and his associates were in no way responsible for the problem. Appropriate steps were taken to ensure that such an error would not be repeated, and the Salk vaccine was again released for use by the public.

CONSEQUENCES

The first reports on the polio epidemic in the United States had occurred on June 27, 1916, when one hundred residents of Brooklyn, New York, were afflicted. Soon, the disease had spread. By August, twenty-seven thousand people had developed polio. Nearly seven thousand afflicted people died, and many survivors of the epidemic were permanently paralyzed to varying extents. In New York City alone, nine thousand people developed polio and two thousand died. Chaos reigned as large numbers of terrified people attempted to leave and were turned back by police. Smaller polio epidemics occurred throughout the nation in the years that followed (for example, the Catawba County, North Carolina, epidemic of 1944). A particularly horrible aspect of polio was the fact that more than 70 percent of polio victims were small children. Adults caught it too; the most famous of these adult polio victims was U.S. President Franklin D. Roosevelt. There was no cure for the disease. The best available treatment was physical therapy.

As of August, 1955, more than four million polio vaccines had been given. The Salk vaccine appeared to work very well. There were only half as many reported cases of polio in 1956 as there had been in 1955. It appeared that polio was being conquered. By 1957, the number of cases reported nationwide had fallen below six thousand. Thus, in two years, its incidence had dropped by about 80 percent.

This was very exciting, and soon other countries clamored for the vaccine. By 1959, ninety other countries had been supplied with the Salk vaccine. Worldwide, the disease was being eradicated. The introduction of an oral polio vaccine by Albert Bruce Sabin supported this progress.

Salk received many honors, including honorary degrees from American and foreign universities, the Lasker Award, a Congressional Medal for Distinguished Civilian Service, and membership in the French Legion of Honor, yet he received neither the Nobel Prize nor membership in the American National Academy of Sciences. It is believed by many that this neglect was a result of the personal antagonism of some of the members of the scientific community who strongly disagreed with his theories of viral inactivation.

See also Antibacterial drugs; Birth control pill; Iron lung; Penicillin; Polio vaccine (Sabin); Reserpine; Salvarsan; Tuberculosis vaccine; Typhus vaccine; Yellow fever vaccine.

FURTHER READING

DeJauregui, Ruth. *100 Medical Milestones That Shaped World History.* San Mateo, Calif.: Bluewood Books, 1998.

Plotkin, Stanley A., and Edward A. Mortimer. *Vaccines.* 2d ed. Philadelphia: W. B. Saunders, 1994.

Seavey, Nina Gilden, Jane S. Smith, and Paul Wagner. *A Paralyzing Fear: The Triumph over Polio in America.* New York: TV Books, 1998.

Smith, Jane S. *Patenting the Sun: Polio and the Salk Vaccine.* New York: Anchor/Doubleday, 1991.

Polyester

THE INVENTION: A synthetic fibrous polymer used especially in fabrics.

THE PEOPLE BEHIND THE INVENTION:
Wallace H. Carothers (1896-1937), an American polymer chemist
Hilaire de Chardonnet (1839-1924), a French polymer chemist
John R. Whinfield (1901-1966), a British polymer chemist

A STORY ABOUT THREADS

Human beings have worn clothing since prehistoric times. At first, clothing consisted of animal skins sewed together. Later, people learned to spin threads from the fibers in plant or animal materials and to weave fabrics from the threads (for example, wool, silk, and cotton). By the end of the nineteenth century, efforts were begun to produce synthetic fibers for use in fabrics. These efforts were motivated by two concerns. First, it seemed likely that natural materials would become too scarce to meet the needs of a rapidly increasing world population. Second, a series of natural disasters—affecting the silk industry in particular—had demonstrated the problems of relying solely on natural fibers for fabrics.

The first efforts to develop synthetic fabric focused on artificial silk, because of the high cost of silk, its beauty, and the fact that silk production had been interrupted by natural disasters more often than the production of any other material. The first synthetic silk was rayon, which was originally patented by a French count, Hilaire de Chardonnet, and was later much improved by other polymer chemists. Rayon is a semisynthetic material that is made from wood pulp or cotton.

Because there was a need for synthetic fabrics whose manufacture did not require natural materials, other avenues were explored. One of these avenues led to the development of totally synthetic polyester fibers. In the United States, the best-known of these is Dacron, which is manufactured by E. I. Du Pont de Nemours. Easily made into threads, Dacron is widely used in clothing. It is also used to make audiotapes and videotapes and in automobile and boat bodies.

From Polymers to Polyester

Dacron belongs to a group of chemicals known as "synthetic polymers." All polymers are made of giant molecules, each of which is composed of a large number of simpler molecules ("monomers") that have been linked, chemically, to form long strings. Efforts by industrial chemists to prepare synthetic polymers developed in the twentieth century after it was discovered that many natural building materials and fabrics (such as rubber, wood, wool, silk, and cotton) were polymers, and as the ways in which monomers could be joined to make polymers became better understood. One group of chemists who studied polymers sought to make inexpensive synthetic fibers to replace expensive silk and wool. Their efforts led to the development of well-known synthetic fibers such as nylon and Dacron.

Wallace H. Carothers of Du Pont pioneered the development of polyamide polymers, collectively called "nylon," and was the first researcher to attempt to make polyester. It was British polymer chemists John R. Whinfield and J. T. Dickson of Calico Printers Association (CPA) Limited, however, who in 1941 perfected and patented polyester that could be used to manufacture clothing. The first polyester fiber products were produced in 1950 in Great Britain by London's British Imperial Chemical Industries, which had secured the British patent rights from CPA. This polyester, which was made of two monomers, terphthalic acid and ethylene glycol, was called Terylene. In 1951, Du Pont, which had acquired Terylene patent rights for the Western Hemisphere, began to market its own version of this polyester, which was called Dacron. Soon, other companies around the world were selling polyester materials of similar composition.

Dacron and other polyesters are used in many items in the United States. Made into fibers and woven, Dacron becomes cloth. When pressed into thin sheets, it becomes Mylar, which is used in videotapes and audiotapes. Dacron polyester, mixed with other materials, is also used in many industrial items, including motor vehicle and boat bodies. Terylene and similar polyester preparations serve the same purposes in other countries.

The production of polyester begins when monomers are mixed in huge reactor tanks and heated, which causes them to form giant polymer chains composed of thousands of alternating monomer units. If T represents terphthalic acid and E represents ethylene glycol, a small part of a necklace-like polymer can be shown in the following way: (TETETETETE). Once each batch of polyester polymer has the desired composition, it is processed for storage until it is needed. In this procedure, the material, in liquid form in the high-temperature reactor, is passed through a device that cools it and forms solid strips. These strips are then diced, dried, and stored.

When polyester fiber is desired, the diced polyester is melted and then forced through tiny holes in a "spinneret" device; this process is called "extruding." The extruded polyester cools again, while passing through the spinneret holes, and becomes fine fibers called "filaments." The filaments are immediately wound into threads that are collected in rolls. These rolls of thread are then dyed and used to weave various fabrics. If polyester sheets or other forms of polyester are desired, the melted, diced polyester is processed in other ways. Polyester preparations are often mixed with cotton, glass fibers, or other synthetic polymers to produce various products.

Impact

The development of polyester was a natural consequence of the search for synthetic fibers that developed from work on rayon. Once polyester had been developed, its great utility led to its widespread use in industry. In addition, the profitability of the material spurred efforts to produce better synthetic fibers for specific uses. One example is that of stretchy polymers such as Helance, which is a form of nylon. In addition, new chemical types of polymer fibers were developed, including the polyurethane materials known collectively as "spandex" (for example, Lycra and Vyrenet).

The wide variety of uses for polyester is amazing. Mixed with cotton, it becomes wash-and-wear clothing; mixed with glass, it is used to make boat and motor vehicle bodies; combined with other materials, it is used to make roofing materials, conveyor belts, hoses, and tire cords. In Europe, polyester has become the main packaging material for consumer goods, and the United States does not lag far behind in this area.

The future is sure to hold more uses for polyester and the invention of new polymers. These spinoffs of polyester will be essential in the development of high technology.

See also Buna rubber; Neoprene; Nylon; Orlon; Plastic; Polyethylene; Polystyrene.

FURTHER READING

Furukawa, Yasu. *Inventing Polymer Science: Staudinger, Carothers, and the Emergence of Macromolecular Chemistry*. Philadelphia: University of Pennsylvania Press, 1998.

Handley, Susannah. *Nylon: The Story of a Fashion Revolution, A Celebration of Design from Art Silk to Nylon and Thinking Fibres*. Baltimore: Johns Hopkins University Press, 1999.

Hermes, Matthew E. *Enough for One Lifetime: Wallace Carothers, Inventor of Nylon*. Washington, D.C.: American Chemical Society and the Chemical Heritage Foundation, 1996.

Smith, Matthew Boyd. *Polyester: The Indestructible Fashion*. Atglen, Pa.: Schiffer, 1998.

Polyethylene

THE INVENTION: An artificial polymer with strong insulating properties and many other applications.

THE PEOPLE BEHIND THE INVENTION:
Karl Ziegler (1898-1973), a German chemist
Giulio Natta (1903-1979), an Italian chemist
August Wilhelm von Hofmann (1818-1892), a German chemist

The Development of Synthetic Polymers

In 1841, August Hofmann completed his Ph.D. with Justus von Liebig, a German chemist and founding father of organic chemistry. One of Hofmann's students, William Henry Perkin, discovered that coal tars could be used to produce brilliant dyes. The German chemical industry, under Hofmann's leadership, soon took the lead in this field, primarily because the discipline of organic chemistry was much more developed in Germany than elsewhere.

The realities of the early twentieth century found the chemical industry struggling to produce synthetic substitutes for natural materials that were in short supply, particularly rubber. Rubber is a natural polymer, a material composed of a long chain of small molecules that are linked chemically. An early synthetic rubber, neoprene, was one of many synthetic polymers (some others were Bakelite, polyvinyl chloride, and polystyrene) developed in the 1920's and 1930's. Another polymer, polyethylene, was developed in 1936 by Imperial Chemical Industries. Polyethylene was a tough, waxy material that was produced at high temperature and at pressures of about one thousand atmospheres. Its method of production made the material expensive, but it was useful as an insulating material.

World War II and the material shortages associated with it brought synthetic materials into the limelight. Many new uses for polymers were discovered, and after the war they were in demand for the production of a variety of consumer goods, although polyethylene was still too expensive to be used widely.

Organometallics Provide the Key

Karl Ziegler, an organic chemist with an excellent international reputation, spent most of his career in Germany. With his international reputation and lack of political connections, he was a natural candidate to take charge of the Kaiser Wilhelm Institute for Coal Research (later renamed the Max Planck Institute) in 1943. Wise planners saw him as a director who would be favored by the conquering Allies. His appointment was a shrewd one, since he was allowed to retain his position after World War II ended. Ziegler thus played a key role in the resurgence of German chemical research after the war.

Before accepting the position at the Kaiser Wilhelm Institute, Ziegler made it clear that he would take the job only if he could pursue his own research interests in addition to conducting coal research. The location of the institute in the Ruhr Valley meant that abundant supplies of ethylene were available from the local coal industry, so it is not surprising that Ziegler began experimenting with that material.

Although Ziegler's placement as head of the institute was an important factor in his scientific breakthrough, his previous research was no less significant. Ziegler devoted much time to the field of organometallic compounds, which are compounds that contain a metal atom that is bonded to one or more carbon atoms. Ziegler was interested in organoaluminum compounds, which are compounds that contain aluminum-carbon bonds.

Ziegler was also interested in polymerization reactions, which involve the linking of thousands of smaller molecules into the single long chain of a polymer. Several synthetic polymers were known, but chemists could exert little control on the actual process. It was impossible to regulate the length of the polymer chain, and the extent of branching in the chain was unpredictable. It was as a result of studying the effect of organoaluminum compounds on these chain formation reactions that the key discovery was made.

Ziegler and his coworkers already knew that ethylene would react with organoaluminum compounds to produce hydrocarbons, which are compounds that contain only carbon and hydrogen and that have varying chain lengths. Regulating the product chain length continued to be a problem.

At this point, fate intervened in the form of a trace of nickel left in a reactor from a previous experiment. The nickel caused the chain lengthening to stop after two ethylene molecules had been linked. Ziegler and his colleagues then tried to determine whether metals other than nickel caused a similar effect with a longer polymeric chain. Several metals were tested, and the most important finding was that a trace of titanium chloride in the reactor caused the deposition of large quantities of high-density polyethylene at low pressures.

Ziegler licensed the procedure, and within a year, Giulio Natta had modified the catalysts to give high yields of polymers with highly ordered side chains branching from the main chain. This opened the door for the easy production of synthetic rubber. For their discovery of Ziegler-Natta catalysts, Ziegler and Natta shared the 1963 Nobel Prize in Chemistry.

CONSEQUENCES

Ziegler's process produced polyethylene that was much more rigid than the material produced at high pressure. His product also had a higher density and a higher softening temperature. Industrial exploitation of the process was unusually rapid, and within ten years more than twenty plants utilizing the process had been built throughout Europe, producing more than 120,000 metric tons of polyethylene. This rapid exploitation was one reason Ziegler and Natta were awarded the Nobel Prize after such a relatively short time.

By the late 1980's, total production stood at roughly 18 billion pounds worldwide. Other polymeric materials, including polypropylene, can be produced by similar means. The ready availability and low cost of these versatile materials have radically transformed the packaging industry. Polyethylene bottles are far lighter than their glass counterparts; in addition, gases and liquids do not diffuse into polyethylene very easily, and it does not break easily. As a result, more and more products are bottled in containers made of polyethylene or other polymers. Other novel materials possessing properties unparalleled by any naturally occurring material (Kevlar, for example, which is used to make bullet-resistant vests) have also been an outgrowth of the availability of low-cost polymeric materials.

See also Buna rubber; Neoprene; Nylon; Orlon; Plastic; Polyester; Polystyrene.

FURTHER READING

Boor, John. *Ziegler-Natta Catalysts and Polymerizations.* New York: Academic Press, 1979.

Clarke, Alison J. *Tupperware: The Promise of Plastic in 1950s America.* Washington, D.C.: Smithsonian Institution Press, 1999.

Natta, Giulio. "From Stereospecific Polymerization to Asymmetric Autocatalytic Synthesis of Macromolecules." In *Chemistry, 1963-1970.* River Edge, N.J.: World Scientific, 1999.

Ziegler, Karl. "Consequences and Development of an Invention." In *Chemistry, 1963-1970.* River Edge, N.J.: World Scientific, 1999.

Polystyrene

THE INVENTION: A clear, moldable polymer with many industrial uses whose overuse has also threatened the environment.

THE PEOPLE BEHIND THE INVENTION:
Edward Simon, an American chemist
Charles Gerhardt (1816-1856), a French chemist
Marcellin Pierre Berthelot (1827-1907), a French chemist

POLYSTYRENE IS CHARACTERIZED

In the late eighteenth century, a scientist by the name of Casper Neuman described the isolation of a chemical called "storax" from a balsam tree that grew in Asia Minor. This isolation led to the first report on the physical properties of the substance later known as "styrene." The work of Neuman was confirmed and expanded upon years later, first in 1839 by Edward Simon, who evaluated the temperature dependence of styrene, and later by Charles Gerhardt, who proposed its molecular formula. The work of these two men sparked an interest in styrene and its derivatives.

Polystyrene belongs to a special class of molecules known as *polymers*. A polymer (the name means "many parts") is a giant molecule formed by combining small molecular units, called "monomers." This combination results in a macromolecule whose physical properties—especially its strength and flexibility—are significantly different from those of its monomer components. Such polymers are often simply called "plastics."

Polystyrene has become an important material in modern society because it exhibits a variety of physical characteristics that can be manipulated for the production of consumer products. Polystyrene is a "thermoplastic," which means that it can be softened by heat and then reformed, after which it can be cooled to form a durable and resilient product.

At 94 degrees Celsius, polystyrene softens; at room temperature, however, it rings like a metal when struck. Because of the glasslike nature and high refractive index of polystyrene, products made from it are known for their shine and attractive texture. In addition, the material is characterized by a high level of water resistance and by electrical insulating qualities. It is also flammable, can by dissolved or softened by many solvents, and is sensitive to light. These qualities make polystyrene a valuable material in the manufacture of consumer products.

PLASTICS ON THE MARKET

In 1866, Marcellin Pierre Berthelot prepared styrene from ethylene and benzene mixtures in a heated reaction flask. This was the first synthetic preparation of polystyrene. In 1925, the Naugatuck Chemical Company began to operate the first commercial styrene/polystyrene manufacturing plant. In the 1930's, the Dow Chemical Company became involved in the manufacturing and marketing of styrene/polystyrene products. Dow's Styron 666 was first marketed as a general-purpose polystyrene in 1938. This material was the first plastic product to demonstrate polystyrene's excellent mechanical properties and ease of fabrication.

The advent of World War II increased the need for plastics. When the Allies' supply of natural rubber was interrupted, chemists sought to develop synthetic substitutes. The use of additives with polymer species was found to alter some of the physical properties of those species. Adding substances called "elastomers" during the polymerization process was shown to give a rubberlike quality to a normally brittle species. An example of this is Dow's Styron 475, which was marketed in 1948 as the first "impact" polystyrene. It is called an impact polystyrene because it also contains butadiene, which increases the product's resistance to breakage. The continued characterization of polystyrene products has led to the development of a worldwide industry that fills a wide range of consumer needs.

Following World War II, the plastics industry revolutionized many aspects of modern society. Polystyrene is only one of the many plastics involved in this process, but it has found its way into a multitude of consumer products. Disposable kitchen utensils, trays and packages, cups, videocassettes, insulating foams, egg cartons, food wrappings, paints, and appliance parts are only a few of the typical applications of polystyrenes. In fact, the production of polystyrene has grown to exceed 5 billion pounds per year.

The tremendous growth of this industry in the postwar era has been fueled by a variety of factors. Having studied the physical and chemical properties of polystyrene, chemists and engineers were able to envision particular uses and to tailor the manufacture of the product to fit those uses precisely. Because of its low cost of production, superior performance, and light weight, polystyrene has become the material of choice for the packaging industry. The automobile industry also enjoys its benefits. Polystyrene's lower density compared to those of glass and steel makes it appropriate for use in automobiles, since its light weight means that using it can reduce the weight of automobiles, thereby increasing gas efficiency.

IMPACT

There is no doubt that the marketing of polystyrene has greatly affected almost every aspect of modern society. From computer keyboards to food packaging, the use of polystyrene has had a powerful impact on both the quality and the prices of products. Its use is not, however, without drawbacks; it has also presented humankind with a dilemma. The wholesale use of polystyrene has created an environmental problem that represents a danger to wildlife, adds to roadside pollution, and greatly contributes to the volume of solid waste in landfills.

Polystyrene has become a household commodity because it lasts. The reciprocal effect of this fact is that it may last forever. Unlike natural products, which decompose upon burial, polystyrene is very difficult to convert into degradable forms. The newest challenge facing engineers and chemists is to provide for the safe and efficient disposal of plastic products. Thermoplastics such as polystyrene can be melted down and remolded into new products, which makes recycling and reuse of polystyrene a viable option, but this option requires the cooperation of the same consumers who have benefited from the production of polystyrene products.

See also Food freezing; Nylon; Orlon; Plastic; Polyester; Polyethylene; Pyrex glass; Teflon; Tupperware.

FURTHER READING

Fenichell, Stephen. *Plastic: The Making of a Synthetic Century.* New York: HarperBusiness, 1997.

Mossman, S. T. I. *Early Plastics: Perspectives, 1850-1950.* London: Science Museum, 1997.

Wünsch, J. R. *Polystyrene: Synthesis, Production and Applications.* Shropshire, England: Rapra Technology, 2000.

PROPELLER-COORDINATED MACHINE GUN

THE INVENTION: A mechanism that synchronized machine gun fire with propeller movement to prevent World War I fighter plane pilots from shooting off their own propellers during combat.

THE PEOPLE BEHIND THE INVENTION:

Anthony Herman Gerard Fokker (1890-1939), a Dutch-born American entrepreneur, pilot, aircraft designer, and manufacturer

Roland Garros (1888-1918), a French aviator

Max Immelmann (1890-1916), a German aviator

Raymond Saulnier (1881-1964), a French aircraft designer and manufacturer

FRENCH INNOVATION

The first true aerial combat of World War I took place in 1915. Before then, weapons attached to airplanes were inadequate for any real combat work. Hand-held weapons and clumsily mounted machine guns were used by pilots and crew members in attempts to convert their observation planes into fighters. On April 1, 1915, this situation changed. From an airfield near Dunkerque, France, a French airman, Lieutenant Roland Garros, took off in an airplane equipped with a device that would make his plane the most feared weapon in the air at that time.

During a visit to Paris, Garros met with Raymond Saulnier, a French aircraft designer. In April of 1914, Saulnier had applied for a patent on a device that mechanically linked the trigger of a machine gun to a cam on the engine shaft. Theoretically, such an assembly would allow the gun to fire between the moving blades of the propeller. Unfortunately, the available machine gun Saulnier used to test his device was a Hotchkiss gun, which tended to fire at an uneven rate. On Garros's arrival, Saulnier showed him a new invention: a steel deflector shield that, when fastened to the propeller, would deflect the small percentage of mistimed bullets that would otherwise destroy the blade.

The first test-firing was a disaster, shooting the propeller off and destroying the fuselage. Modifications were made to the deflector braces, streamlining its form into a wedge shape with gutter-channels for deflected bullets. The invention was attached to a Morane-Saulnier monoplane, and on April 1, Garros took off alone toward the German lines. Success was immediate. Garros shot down a German observation plane that morning. During the next two weeks, Garros shot down five more German aircraft.

GERMAN LUCK

The German high command, frantic over the effectiveness of the French "secret weapon," sent out spies to try to steal the secret and also ordered engineers to develop a similar weapon. Luck was with them. On April 18, 1915, despite warnings by his superiors not to fly over enemy-held territory, Garros was forced to crash-land behind German lines with engine trouble. Before he could destroy his aircraft, Garros and his plane were captured by German troops. The secret weapon was revealed.

The Germans were ecstatic about the opportunity to examine the new French weapon. Unlike the French, the Germans had the first air-cooled machine gun, the Parabellum, which shot continuous bands of one hundred bullets and was reliable enough to be adapted to a timing mechanism.

In May of 1915, Anthony Herman Gerard Fokker was shown Garros's captured plane and was ordered to copy the idea. Instead, Fokker and his assistant designed a new firing system. It is unclear whether Fokker and his team were already working on a synchronizer or to what extent they knew of Saulnier's previous work in France. Within several days, however, they had constructed a working prototype and attached it to a Fokker *Eindecker 1* airplane. The design consisted of a simple linkage of cams and push-rods connected to the oil-pump drive of an Oberursel engine and the trigger of a Parabellum machine gun. The firing of the gun had to be timed precisely to fire its six hundred rounds per minute between the twelve-hundred-revolutions-per-minute propeller blades.

Fokker took his invention to Doberitz air base, and after a series

ANTHONY HERMAN GERARD FOKKER

Anthony Fokker was born on the island of Java in the Dutch East Indies (now Indonesia) in 1890. He returned to his parent's home country, the Netherlands, to attend school and then studied aeronautics in Germany. He built his first plane in 1910 and established Fokker Aeroplanbau near Berlin in 1912.

His monoplanes were highly esteemed when World War I erupted in 1914, and he offered his designs to both the German and the French governments. The Germans hired him. By the end of the war his fighters, especially the Dr I triplane and D VII biplane, were practically synonymous with German air warfare because they had been the scourge of Allied pilots.

In 1922 Fokker moved to the United States and opened the Atlantic Aircraft Corporation in New Jersey. He had lost enthusiasm for military aircraft and turned his skills toward producing advanced designs for civilian use. The planes his company turned out established one first after another. His T-2 monoplane became the first to fly nonstop from coast to coast, New York to San Diego. His ten-seat airliner, the F VII/3m, carried Lieutenant Commander Richard Byrd over the North Pole in 1926 and Charles Kingsford-Smith across the Pacific Ocean in 1928.

By the time Fokker died in New York in 1939, he had become a visionary. He foresaw passenger planes as the means to knit together the far-flung nations of the world into a network of rapid travel and communications.

of exhausting trials before the German high command, both on the ground and in the air, he was allowed to take two prototypes of the machine-gun-mounted airplanes to Douai in German-held France. At Douai, two German pilots crowded into the cockpit with Fokker and were given demonstrations of the plane's capabilities. The airmen were Oswald Boelcke, a test pilot and veteran of forty reconnaissance missions, and Max Immelmann, a young, skillful aviator who was assigned to the front.

When the first combat-ready versions of Fokker's *Eindecker 1* were delivered to the front lines, one was assigned to Boelcke, the other to Immelmann. On August 1, 1915, with their aerodrome under attack from nine English bombers, Boelcke and Immelmann manned their aircraft and attacked. Boelcke's gun jammed, and he was forced to cut off his attack and return to the aerodrome. Immelmann, however, succeeded in shooting down one of the bombers with his synchronized machine gun. It was the first victory credited to the Fokker-designed weapon system.

IMPACT

At the outbreak of World War I, military strategists and commanders on both sides saw the wartime function of airplanes as a means to supply intelligence information behind enemy lines or as airborne artillery spotting platforms. As the war progressed and aircraft flew more or less freely across the trenches, providing vital information to both armies, it became apparent to ground commanders that while it was important to obtain intelligence on enemy movements, it was important also to deny the enemy similar information.

Early in the war, the French used airplanes as strategic bombing platforms. As both armies began to use their air forces for strategic bombing of troops, railways, ports, and airfields, it became evident that aircraft would have to be employed against enemy aircraft to prevent reconnaissance and bombing raids.

With the invention of the synchronized forward-firing machine gun, pilots could use their aircraft as attack weapons. A pilot finally could coordinate control of his aircraft and his armaments with maximum efficiency. This conversion of aircraft from nearly passive observation platforms to attack fighters is the single greatest innovation in the history of aerial warfare. The development of fighter aircraft forced a change in military strategy, tactics, and logistics and ushered in the era of modern warfare. Fighter planes are responsible for the battle-tested military adage: Whoever controls the sky controls the battlefield.

See also Airplane; Radar; Stealth aircraft.

FURTHER READING

Dierikx, M. L. J. *Fokker: A Transatlantic Biography.* Washington: Smithsonian Institution Press, 1997.

Franks, Norman L. R. *Aircraft Versus Aircraft: The Illustrated Story of Fighter Pilot Combat from 1914 to the Present Day.* New York: Barnes & Noble Books, 1999.

Guttman, Jon. *Fighting Firsts: Fighter Aircraft Combat Debuts from 1914 to 1944.* London: Cassell, 2000.

Pyrex Glass

The invention: A superhard and durable glass product with widespread uses in industry and home products.

The people behind the invention:
Jesse T. Littleton (1888-1966), the chief physicist of Corning Glass Works' research department
Eugene G. Sullivan (1872-1962), the founder of Corning's research laboratories
William C. Taylor (1886-1958), an assistant to Sullivan

Cooperating with Science

By the twentieth century, Corning Glass Works had a reputation as a corporation that cooperated with the world of science to improve existing products and develop new ones. In the 1870's, the company had hired university scientists to advise on improving the optical quality of glasses, an early example of today's common practice of academics consulting for industry.

When Eugene G. Sullivan established Corning's research laboratory in 1908 (the first of its kind devoted to glass research), the task that he undertook with William C. Taylor was that of making a heat-resistant glass for railroad lantern lenses. The problem was that ordinary flint glass (the kind in bottles and windows, made by melting together silica sand, soda, and lime) has a fairly high thermal expansion, but a poor heat conductivity. The glass thus expands unevenly when exposed to heat. This condition can cause the glass to break, sometimes violently. Colored lenses for oil or gas railroad signal lanterns sometimes shattered if they were heated too much by the flame that produced the light and were then sprayed by rain or wet snow. This changed a red "stop" light to a clear "proceed" signal and caused many accidents or near misses in railroading in the late nineteenth century.

Two solutions were possible: to improve the thermal conductivity or reduce the thermal expansion. The first is what metals do: When exposed to heat, most metals have an expansion much greater than that of glass, but they conduct heat so quickly that they expand nearly equally throughout and seldom lose structural integrity from uneven expansion. Glass, however, is an inherently poor heat conductor, so this approach was not possible.

Therefore, a formulation had to be found that had little or no thermal expansivity. Pure silica (one example is quartz) fits this description, but it is expensive and, with its high melting point, very difficult to work.

The formulation that Sullivan and Taylor devised was a borosilicate glass—essentially a soda-lime glass with the lime replaced by borax, with a small amount of alumina added. This gave the low thermal expansion needed for signal lenses. It also turned out to have good acid-resistance, which led to its being used for the battery jars required for railway telegraph systems and other applications. The glass was marketed as "Nonex" (for "nonexpansion glass").

FROM THE RAILROAD TO THE KITCHEN

Jesse T. Littleton joined Corning's research laboratory in 1913. The company had a very successful lens and battery jar material, but no one had even considered it for cooking or other heat-transfer applications, because the prevailing opinion was that glass absorbed and conducted heat poorly. This meant that, in glass pans, cakes, pies, and the like would cook on the top, where they were exposed to hot air, but would remain cold and wet (or at least undercooked) next to the glass surface. As a physicist, Littleton knew that glass absorbed radiant energy very well. He thought that the heat-conduction problem could be solved by using the glass vessel itself to absorb and distribute heat. Glass also had a significant advantage over metal in baking. Metal bakeware mostly reflects radiant energy to the walls of the oven, where it is lost ultimately to the surroundings. Glass would absorb this radiation energy and conduct it evenly to the cake or pie, giving a better result than that of the metal bakeware. Moreover, glass would not absorb and carry over flavors from one baking effort to the next, as some metals do.

Littleton took a cut-off battery jar home and asked his wife to bake a cake in it. He took it to the laboratory the next day, handing pieces around and not disclosing the method of baking until all had

JESSE T. LITTLETON

To prove that glass is good for baking, place an uncooked pie in a pie tin and place another pie pan under it, made half of tin and half of non-expanding glass. Place it in all the oven. That is the experiment Jesse Talbot Littleton, Jr., used at Corning Glass Works soon after he hired on in 1913. The story behind it began with a ceramic dish that cracked when his wife baked a cake. That would not happen, he realized, with the right kind of glass. Although his wife baked a cake successfully in a glass battery jar bottom at his request, Littleton had to demonstrate the feat for his superiors scientifically. The half of the pie over the glass, it turned out, cooked faster and more evenly. Kitchen glassware was born.

Littleton was born in Belle Haven, Virginia, in 1888. After taking degrees from Southern University and Tulane University, he earned a doctorate in physics from the University of Wisconsin in 1911. He briefly vowed to remain a bachelor and dedicate his life to physics, but Besse Cook, a pretty Mississippi school teacher, turned his head, and so he got married instead. He was the first physicist added to the newly organized research laboratories at Corning in New York. There he studied practical problems involved in the industrial applications of glass, including tempering, and helped invent a gas pressure meter to measure the flow of air in blowing glass and a sensitive, faster thermometer. He rose rapidly in the organization. In 1920 he became chief of the physical lab, assistant director of research in 1940, vice president in 1943, director of all Corning research and development in 1946, and general technical adviser in 1951.

Littleton retired a year later and, a passionate outdoorsman, devoted himself to hunting and fishing. A leading figure in the ceramics industry, he belonged to the American Academy for the Advancement of Science, American Physical Society, and the American Institute of Engineers and was an editor for the *Journal of Applied Physics*. He died in 1966.

agreed that the results were excellent. With this agreement, he was able to commit laboratory time to developing variations on the Nonex formula that were more suitable for cooking. The result was Pyrex, patented and trademarked in May of 1915.

IMPACT

In the 1930's, Pyrex "Flameware" was introduced, with a new glass formulation that could resist the increased heat of stove-top cooking. In the half century since Flameware was introduced, Corning went on to produce a variety of other products and materials: tableware in tempered opal glass; cookware in Pyroceram, a glass product that during heat treatment gained such mechanical strength as to be virtually unbreakable; even hot plates and stoves topped with Pyroceram.

In the same year that Pyrex was marketed for cooking, it was also introduced for laboratory apparatus. Laboratory glassware had been coming from Germany at the beginning of the twentieth century; World War I cut off the supply. Corning filled the gap with Pyrex beakers, flasks, and other items. The delicate blown-glass equipment that came from Germany was completely displaced by the more rugged and heat-resistant machine-made Pyrex ware.

Any number of operations are possible with Pyrex that cannot be performed safely in flint glass: Test tubes can be thrust directly into burner flames, with no preliminary warming; beakers and flasks can be heated on hot plates; and materials that dissolve when exposed to heat can be made into solutions directly in Pyrex storage bottles, a process that cannot be performed in regular glass. The list of such applications is almost endless.

Pyrex has also proved to be the material of choice for lenses in the great reflector telescopes, beginning in 1934 with that at Mount Palomar. By its nature, astronomical observation must be done with the scope open to the weather. This means that the mirror must not change shape with temperature variations, which rules out metal mirrors. Silvered (or aluminized) Pyrex serves very well, and Corning has developed great expertise in casting and machining Pyrex blanks for mirrors of all sizes.

See also Laminated glass; Microwave cooking; Plastic; Polystyrene; Teflon; Tupperware.

FURTHER READING

Blaszczyk, Regina Lee. *Imagining Consumers: Design and Innovation from Wedgwood to Corning.* Baltimore: Johns Hopkins University Press, 2000.

Graham, Margaret B. W., and Alec T. Shuldiner. *Corning and the Craft of Innovation.* New York: Oxford University Press, 2001.

Stage, Sarah, and Virginia Bramble Vincenti. *Rethinking Home Economics: Women and the History of a Profession.* Ithaca, N.Y.: Cornell University Press, 1997.

Rogove, Susan Tobier, and Marcia B. Steinhauer. *Pyrex by Corning: A Collector's Guide.* Marietta, Ohio: Antique Publications, 1993.

Radar

THE INVENTION: An electronic system for detecting objects at great distances, radar was a major factor in the Allied victory of World War II and now pervades modern life, including scientific research.

THE PEOPLE BEHIND THE INVENTION:

Sir Robert Watson-Watt (1892-1973), the father of radar who proposed the chain air-warning system

Arnold F. Wilkins, the person who first calculated the intensity of a radio wave

William C. Curtis (1914-1976), an American engineer

LOOKING FOR THUNDER

Sir Robert Watson-Watt, a scientist with twenty years of experience in government, led the development of the first radar, an acronym for *radio detection and ranging*. "Radar" refers to any instrument that uses the reflection of radio waves to determine the distance, direction, and speed of an object.

In 1915, during World War I (1914-1918), Watson-Watt joined Great Britain's Meteorological Office. He began work on the detection and location of thunderstorms at the Royal Aircraft Establishment in Farnborough and remained there throughout the war. Thunderstorms were known to be a prolific source of "atmospherics" (audible disturbances produced in radio receiving apparatus by atmospheric electrical phenomena), and Watson-Watt began the design of an elementary radio direction finder that gave the general position of such storms. Research continued after the war and reached a high point in 1922 when sealed-off cathode-ray tubes first became available. With assistance from J. F. Herd, a fellow Scot who had joined him at Farnborough, he constructed an instantaneous direction finder, using the new cathode-ray tubes, that gave the direction of thunderstorm activity. It was admittedly of low sensitivity, but it worked, and it was the first of its kind.

WILLIAM C. CURTIS

In addition to radar's applications in navigation, civil aviation, and science, it rapidly became an integral part of military aircraft by guiding weaponry and detecting enemy aircraft and missiles. The research and development industry that grew to provide offensive and defensive systems greatly expanded the opportunities for young scientists during the Cold War. Among them was William C. Curtis (1914-1976), one of the most influential African Americans in defense research.

Curtis graduated from the Tuskegee Institute (later Tuskegee University), where he later served as its first dean of engineering. While there, he helped form and train the Tuskegee Airmen, a famous squadron of African American fighter pilots during World War II. He also worked for the Radio Corporation of American (RCA) for twenty-three years. It was while at RCA that he contributed innovations to military radar. These include the Black Cat weapons system, MG-3 fire control system, 300-A weapon radar system, and Airborne Interceptor Data Link.

Watson-Watt did much of this work at a new site at Ditton Park, near Slough, where the National Physical Laboratory had a field station devoted to radio research. In 1927, the two endeavors were combined as the Radio Research Station; it came under the general supervision of the National Physical Laboratory, with Watson-Watt as the first superintendent. This became a center with unrivaled expertise in direction finding using the cathode-ray tube and in studying the ionosphere using radio waves. No doubt these facilities were a factor when Watson-Watt invented radar in 1935.

As radar developed, its practical uses expanded. Meteorological services around the world, using ground-based radar, gave warning of approaching rainstorms. Airborne radars proved to be a great help to aircraft by allowing them to recognize potentially hazardous storm areas. This type of radar was used also to assist research into cloud and rain physics. In this type of research, radar-equipped research aircraft observe the radar echoes inside a cloud as rain develops, and then fly through the cloud, using on-board instruments to measure the water content.

Technician at a modern radar display. (PhotoDisc)

AIMING RADAR AT THE MOON

The principles of radar were further developed through the discipline of radio astronomy. This field began with certain observations made by the American electrical engineer Karl Jansky in 1933 at the Bell Laboratories at Holmdell, New Jersey. Radio astronomers learn about objects in space by intercepting the radio waves that these objects emit.

Jansky found that radio signals were coming to Earth from space. He called these mysterious pulses "cosmic noise." In particular, there was an unusual amount of radar noise when the radio antennas were pointed at the Sun, which increased at the time of sun-spot activity.

All this information lay dormant until after World War II (1939-1945), at which time many investigators turned their attention to interpreting the cosmic noise. The pioneers were Sir Bernard Lovell at Manchester, England, Sir Martin Ryle at Cambridge, England, and Joseph Pawsey of the Commonwealth of Science Industrial Research Organization, in Australia. The intensity of these radio waves was first calculated by Arnold F. Wilkins.

As more powerful tools became available toward the end of World War II, curiosity caused experimenters to try to detect radio signals from the Moon. This was accomplished successfully in the late 1940's and led to experiments on other objects in the solar system: planets, satellites, comets, and asteroids.

IMPACT

Radar introduced some new and revolutionary concepts into warfare, and in doing so gave birth to entirely new branches of technology.

In the application of radar to marine navigation, the long-range navigation system developed during the war was taken up at once by the merchant fleets that used military-style radar equipment without modification. In addition, radar systems that could detect buoys and other ships and obstructions in closed waters, particularly under conditions of low visibility, proved particularly useful to peacetime marine navigation.

In the same way, radar was adopted to assist in the navigation of civil aircraft. The various types of track guidance systems developed after the war were aimed at guiding aircraft in the critical last hundred kilometers or so of their run into an airport. Subsequent improvements in the system meant that an aircraft could place itself on an approach or landing path with great accuracy.

The ability of radar to measure distance to an extraordinary degree of accuracy resulted in the development of an instrument that provided pilots with a direct measurement of the distances between airports. Along with these aids, ground-based radars were developed for the control of aircraft along the air routes or in the airport control area.

The development of electronic computers can be traced back to the enormous advances in circuit design, which were an integral part of radar research during the war. During that time, some elements of electronic computing had been built into bombsights and other weaponry; later, it was realized that a whole range of computing operations could be performed electronically. By the end of the war, many pulse-forming networks, pulse-counting circuits, and memory circuits existed in the form needed for an electronic computer.

Finally, the developing radio technology has continued to help astronomers explore the universe. Large radio telescopes exist in almost every country and enable scientists to study the solar system in great detail. Radar-assisted cosmic background radiation studies have been a building block for the big bang theory of the origin of the universe.

See also Airplane; Cruise missile; Radio interferometer; Sonar; Stealth aircraft.

FURTHER READING

Brown, Louis. *A Radar History of World War II: Technical and Military Imperatives*. Philadelphia: Institute of Physics, 1999.

Latham, Colin, and Anne Stobbs. *Pioneers of Radar*. Gloucestershire: Sutton, 1999.

Rowland, John. *The Radar Man: The Story of Sir Robert Watson-Watt*. New York: Roy Publishers, 1964.

Watson-Watt, Robert Alexander. *The Pulse of Radar: The Autobiography of Sir Robert Watson-Watt*. New York: Dial Press, 1959.

Radio

THE INVENTION: The first radio transmissions of music and voice laid the basis for the modern radio and television industries.

THE PEOPLE BEHIND THE INVENTION:
Guglielmo Marconi (1874-1937), an Italian physicist and inventor
Reginald Aubrey Fessenden (1866-1932), an American radio pioneer

TRUE RADIO

The first major experimenter in the United States to work with wireless radio was Reginald Aubrey Fessenden. This transplanted Canadian was a skilled, self-made scientist, but unlike American inventor Thomas Alva Edison, he lacked the business skills to gain the full credit and wealth that such pathbreaking work might have merited. Guglielmo Marconi, in contrast, is most often remembered as the person who invented wireless (as opposed to telegraphic) radio.

There was a great difference between the contributions of Marconi and Fessenden. Marconi limited himself to experiments with radio telegraphy; that is, he sought to send through the air messages that were currently being sent by wire—signals consisting of dots and dashes. Fessenden sought to perfect radio telephony, or voice communication by wireless transmission. Fessenden thus pioneered the essential precursor of modern radio broadcasting. At the beginning of the twentieth century, Fessenden spent much time and energy publicizing his experiments, thus promoting interest in the new science of radio broadcasting.

Fessenden began his career as an inventor while working for the U.S. Weather Bureau. He set out to invent a radio system by which to broadcast weather forecasts to users on land and at sea. Fessenden believed that his technique of using continuous waves in the radio frequency range (rather than interrupted waves Marconi had used to produce the dots and dashes of Morse code) would provide the power necessary to carry Morse telegraph code yet be effective enough to handle voice communication. He would turn out to be correct. He conducted experiments as early as 1900 at Rock Point, Maryland, about 80 kilometers south of Washington, D.C., and registered his first patent in the area of radio research in 1902.

FAME AND GLORY

In 1900, Fessenden asked the General Electric Company to produce a high-speed generator of alternating current—or alternator—to use as the basis of his radio transmitter. This proved to be the first major request for wireless radio apparatus that could project voices and music. It took the engineers three years to design and deliver the alternator. Meanwhile, Fessenden worked on an improved radio receiver. To fund his experiments, Fessenden aroused the interest of financial backers, who put up one million dollars to create the National Electric Signalling Company in 1902.

Fessenden, along with a small group of handpicked scientists, worked at Brant Rock on the Massachusetts coast south of Boston. Working outside the corporate system, Fessenden sought fame and glory based on his own work, rather than on something owned by a corporate patron.

Fessenden's moment of glory came on December 24, 1906, with the first announced broadcast of his radio telephone. Using an ordinary telephone microphone and his special alternator to generate the necessary radio energy, Fessenden alerted ships up and down the Atlantic coast with his wireless telegraph and arranged for newspaper reporters to listen in from New York City. Fessenden made himself the center of the show. He played the violin, sang, and read from the Bible. Anticipating what would become standard practice fifty years later, Fessenden also transmitted the sounds of a phonograph recording. He ended his first broadcast by wishing those listening "a Merry Christmas." A similar, equally well-publicized demonstration came on December 31.

Although Fessenden was skilled at drawing attention to his invention and must be credited, among others, as one of the engineering founders of the principles of radio, he was far less skilled at making money with his experiments, and thus his long-term impact was limited. The National Electric Signalling Company had a fine beginning and for a time was a supplier of equipment to the United

Fruit Company. The financial panic of 1907, however, wiped out an opportunity to sell the Fessenden patents—at a vast profit—to a corporate giant, the American Telephone and Telegraph Corporation.

IMPACT

Had there been more receiving equipment available and in place, a massive audience could have heard Fessenden's first broadcast. He had the correct idea, even to the point of playing a crude phonograph record. Yet Fessenden, Marconi, and their rivals were unable to establish a regular series of broadcasts. Their "stations" were experimental and promotional.

It took the stresses of World War I to encourage broader use of wireless radio based on Fessenden's experiments. Suddenly, communicating from ship to ship or from a ship to shore became a frequent matter of life or death. Generating publicity was no longer necessary. Governments fought over crucial patent rights. The Radio Corporation of America (RCA) pooled vital knowledge. Ultimately, RCA came to acquire the Fessenden patents. Radio broadcasting commenced, and the radio industry, with its multiple uses for mass communication, was off and running.

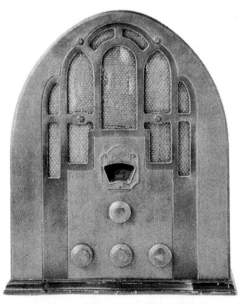

Antique tabletop radio. (PhotoDisc)

GUGLIELMO MARCONI

Guglielmo Marconi failed his entrance examinations to the University of Bologna in 1894. He had a weak educational background, particularly in science, but he was not about to let that—or his father's disapproval—stop him after he conceived a deep interest in wireless telegraphy during his teenage years.

Marconi was born in 1874 to a wealthy Italian landowner and an Irish whiskey distiller's daughter and grew up both in Italy and England. His parents provided tutors for him, but he and his brother often accompanied their mother, a socialite, on extensive travels. He acquired considerable social skills, easy self-confidence, and determination from the experience.

Thus, when he failed his exams, he simply tried another route for his ambitions. He and his mother persuaded a science professor to let Marconi use a university laboratory unofficially. His father thought it a waste of time. However, he changed his mind when his son succeeded in building equipment that could transmit electronic signals around their house without wires, an achievement right at the vanguard of technology.

Now supported by his father's money, Marconi and his brother built an elaborate set of equipment—including an oscillator, coherer, galvanometer, and antennas—that they hoped would send a signal outside over a long distance. His brother walked off a mile and a half, out of sight, with the galvanometer and a rifle. When the galvanometer moved, indicating a signal had arrived from the oscillator, he fired the rifle to let Marconi know he had succeeded. The incident is widely cited as the first radio transmission.

Marconi went on to send signals over greater and greater distances. He patented a tuner to permit transmissions at specific frequencies, and he started the Wireless Telegraph and Signal Company to bring his inventions to the public; its American branch was the Radio Corporation of America (RCA). He not only grew wealthy at a young age; he also was awarded half of the 1909 Nobel Prize in Physics for his work. He died in Rome in 1937, one of the most famous inventors in the world.

(Library of Congress)

See also Communications satellite; Compact disc; Dolby noise reduction; FM radio; Long-distance radiotelephony; Radio crystal sets; Television; Transistor; Transistor radio.

Further Reading

Fessenden, and Helen May Trott. *Fessenden: Builder of Tomorrows.* New York: Arno Press, 1974.

Lewis, Tom. *Empire of the Air: The Men Who Made Radio.* New York: HarperPerennial, 1993.

Masini, Giancarlo. *Marconi.* New York: Marsilio Publishers, 1995.

Seitz. Frederick. *The Cosmic Inventor: Reginald Aubrey Fessenden, 1866-1932.* Philadelphia: American Philosophical Society, 1999.

Radio crystal sets

THE INVENTION: The first primitive radio receivers, crystal sets led to the development of the modern radio.

THE PEOPLE BEHIND THE INVENTION:
H. H. Dunwoody (1842-1933), an American inventor
Sir John A. Fleming (1849-1945), a British scientist-inventor
Heinrich Rudolph Hertz (1857-1894), a German physicist
Guglielmo Marconi (1874-1937), an Italian engineer-inventor
James Clerk Maxwell (1831-1879), a Scottish physicist
Greenleaf W. Pickard (1877-1956), an American inventor

FROM MORSE CODE TO MUSIC

In the 1860's, James Clerk Maxwell demonstrated that electricity and light had electromagnetic and wave properties. The conceptualization of electromagnetic waves led Maxwell to propose that such waves, made by an electrical discharge, would eventually be sent long distances through space and used for communication purposes. Then, near the end of the nineteenth century, the technology that produced and transmitted the needed Hertzian (or radio) waves was devised by Heinrich Rudolph Hertz, Guglielmo Marconi (inventor of the wireless telegraph), and many others. The resultant radio broadcasts, however, were limited to the dots and dashes of the Morse code.

Then, in 1901, H. H. Dunwoody and Greenleaf W. Pickard invented the crystal set. Crystal sets were the first radio receivers that made it possible to hear music and the many other types of now-familiar radio programs. In addition, the simple construction of the crystal set enabled countless amateur radio enthusiasts to build "wireless receivers" (the name for early radios) and to modify them. Although, except as curiosities, crystal sets were long ago replaced by more effective radios, they are where it all began.

CRYSTALS, DIODES, TRANSISTORS, AND CHIPS

Radio broadcasting works by means of electromagnetic radio waves, which are low-energy cousins of light waves. All electromagnetic waves have characteristic vibration frequencies and wavelengths. This article will deal mostly with long radio waves of frequencies from 550 to 1,600 kilocycles (kilohertz), which can be seen on amplitude-modulation (AM) radio dials. Frequency-modulation (FM), shortwave, and microwave radio transmission use higher-energy radio frequencies.

The broadcasting of radio programs begins with the conversion of sound to electrical impulses by means of microphones. Then, radio transmitters turn the electrical impulses into radio waves that are broadcast together with higher-energy carrier waves. The combined waves travel at the speed of light to listeners. Listeners hear radio programs by using radio receivers that pick up broadcast waves through antenna wires and reverse the steps used in broadcasting. This is done by converting those waves to electrical impulses and then into sound waves. The two main types of radio broadcasting are AM and FM, which allow the selection (modulation) of the power (amplitude) or energy (frequency) of the broadcast waves.

The crystal set radio receiver of Dunwoody and Pickard had many shortcomings. These led to the major modifications that produced modern radios. Crystal sets, however, began the radio industry and fostered its development. Today, it is possible to purchase somewhat modified forms of crystal sets, as curiosity items. All crystal sets, original or modern versions, are crude AM radio receivers that are composed of four components: an antenna wire, a crystal detector, a tuning circuit, and a headphone or loudspeaker.

Antenna wires (aerials) pick up radio waves broadcast by external sources. Originally simple wires, today's aerials are made to work better by means of insulation and grounding. The crystal detector of a crystal set is a mineral crystal that allows radio waves to be selected (tuned). The original detectors were crystals of a lead-sulfur mineral, galena. Later, other minerals (such as silicon and carborundum) were also found to work. The tuning circuit is composed of 80 to 100 turns of insulated wire, wound on a 0.33-inch support. Some surprising supports used in homemade tuning circuits include cardboard toilet-paper-roll centers and Quaker Oats cereal boxes. When realism is desired in collector crystal sets, the coil is usually connected to a wire probe selector called a "cat's whisker." In some such crystal sets, a condenser (capacitor) and additional components are used to extend the range of tunable signals. Headphones convert chosen radio signals to sound waves that are heard by only one listener. If desired, loudspeakers can be used to enable a roomful of listeners to hear chosen programs.

An interesting characteristic of the crystal set is the fact that its operation does not require an external power supply. Offsetting this are its short reception range and a great difficulty in tuning or maintaining tuned-in radio signals. The short range of these radio receivers led to, among other things, the use of power supplies (house current or batteries) in more sophisticated radios. Modern solutions to tuning problems include using manufactured diode vacuum tubes to replace crystal detectors, which are a kind of natural diode. The first manufactured diodes, used in later crystal sets and other radios, were invented by John Ambrose Fleming, a colleague of Marconi's. Other modifications of crystal sets that led to more sophisticated modern radios include more powerful aerials, better circuits, and vacuum tubes. Then came miniaturization, which was made possible by the use of transistors and silicon chips.

IMPACT

The impact of the invention of crystal sets is almost incalculable, since they began the modern radio industry. These early radio receivers enabled countless radio enthusiasts to build radios, to receive radio messages, and to become interested in developing radio communication systems. Crystal sets can be viewed as having spawned all the variant modern radios. These include boom boxes and other portable radios; navigational radios used in ships and supersonic jet airplanes; and the shortwave, microwave, and satellite networks used in the various aspects of modern communication.

The later miniaturization of radios and the development of sophisticated radio system components (for example, transistors and silicon chips) set the stage for both television and computers.

Certainly, if one tried to assess the ultimate impact of crystal sets by simply counting the number of modern radios in the United States, one would find that few Americans more than ten years old own fewer than two radios. Typically, one of these is run by house electric current and the other is a portable set that is carried almost everywhere.

See also FM radio; Long-distance radiotelephony; Radio; Television; Transistor radio.

FURTHER READING

Masini, Giancarlo. *Marconi*. New York: Marsilio, 1995.
Sievers, Maurice L. *Crystal Clear: Vintage American Crystal Sets, Crystal Detectors, and Crystals*. Vestal, N.Y.: Vestal Press, 1991.
Tolstoy, Ivan. *James Clerk Maxwell: A Biography*. Chicago: University of Chicago Press, 1982.

Radio Interferometer

The invention: An astronomical instrument that combines multiple radio telescopes into a single system that makes possible the exploration of distant space.

The people behind the invention:
Sir Martin Ryle (1918-1984), an English astronomer
Karl Jansky (1905-1950), an American radio engineer
Hendrik Christoffel van de Hulst (1918-   ), a Dutch radio astronomer
Harold Irving Ewan (1922-   ), an American astrophysicist
Edward Mills Purcell (1912-1997), an American physicist

Seeing with Radio

Since the early 1600's, astronomers have relied on optical telescopes for viewing stellar objects. Optical telescopes detect the visible light from stars, galaxies, quasars, and other astronomical objects. Throughout the late twentieth century, astronomers developed more powerful optical telescopes for peering deeper into the cosmos and viewing objects located hundreds of millions of light-years away from the earth.

In 1933, Karl Jansky, an American radio engineer with Bell Telephone Laboratories, constructed a radio antenna receiver for locating sources of telephone interference. Jansky discovered a daily radio burst that he was able to trace to the center of the Milky Way galaxy. In 1935, Grote Reber, another American radio engineer, followed up Jansky's work with the construction of the first dish-shaped "radio" telescope. Reber used his 9-meter-diameter radio telescope to repeat Jansky's experiments and to locate other radio sources in space. He was able to map precisely the locations of various radio sources in space, some of which later were identified as galaxies and quasars.

Following World War II (that is, after 1945), radio astronomy blossomed with the help of surplus radar equipment. Radio astronomy tries to locate objects in space by picking up the radio waves that they emit. In 1944, the Dutch astronomer Hendrik Christoffel van de Hulst had proposed that hydrogen atoms emit radio waves with a 21-centimeter wavelength. Because hydrogen is the most abundant element in the universe, van de Hulst's discovery had explained the nature of extraterrestrial radio waves. His theory later was confirmed by the American radio astronomers Harold Irving Ewen and Edward Mills Purcell of Harvard University.

By coupling the newly invented computer technology with radio telescopes, astronomers were able to generate a radio image of a star almost identical to the star's optical image. A major advantage of radio telescopes over optical telescopes is the ability of radio telescopes to detect extraterrestrial radio emissions day or night, as well as their ability to bypass the cosmic dust that dims or blocks visible light.

MORE WITH LESS

After 1945, major research groups were formed in England, Australia, and The Netherlands. Sir Martin Ryle was head of the Mullard Radio Astronomy Observatory of the Cavendish Laboratory, University of Cambridge. He had worked with radar for the Telecommunications Research Establishment during World War II.

The radio telescopes developed by Ryle and other astronomers operate on the same basic principle as satellite television receivers. A constant stream of radio waves strikes the parabolic-shaped reflector dish, which aims all the radio waves at a focusing point above the dish. The focusing point directs the concentrated radio beam to the center of the dish, where it is sent to a radio receiver, then an amplifier, and finally to a chart recorder or computer.

With large-diameter radio telescopes, astronomers can locate stars and galaxies that cannot be seen with optical telescopes. This ability to detect more distant objects is called "resolution." Like optical telescopes, large-diameter radio telescopes have better resolution than smaller ones. Very large radio telescopes were constructed in the late 1950's and early 1960's (Jodrell Bank, England; Green Bank, West Virginia; Arecibo, Puerto Rico). Instead of just building larger radio telescopes to achieve greater resolution, however, Ryle developed a method called "interferometry." In Ryle's method, a computer is used to combine the incoming radio waves

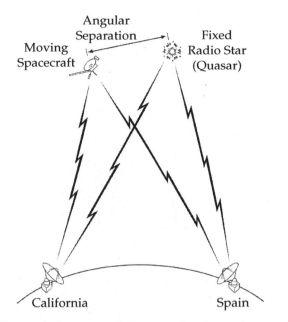

One use of VLBI is to navigate a spacecraft: By measuring the angular separation between a fixed radio star, such as a quasar, and a moving spacecraft, the craft's location, orientation, and path can be precisely monitored and adjusted.

of two or more movable radio telescopes pointed at the same stellar object.

Suppose that one had a 30-meter-diameter radio telescope. Its radio wave-collecting area would be limited to its diameter. If a second identical 30-meter-diameter radio telescope was linked with the first, then one would have an interferometer. The two radio telescopes would point exactly at the same stellar object, and the radio emissions from this object captured by the two telescopes would be combined by computer to produce a higher-resolution image. If the two radio telescopes were located 1.6 kilometers apart, then their combined resolution would be equivalent to that of a single radio telescope dish 1.6 kilometers in diameter.

Ryle constructed the first true radio telescope interferometer at the Mullard Radio Astronomy Observatory in 1955. He used combinations of radio telescopes to produce interferometers containing about twelve radio receivers. Ryle's interferometer greatly improved radio telescope resolution for detecting stellar radio sources, mapping the locations of stars and galaxies, assisting in the discovery of "quasars" (quasi-stellar radio sources), measuring the earth's rotation around the Sun, and measuring the motion of the solar system through space.

CONSEQUENCES

Following Ryle's discovery, interferometers were constructed at radio astronomy observatories throughout the world. The United States established the National Radio Astronomy Observatory (NRAO) in rural Green Bank, West Virginia. The NRAO is operated by nine eastern universities and is funded by the National Science Foundation. At Green Bank, a three-telescope interferometer was constructed, with each radio telescope having a 26-meter-diameter dish. During the late 1970's, the NRAO constructed the largest radio interferometer in the world, the Very Large Array (VLA). The VLA, located approximately 80 kilometers west of Socorro, New Mexico, consists of twenty-seven 25-meter-diameter radio telescopes linked by a supercomputer. The VLA has a resolution equivalent to that of a single radio telescope 32 kilometers in diameter.

Even larger radio telescope interferometers can be created with a technique known as "very long baseline interferometry" (VLBI). VLBI has been used to construct a radio telescope having an effective diameter of several thousand kilometers. Such an arrangement involves the precise synchronization of radio telescopes located in several different parts of the world. Supernova 1987A in the Large Magellanic Cloud was studied using a VLBI arrangement between observatories located in Australia, South America, and South Africa.

Launching radio telescopes into orbit and linking them with ground-based radio telescopes could produce a radio telescope whose effective diameter would be larger than that of the earth. Such instruments will enable astronomers to map the distribution of galaxies, quasars, and other cosmic objects, to understand the origin and evolution of the universe, and possibly to detect meaningful radio signals from extraterrestrial civilizations.

See also Neutrino detector; Weather satellite; Artificial satellite; Communications satellite; Radar; Rocket; Weather satellite.

FURTHER READING

Graham-Smith, Francis. *Sir Martin Ryle: A Biographical Memoir.* London: Royal Society, 1987.

Malphrus, Benjamin K. *The History of Radio Astronomy and the National Radio Astronomy Observatory: Evolution Toward Big Science.* Malabar, Fla.: Krieger, 1996.

Pound, Robert V. *Edward Mills Purcell: August 30, 1912-March 7, 1997.* Washington, D.C.: National Academy Press, 2000.

Refrigerant Gas

The invention: A safe refrigerant gas for domestic refrigerators, dichlorodifluoromethane helped promote a rapid growth in the acceptance of electrical refrigerators in homes.

The people behind the invention:
Thomas Midgley, Jr. (1889-1944), an American engineer and chemist
Charles F. Kettering (1876-1958), an American engineer and inventor who was the head of research for General Motors
Albert Henne (1901-1967), an American chemist who was Midgley's chief assistant
Frédéric Swarts (1866-1940), a Belgian chemist

Toxic Gases

Refrigerators, freezers, and air conditioners have had a major impact on the way people live and work in the twentieth century. With them, people can live more comfortably in hot and humid areas, and a great variety of perishable foods can be transported and stored for extended periods. As recently as the early nineteenth century, the foods most regularly available to Americans were bread and salted meats. Items now considered essential to a balanced diet, such as vegetables, fruits, and dairy products, were produced and consumed only in small amounts.

Through the early part of the twentieth century, the pattern of food storage and distribution evolved to make perishable foods more available. Farmers shipped dairy products and frozen meats to mechanically refrigerated warehouses. Smaller stores and most American households used iceboxes to keep perishable foods fresh. The iceman was a familiar figure on the streets of American towns, delivering large blocks of ice regularly.

In 1930, domestic mechanical refrigerators were being produced in increasing numbers. Most of them were vapor compression machines, in which a gas was compressed in a closed system of pipes outside the refrigerator by a mechanical pump and condensed to a liquid. The liquid was pumped into a sealed chamber in the refrigerator and allowed to evaporate to a gas. The process of evaporation removes heat from the environment, thus cooling the interior of the refrigerator.

The major drawback of early home refrigerators involved the types of gases used. In 1930, these included ammonia, sulfur dioxide, and methyl chloride. These gases were acceptable if the refrigerator's gas pipes never sprang a leak. Unfortunately, leaks sometimes occurred, and all these gases are toxic. Ammonia and sulfur dioxide both have unpleasant odors; if they leaked, at least they would be detected rapidly. Methyl chloride however, can form a dangerously explosive mixture with air, and it has only a very faint, and not unpleasant, odor. In a hospital in Cleveland during the 1920's, a refrigerator with methyl chloride leaked, and there was a disastrous explosion of the methyl chloride-air mixture. After that, methyl chloride for use in refrigerators was mixed with a small amount of a very bad-smelling compound to make leaks detectable. (The same tactic is used with natural gas.)

THREE-DAY SUCCESS

General Motors, through its Frigidaire division, had a substantial interest in the domestic refrigerator market. Frigidaire refrigerators used sulfur dioxide as the refrigerant gas. Charles F. Kettering, director of research for General Motors, decided that Frigidaire needed a new refrigerant gas that would have good thermal properties but would be nontoxic and nonexplosive. In early 1930, he sent Lester S. Keilholtz, chief engineer of General Motors' Frigidaire division, to Thomas Midgley, Jr., a mechanical engineer and self-taught chemist. He challenged them to develop such a new gas.

Midgley's associates, Albert Henne and Robert McNary, researched what types of compounds might already fit Kettering's specifications. Working with research that had been done by the Belgian chemist Frédéric Swarts in the late nineteenth and early twentieth centuries, Midgley, Henne, and McNary realized that dichlorodifluoromethane would have ideal thermal properties and the right boiling point for a refrigerant gas. The only question left to be answered was whether the compound was toxic.

The chemists prepared a few grams of dichlorodifluoromethane and put it, along with a guinea pig, into a closed chamber. They were delighted to see that the animal seemed to suffer no ill effects at all and was able to breathe and move normally. They were briefly puzzled when a second batch of the compound killed a guinea pig almost instantly. Soon, they discovered that an impurity in one of the ingredients had produced a potent poison in their refrigerant gas. A simple washing procedure completely removed the poisonous contaminant.

This astonishingly successful research project was completed in three days. The boiling point of dichlorodifluoromethane is –5.6 degrees Celsius. It is nontoxic and nonflammable and possesses excellent thermal properties. When Midgley was awarded the Perkin Medal for industrial chemistry in 1937, he gave the audience a graphic demonstration of the properties of dichlorodifluoromethane: He inhaled deeply of its vapors and exhaled gently into a jar containing a burning candle. The candle flame promptly went out. This visual evidence proved that dichlorodifluoromethane was not poisonous and would not burn.

IMPACT

The availability of this safe refrigerant gas, which was renamed Freon, led to drastic changes in the United States. The current patterns of food production, distribution, and consumption are a direct result, as is air conditioning. Air conditioning was developed early in the twentieth century; by the late 1970's, most American cars and residences were equipped with air conditioning, and other countries with hot climates followed suit. Consequently, major relocations of populations and businesses have become possible. Since World War II, there have been steady migrations to the "Sun Belt," the states spanning the United States from southeast to southwest, because air conditioners have made these areas much more livable.

Freon is a member of a family of chemicals called "chlorofluorocarbons." In addition to refrigeration, it is also used as a propellant in aerosols and in the production of polystyrene plastics. In 1974, scientists began to suspect that chlorofluorocarbons, when released into the air, might have a serious effect on the environment. They speculated that the compounds might migrate into the stratosphere, where they could be decomposed by the intense ultraviolet light from the sunlight that is prevented from reaching the earth's surface by the thin but vital layer of ozone in the stratosphere. In the process, large amounts of the ozone layer might also be destroyed— letting in the dangerous ultraviolet light. In addition to possible climatic effects, the resulting increase in ultraviolet light reaching the earth's surface would raise the incidence of skin cancers. As a result, chemical manufacturers are trying to develop alternative refrigerant gases that will not harm the ozone layer.

See also Electric refrigerator; Electric refrigerator; Food freezing; Microwave cooking.

FURTHER READING

Leslie, Stuart W. *Boss Kettering.* New York: Columbia University Press, 1983.
Mahoney, Thomas A. "The Seventy-one-year Saga of CFC's." *Air Conditioning, Heating and Refrigeration News* (March 15, 1999).
Preville, Cherie R., and Chris King. "Cooling Takes Off in the Roaring Twenties." *Air Conditioning, Heating and Refrigeration News* (April 30, 2001).

RESERPINE

THE INVENTION: A drug with unique hypertension-decreasing effects that provides clinical medicine with a versatile and effective tool.

THE PEOPLE BEHIND THE INVENTION:
Robert Wallace Wilkins (1906-   ), an American physician and clinical researcher
Walter E. Judson (1916-   ) , an American clinical researcher

TREATING HYPERTENSION

Excessively elevated blood pressure, clinically known as "hypertension," has long been recognized as a pervasive and serious human malady. In a few cases, hypertension is recognized as an effect brought about by particular pathologies (diseases or disorders). Often, however, hypertension occurs as the result of unknown causes. Despite the uncertainty about its origins, unattended hypertension leads to potentially dramatic health problems, including increased risk of kidney disease, heart disease, and stroke.

Recognizing the need to treat hypertension in a relatively straightforward and effective way, Robert Wallace Wilkins, a clinical researcher at Boston University's School of Medicine and the head of Massachusetts Memorial Hospital's Hypertension Clinic, began to experiment with reserpine in the early 1950's. Initially, the samples that were made available to Wilkins were crude and unpurified. Eventually, however, a purified version was used.

Reserpine has a long and fascinating history of use—both clinically and in folk medicine—in India. The source of reserpine is the root of the shrub *Rauwolfia serpentina*, first mentioned in Western medical literature in the 1500's but virtually unknown, or at least unaccepted, outside India until the mid-twentieth century. Crude preparations of the shrub had been used for a variety of ailments in India for centuries prior to its use in the West.

Wilkins's work with the drug did not begin on an encouraging note, because reserpine does not act rapidly—a fact that had been noted in Indian medical literature. The standard observation in Western pharmacotherapy, however, was that most drugs work rapidly; if a week has elapsed without positive effects being shown by a drug, the conventional Western wisdom is that it is unlikely to work at all. Additionally, physicians and patients alike tend to look for rapid improvement or at least positive indications. Reserpine is deceptive in this temporal context, and Wilkins and his coworkers were nearly deceived. In working with crude preparations of *Rauwolfia serpentina*, they were becoming very pessimistic, when a patient who had been treated for many consecutive days began to show symptomatic relief. Nevertheless, only after months of treatment did Wilkins become a believer in the drug's beneficial effects.

THE ACTION OF RESERPINE

When preparations of pure reserpine became available in 1952, the drug did not at first appear to be the active ingredient in the crude preparations. When patients' heart rate and blood pressure began to drop after weeks of treatment, however, the investigators saw that reserpine was indeed responsible for the improvements.

Once reserpine's activity began, Wilkins observed a number of important and unique consequences. Both the crude preparations and pure reserpine significantly reduced the two most meaningful measures of blood pressure. These two measures are systolic blood pressure and diastolic blood pressure. Systolic pressure represents the peak of pressure produced in the arteries following a contraction of the heart. Diastolic pressure is the low point that occurs when the heart is resting. To lower the mean blood pressure in the system significantly, both of these pressures must be reduced. The administration of low doses of reserpine produced an average drop in pressure of about 15 percent, a figure that was considered less than dramatic but still highly significant. The complex phenomenon of blood pressure is determined by a multitude of factors, including the resistance of the arteries, the force of contraction of the heart, and the heartbeat rate. In addition to lowering the blood pressure, reserpine reduced the heartbeat rate by about 15 percent, providing an important auxiliary action.

In the early 1950's, various therapeutic drugs were used to treat hypertension. Wilkins recognized that reserpine's major contribution would be as a drug that could be used in combination with drugs that were already in use. His studies established that reserpine, combined with at least one of the drugs already in use, produced an additive effect in lowering blood pressure. Indeed, at times, the drug combinations produced a "synergistic effect," which means that the combination of drugs created an effect that was more effective than the sum of the effects of the drugs when they were administered alone. Wilkins also discovered that reserpine was most effective when administered in low dosages. Increasing the dosage did not increase the drug's effect significantly, but it did increase the likelihood of unwanted side effects. This fact meant that reserpine was indeed most effective when administered in low dosages along with other drugs.

Wilkins believed that reserpine's most unique effects were not those found directly in the cardiovascular system but those produced indirectly by the brain. Hypertension is often accompanied by neurotic anxiety, which is both a consequence of the justifiable fears of future negative health changes brought on by prolonged hypertension and contributory to the hypertension itself. Wilkins's patients invariably felt better mentally, were less anxious, and were sedated, but in an unusual way. Reserpine made patients drowsy but did not generally cause sleep, and if sleep did occur, patients could be awakened easily. Such effects are now recognized as characteristic of tranquilizing drugs, or antipsychotics. In effect, Wilkins had discovered a new and important category of drugs: tranquilizers.

IMPACT

Reserpine holds a vital position in the historical development of antihypertensive drugs for two reasons. First, it was the first drug that was discovered to block activity in areas of the nervous system that use norepinephrine or its close relative dopamine as transmitter substances. Second, it was the first hypertension drug to be widely accepted and used. Its unusual combination of characteristics made it effective in most patients.

Since the 1950's, medical science has rigorously examined cardiovascular functioning and diseases such as hypertension. Many new factors, such as diet and stress, have been recognized as factors in hypertension. Controlling diet and life-style help tremendously in treating hypertension, but if the nervous system could not be partially controlled, many cases of hypertension would continue to be problematic. Reserpine has made that control possible.

See also Abortion pill; Antibacterial drugs; Artificial kidney; Birth control pill; Salvarsan.

FURTHER READING

MacGregor, G. A., and Norman M. Kaplan. *Hypertension*. 2d ed. Abingdon: Health Press, 2001.
"Reconsidering Reserpine." *American Family Physician* 45 (March, 1992).
Weber, Michael A. *Hypertension Medicine*. Totowa, N.J.: Humana, 2001.

Rice and Wheat Strains

The invention: Artificially created high-yielding wheat and rice
varieties that are helping food producers in developing countries
keep pace with population growth

The people behind the invention:
Orville A. Vogel (1907-1991), an agronomist who developed
high-yielding semidwarf winter wheats and equipment for
wheat research
Norman E. Borlaug (1914-   ), a distinguished agricultural
scientist
Robert F. Chandler, Jr. (1907-1999), an international agricultural
consultant and director of the International Rice Research
Institute, 1959-1972
William S. Gaud (1907-1977), a lawyer and the administrator of
the U.S. Agency for International Development, 1966-1969

The Problem of Hunger

In the 1960's, agricultural scientists created new, high-yielding
strains of rice and wheat designed to fight hunger in developing
countries. Although the introduction of these new grains raised levels of food production in poor countries, population growth and
other factors limited the success of the so-called "Green Revolution."

Before World War II, many countries of Asia, Africa, and Latin
America exported grain to Western Europe. After the war, however,
these countries began importing food, especially from the United
States. By 1960, they were importing about nineteen million tons of
grain a year; that level nearly doubled to thirty-six million tons in
1966. Rapidly growing populations forced the largest developing
countries—China, India, and Brazil in particular—to import huge
amounts of grain. Famine was averted on the Indian subcontinent
in 1966 and 1967 only by the United States shipping wheat to the region. The United States then changed its food policy. Instead of contributing food aid directly to hungry countries, the U.S. began

working to help such countries feed themselves.

The new rice and wheat strains were introduced just as countries in Africa and Asia were gaining their independence from the European nations that had colonized them. The Cold War was still going strong, and Washington and other Western capitals feared that the Soviet Union was gaining influence in the emerging countries. To help counter this threat, the U.S. Agency for International Development (USAID) was active in the Third World in the 1960's, directing or contributing to dozens of agricultural projects, including building rural infrastructure (farm-to-market roads, irrigation projects, and rural electric systems), introducing modern agricultural techniques, and importing fertilizer or constructing fertilizer factories in other countries. By raising the standard of living of impoverished people in developing countries through applying technology to agriculture, policymakers hoped to eliminate the socioeconomic conditions that would support communism.

THE GREEN REVOLUTION

It was against this background that William S. Gaud, administrator of USAID from 1966 to 1969, first talked about a "green revolution" in a 1968 speech before the Society for International Development in Washington, D.C. The term "green revolution" has been used to refer to both the scientific development of high-yielding food crops and the broader socioeconomic changes in a country's agricultural sector stemming from farmers' adoption of these crops.

In 1947, S. C. Salmon, a United States Department of Agriculture (USDA) scientist, brought a wheat-dwarfing gene to the United States. Developed in Japan, the gene produced wheat on a short stalk that was strong enough to bear a heavy head of grain. Orville Vogel, another USDA scientist, then introduced the gene into local wheat strains, creating a successful dwarf variety known as Gaines wheat. Under irrigation, Gaines wheat produced record yields. After hearing about Vogel's work, Norman Borlaug, who headed the Rockefeller Foundation's wheat-breeding program in Mexico, adapted Gaines wheat, later called "miracle wheat," to a variety of growing conditions in Mexico.

Workers in an Asian rice field. (PhotoDisc)

Success with the development of high-yielding wheat varieties persuaded the Rockefeller and Ford foundations to pursue similar ends in rice culture. The foundations funded the International Rice Research Institute (IRRI) in Los Banos, Philippines, appointing as director Robert F. Chandler, Jr., an international agricultural consultant. Under his leadership, IRRI researchers cross-bred Peta, a tall variety of rice from Indonesia, with Deo-geo-woo-gen, a dwarf rice from Taiwan, to produce a new strain, IR-8. Released in 1966 and dubbed "miracle rice," IR-8 produced yields double those of other Asian rice varieties and in a shorter time, 120 days in contrast to 150 to 180 days.

Statistics from India illustrate the expansion of the new grain varieties. During the 1966-1967 growing season, Indian farmers planted improved rice strains on 900,000 hectares, or 2.5 percent of the total area planted in rice. By 1984-1985, the surface area planted in improved rice varieties stood at 23.4 million hectares, or 56.9 percent of the total. The rate of adoption was even faster for wheat. In 1966-1967, improved varieties covered 500,000 hectares, comprising 4.2 percent of the total wheat crop. By the 1984-1985 growing season, the surface area had expanded to 19.6 million hectares, or 82.9 percent of the total wheat crop.

To produce such high yields, IR-8 and other genetically engineered varieties of rice and wheat required the use of irrigation, fertilizers, and pesticides. Irrigation further increased food production by allowing year-round farming and the planting of multiple crops on the same plot of land, either two crops of high-yielding grain varieties or one grain crop and another food crop.

EXPECTATIONS

The rationale behind the introduction of high-yielding grains in developing countries was that it would start a cycle of improvement in the lives of the rural poor. High-yielding grains would lead to bigger harvests and better-nourished and healthier families. If better nutrition enabled more children to survive, the need to have large families to ensure care for elderly parents would ease. A higher survival rate of children would lead couples to use family planning, slowing overall population growth and allowing per capita food intake to rise.

The greatest impact of the Green Revolution has been seen in Asia, which experienced dramatic increases in rice production, and on the Indian subcontinent, with increases in rice and wheat yields. Latin America, especially Mexico, enjoyed increases in wheat harvests. Subsaharan Africa initially was left out of the revolution, as scientists paid scant attention to increasing the yields of such staple food crops as yams, cassava, millet, and sorghum. By the 1980's, however, this situation was being remedied with new research directed toward millet and sorghum.

Research is conducted by a network of international agricultural research centers. Backed by both public and private funds, these centers cooperate with international assistance agencies, private foundations, universities, multinational corporations, and government agencies to pursue and disseminate research into improved crop varieties to farmers in the Third World. IRRI and the International Maize and Wheat Improvement Center (IMMYT) in Mexico City are two of these agencies.

IMPACT

Expectations went unrealized in the first few decades following the green revolution. Despite the higher yields from millions of tons of improved grain seeds imported into the developing world, lower-yielding grains still accounted for much of the surface area planted in grain. The reasons for this explain the limits and impact of the Green Revolution.

The subsistence mentality dies hard. The main targets of Green Revolution programs were small farmers, people whose crops provide barely enough to feed their families and provide seed for the next crop. If an experimental grain failed, they faced starvation. Such farmers hedged their bets when faced with a new proposition, for example, by intercropping, alternating rows of different grains in the same field. In this way, even if one crop failed, another might feed the family.

Poor farmers in developing countries also were likely to be illiterate and not eager to try something they did not fully understand. Also, by definition, poor farmers often did not have the means to purchase the inputs—irrigation, fertilizer, and pesticides—required to grow the improved varieties.

In many developing countries, therefore, rich farmers tended to be the innovators. More likely than poor farmers to be literate, they also had the money to exploit fully the improved grain varieties. They also were more likely than subsistence-level farmers to be in touch with the monetary economy, making purchases from the agricultural supply industry and arranging sales through established marketing channels, rather than producing primarily for personal or family use.

Once wealthy farmers adopted the new grains, it often became more difficult for poor farmers to do so. Increased demand for limited supplies, such as pesticides and fertilizers, raised costs, while bigger-than-usual harvests depressed market prices. With high sales volumes, owners of large farms could withstand the higher costs and lower-per-unit profits, but smaller farmers often could not.

Often, the result of adopting improved grains was that small farmers could no longer make ends meet solely by farming. Instead, they were forced to hire themselves out as laborers on large farms. Surges of laborers into a limited market depressed rural wages,

ORVILLE A. VOGEL

Born in 1907, Orville Vogel grew up on a farm in eastern Nebraska, and farming remained his passion for his entire life. He earned bachelor's and master's degrees in agriculture from the University of Nebraska, and then a doctorate in agronomy from Washington State University (WSU) in 1939.

Eastern Washington agreed with him, and he stayed there. He began his career as a wheat breeder 1931 for the U.S. Department of Agriculture, stationed at WSU. During the next forty-two years, he also took on the responsibilities of associate agronomist for the university's Division of Agronomy and from 1960 until his retirement in 1973 was professor of agronomy.

At heart Vogel was an experimenter and tinkerer, renowned among his peers for his keen powers of observation and his unselfishness. In addition to the wheat strains he bred that helped launch the Green Revolution, he took part in the search for plant varieties resistant to snow mold and foot rot. However, according to the father of the Green Revolution, Nobel laureate Norman Borlaug, Vogel's greatest contribution may not have been semi-dwarf wheat varieties but the many innovations in farming equipment he built as a sideline. These unheralded inventions automated the planting and harvesting of research plots, and so made research much easier to carry out and faster.

In recognition of his achievements, Vogel received the U.S. National Medal of Science in 1975 and entered the Agricultural Research Service's Science Hall of Fame in 1987. Vogel died in Washington in 1991.

making it even more difficult for small farmers to eke out a living. The result was that rich farmers got richer and poor farmers got poorer. Often, small farmers who could no longer support their families would leave rural areas and migrate to the cities, seeking work and swelling the ranks of the urban poor.

MIXED RESULTS

The effects of the Green Revolution were thus mixed. The dissemination of improved grain varieties unquestionably increased grain harvests in some of the poorest countries of the world. Seed companies developed, produced, and sold commercial quantities of improved grains, and fertilizer and pesticide manufacturers logged sales to developing countries thanks to USAID-sponsored projects.

Along with disrupting the rural social structure and encouraging rural flight to the cities, the Green Revolution has had other negative effects. For example, the millions of tube wells sunk in India to irrigate crops reduced groundwater levels in some regions faster than they could be recharged. In other areas, excessive use of pesticides created health hazards, and fertilizer use led to streams and ponds being clogged by weeds. The scientific community became concerned that the use of improved varieties of grain, many of which were developed from the same mother variety, reduced the genetic diversity of the world's food crops, making them especially vulnerable to attack by disease or pests.

Perhaps the most significant impact of the Green Revolution is the change it wrought in the income and class structure of rural areas; often, malnutrition was not eliminated in either the countryside or the cities. Almost without exception, the relative position of peasants deteriorated. Many analysts admit that the Green Revolution did not end world hunger, but they argue that it did buy time. The poorest of the poor would be even worse off without it.

See also Artificial chromosome; Cloning; Genetic "fingerprinting"; Genetically engineered insulin; In vitro plant culture.

FURTHER READING

Glaeser, Bernhard, ed. *The Green Revolution Revisited: Critique and Alternatives.* London: Allen & Unwin, 1987.

Hayami, Yujiro, and Masao Kikuchi. *A Rice Village Saga: Three Decades of Green Revolution in the Philippines.* Lanham, Md.: Barnes and Noble, 2000.

Karim, M. Bazlul. *The Green Revolution: An International Bibliography.* New York: Greenwood Press, 1986.

Lipton, Michael, and Richard Longhurst. *New Seeds and Poor People.* Baltimore: Johns Hopkins University Press, 1989.

Perkins, John H. *Geopolitics and the Green Revolution: Wheat, Genes, and the Cold War.* New York: Oxford University Press, 1997.

Richter scale

The invention: A scale for measuring the strength of earthquakes based on their seismograph recordings.

The people behind the invention:
Charles F. Richter (1900-1985), an American seismologist
Beno Gutenberg (1889-1960), a German American seismologist
Kiyoo Wadati (1902-   ), a pioneering Japanese seismologist
Giuseppe Mercalli (1850-1914), an Italian physicist, volcanologist, and meteorologist

Earthquake Study by Eyewitness Report

Earthquakes range in strength from barely detectable tremors to catastrophes that devastate large regions and take hundreds of thousands of lives. Yet the human impact of earthquakes is not an accurate measure of their power; minor earthquakes in heavily populated regions may cause great destruction, whereas powerful earthquakes in remote areas may go unnoticed. To study earthquakes, it is essential to have an accurate means of measuring their power.

The first attempt to measure the power of earthquakes was the development of intensity scales, which relied on damage effects and reports by witnesses to measure the force of vibration. The first such scale was devised by geologists Michele Stefano de Rossi and François-Alphonse Forel in 1883. It ranked earthquakes on a scale of 1 to 10. The de Rossi-Forel scale proved to have two serious limitations: Its level 10 encompassed a great range of effects, and its description of effects on human-made and natural objects was so specifically European that it was difficult to apply the scale elsewhere.

To remedy these problems, Giuseppe Mercalli published a revised intensity scale in 1902. The Mercalli scale, as it came to be called, added two levels to the high end of the de Rossi-Forel scale, making its highest level 12. It also was rewritten to make it more globally applicable. With later modifications by Charles F. Richter, the Mercalli scale is still in use.

Intensity measurements, even though they are somewhat subjec-

CHARLES F. RICHTER

Charles Francis Richter was born in Ohio in 1900. After his mother divorced his father, she moved the family to Los Angles in 1909. A precocious student, Richter entered the University of Southern California at sixteen and transferred to Stanford University a year later, majoring in physics. He graduated in 1920 and finished a doctorate in theoretical physics at the California Institute of Technology in 1928.

While Richter was a graduate student at Caltech, Noble laureate Robert A. Millikan lured him away from his original interest, astronomy, to become an assistant at the seismology laboratory. Richter realized that seismology was then a relatively new discipline and that he could help it mature. He stayed with it— and Caltech—for the rest of his university career, retiring as professor emeritus in 1970. In 1971 he opened a consulting firm—Lindvall, Richter and Associates—to assess the earthquake readiness of structures.

Richter published more than two hundred articles about earthquakes and earthquake engineering and two influential books, *Elementary Seismology* and *Seismicity of the Earth* (with Beno Gutenberg). These works, together with his teaching, trained a generation of earthquake researchers and gave them a basic tool, the Richter scale, to work with. He died in California in 1985.

tive, are very useful in mapping the extent of earthquake effects. Nevertheless, intensity measurements are still not ideal measuring techniques. Intensity varies from place to place and is strongly influenced by geologic features, and different observers frequently report different intensities. There is a need for an objective method of describing the strength of earthquakes with a single measurement.

MEASURING EARTHQUAKES ONE HUNDRED KILOMETERS AWAY

An objective technique for determining the power of earthquakes was devised in the early 1930's by Richter at the California Institute of Technology in Pasadena, California. The eventual usefulness of the scale that came to be called the "Richter scale" was completely unforeseen at first.

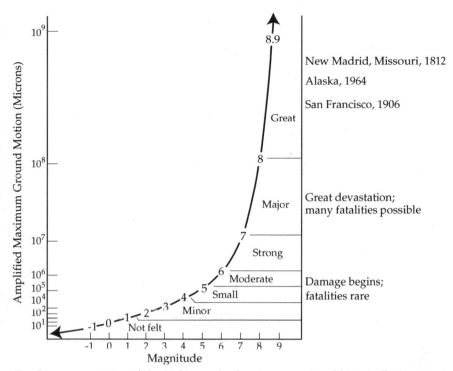

Graphic representation of the Richter scale showing examples of historically important earthquakes.

In 1931, the California Institute of Technology was preparing to issue a catalog of all earthquakes detected by its seismographs in the preceding three years. Several hundred earthquakes were listed, most of which had not been felt by humans, but detected only by instruments. Richter was concerned about the possible misinterpretations of the listing. With no indication of the strength of the earthquakes, the public might overestimate the risk of earthquakes in areas where seismographs were numerous and underestimate the risk in areas where seismographs were few.

To remedy the lack of a measuring method, Richter devised the scale that now bears his name. On this scale, earthquake force is expressed in magnitudes, which in turn are expressed in whole numbers and decimals. Each increase of one magnitude indicates a tenfold jump in the earthquake's force. These measurements were defined for a standard seismograph located one hundred kilometers from the earthquake. By comparing records for earthquakes recorded on different devices at different distances, Richter was able to create conversion tables for measuring magnitudes for any instrument at any distance.

IMPACT

Richter had hoped to create a rough means of separating small, medium, and large earthquakes, but he found that the scale was capable of making much finer distinctions. Most magnitude estimates made with a variety of instruments at various distances from earthquakes agreed to within a few tenths of a magnitude. Richter formally published a description of his scale in January, 1935, in the *Bulletin of the Seismological Society of America*. Other systems of estimating magnitude had been attempted, notably that of Kiyoo Wadati, published in 1931, but Richter's system proved to be the most workable scale yet devised and rapidly became the standard.

Over the next few years, the scale was refined. One critical refinement was in the way seismic recordings were converted into magnitude. Earthquakes produce many types of waves, but it was not known which type should be the standard for magnitude. So-called surface waves travel along the surface of the earth. It is these waves that produce most of the damage in large earthquakes; therefore, it seemed logical to let these waves be the standard. Earthquakes deep within the earth, however, produce few surface waves. Magnitudes based on surface waves would therefore be too small for these earthquakes. Deep earthquakes produce mostly waves that travel through the solid body of the earth; these are the so-called body waves.

It became apparent that two scales were needed: one based on surface waves and one on body waves. Richter and his colleague Beno Gutenberg developed scales for the two different types of waves, which are still in use. Magnitudes estimated from surface waves are symbolized by a capital M, and those based on body waves are denoted by a lowercase m.

From a knowledge of Earth movements associated with seismic waves, Richter and Gutenberg succeeded in defining the energy output of an earthquake in measurements of magnitude. A magnitude 6 earthquake releases about as much energy as a one-megaton nuclear explosion; a magnitude 0 earthquake releases about as much energy as a small car dropped off a two-story building.

See also Carbon dating; Geiger counter; Gyrocompass; Sonar; Scanning tunneling microscope.

FURTHER READING

Bates, Charles C., Thomas Frohock Gaskell, and Robert B. Rice. *Geophysics in the Affairs of Man: A Personalized History of Exploration Geophysics and Its Allied Sciences of Seismology and Oceanography.* New York: Pergamon Press, 1982.

Davison, Charles. 1927. Reprint. *The Founders of Seismology.* New York: Arno Press, 1978.

Howell, Benjamin F. *An Introduction to Seismological Research: History and Development.* Cambridge, N.Y.: Cambridge University Press, 1990.

Robot (household)

The invention: The first available personal robot, Hero 1 could speak; carry small objects in a gripping arm, and sense light, motion, sound, and time.

The people behind the invention:
Karel Čapek (1890-1938), a Czech playwright
The Health Company, an American electronics manufacturer

Personal Robots

In 1920, the Czech playwright Karel Čapek introduced the term *robot*, which he used to refer to intelligent, humanoid automatons that were subservient to humans. Robots such as those described by Čapek have not yet been developed; their closest counterparts are the nonintelligent automatons used by industry and by private individuals. Most industrial robots are heavy-duty, immobile machines designed to replace humans in routine, undesirable, monotonous jobs. Most often, they use programmed gripping arms to carry out tasks such as spray painting cars, assembling watches, and shearing sheep.

Modern personal robots are smaller, more mobile, less expensive models that serve mostly as toys or teaching tools. In some cases, they can be programmed to carry out activities such as walking dogs or serving mixed drinks. Usually, however, it takes more effort to program a robot to perform such activities than it does to do them oneself.

The Hero 1, which was first manufactured by the Heath Company in 1982, has been a very popular personal robot. Conceived as a toy and a teaching tool, the Hero 1 can be programmed to speak; to sense light, sound, motion, and time; and to carry small objects. The Hero 1 and other personal robots are often viewed as tools that will someday make it possible to produce intelligent robots.

HERO 1 OPERATION

The concept of artificial beings serving humanity has existed since antiquity (for example, it is found in Greek mythology). Such devices, which are now called robots, were first actualized, in a simple form, in the 1960's. Then, in the mid-1970's, the manufacture of personal robots began. One of the first personal robots was the Turtle, which was made by the Terrapin Company of Cambridge, Massachusetts. The Turtle was a toy that entertained owners via remote control, programmable motion, a beeper, and blinking displays. The Turtle was controlled by a computer to which it was linked by a cable.

Among the first significant personal robots was the Hero 1. This robot, which was usually sold in the form of a $1,000 kit that had to be assembled, is a squat, thirty-nine-pound mobile unit containing a head, a body, and a base. The head contains control boards, sensors, and a manipulator arm. The body houses control boards and related electronics, while the base contains a three-wheel-drive unit that renders the robot mobile.

The Heath Company, which produced the Hero 1, viewed it as providing entertainment for and teaching people who are interested in robot applications. To facilitate these uses, the following abilities were incorporated into the Hero 1: independent operation via rechargeable batteries; motion- and distance/position-sensing capability; light, sound, and language use/recognition; a manipulator arm to carry out simple tasks; and easy programmability.

The Hero 1 is powered by four rechargeable batteries arranged as two 12-volt power supplies. Recharging is accomplished by means of a recharging box that is plugged into a home outlet. It takes six to eight hours to recharge depleted batteries, and complete charging is signaled by an indicator light. In the functioning robot, the power supplies provide 5-volt and 12-volt outputs to logic and motor circuits, respectively.

The Hero 1 moves by means of a drive mechanism in its base. The mechanism contains three wheels, two of which are unpowered drones. The third wheel, which is powered for forward and reverse motion, is connected to a stepper motor that makes possible directional steering. Also included in the powered wheel is a metal disk with spaced reflective slots that helps Hero 1 to identify its position. As the robot moves, light is used to count the slots, and the slot count is used to measure the distance the robot has traveled, and therefore its position.

The robot's "senses," located in its head, consist of sound, light, and motion detectors as well as a phoneme synthesizer (phonemes are sounds, or units of speech). All these components are connected with the computer. The Hero 1 can detect sounds between 200 and 5,000 hertz. Its motion sensor detects all movement within a 15-foot radius. The phoneme synthesizer is capable of producing most words by using combinations of 64 phonemes. In addition, the robot keeps track of time by using an internal clock/calendar.

The Hero 1 can carry out various tasks by using a gripper that serves as a hand. The arm on which the gripper is located is connected to the back of the robot's head. The head (and, therefore, the arm) can rotate 350 degrees horizontally. In addition, the arm contains a shoulder motor that allows it to rise or drop 150 degrees vertically, and its forearm can be either extended or retracted. Finally, a wrist motor allows the gripper's tip to rotate by 350 degrees, and the two-fingered gripper can open up to a maximum width of 3.5 inches. The arm is not useful except as an educational tool, since its load-bearing capacity is only about a pound and its gripper can exert a force of only 6 ounces.

The computational capabilities of the robot are much more impressive than its physical capabilities. Programming is accomplished by means of a simple keypad located on the robot's head, which provides an inexpensive, easy-to-use method of operator-computer communication. To make things simpler for users who want entertainment without having to learn robotics, a manual mode is included for programming. In the manual mode, a hand-held teaching pendant is connected to Hero 1 and used to program all the motion capabilities of the robot. The programming of sensory and language abilities, however, must be accomplished by using the keyboard. Using the keyboard and the various options that are available enables Hero 1 owners to program the robot to perform many interesting activities.

CONSEQUENCES

The Hero 1 had a huge impact on robotics; thousands of people purchased it and used it for entertainment, study, and robot design. The Heath Company itself learned from the Hero 1 and later introduced an improved version: Heathkit 2000. This personal robot, which costs between $2,000 and $4,500, has ten times the capabilities of Hero 1, operates via radio-controlled keyboard, contains a voice synthesizer that can be programmed in any language, and plugs itself in for recharging.

Other companies, including the Androbot Company in California, have manufactured personal robots that sell for up to $10,000. One such robot is the Androbot BOB (brains on board). It can guard a home, call the police, walk at 2.5 kilometers per hour, and sing. Androbot has also designed Topo, a personal robot that can serve drinks. Still other robots can sort laundry and/or vacuum-clean houses. Although modern robots lack intelligence and merely have the ability to move when they are directed to by a program or by remote control, there is no doubt that intelligent robots will be developed in the future.

See also Electric refrigerator; Microwave cooking; Robot (industrial); Vacuum cleaner; Washing machine.

FURTHER READING

Aleksander, Igor, and Piers Burnett. *Reinventing Man: The Robot Becomes Reality.* London: Kogan Page, 1983.

Asimov, Isaac. *Robots: Machines in Man's Image.* New York: Harmony Books, 1985.

Bell, Trudy E. "Robots in the Home: Promises, Promises." *IEEE Spectrum* 22, no. 5 (May, 1985).

Whalen, Bernie. "Upscale Consumers Adopt Home Robots, but Widespread Lifestyle Impact Is Years Away." *Marketing News* 17, no. 24 (November 25, 1983).

Robot (industrial)

The invention: The first industrial robots, Unimates were designed to replace humans in undesirable, hazardous, and monotonous jobs.

The people behind the invention:
Karel Čapek (1890-1938), a Czech playwright
George C. Devol, Jr. (1912-   ), an American inventor
Joseph F. Engelberger (1925-   ), an American entrepreneur

Robots, from Concept to Reality

The 1920 play *Rossum's Universal Robots*, by Czech writer Karel Čapek, introduced robots to the world. Čapek's humanoid robots—robot, a word created by Čapek, essentially means slave—revolted and took over the world, which made the concept of robots somewhat frightening. The development of robots, which are now defined as machines that do work that would ordinarily be carried out by humans, has not yet advanced to the stage of being able to produce humanoid robots, however, much less robots capable of carrying out a revolt.

Most modern robots are found in industry, where they perform dangerous or monotonous tasks that previously were done by humans. The first industrial robots were the Unimates (short for "universal automaton"), which were derived from a robot design invented by George C. Devol and patented in 1954. The first Unimate prototypes, developed by Devol and Joseph F. Engelberger, were completed in 1962 by Unimation Incorporated and tested in industry. They were so successful that the company, located in Danbury, Connecticut, manufactured and sold thousands of Unimates to companies in the United States and abroad. Unimates are very versatile at performing routine industrial tasks and are easy to program and reprogram. The tasks they perform include various steps in automobile manufacturing, spray painting, and running lathes. The huge success of the Unimates led companies in other countries to produce their own industrial robots, and advancing technology has improved all industrial robots tremendously.

A New Industrial Revolution

Each of the first Unimate robots, which were priced at $25,000, was almost five feet tall and stood on a four-foot by five-foot base. It has often been said that a Unimate resembles the gun turret of a minitank, set atop a rectangular box. In operation, such a robot will swivel, swing, and/or dip and turn at the wrist of its hydraulically powered arm, which has a steel hand. The precisely articulated hand can pick up an egg without breaking it. At the same time, however, it is powerful enough to lift a hundred-pound weight.

The Unimate is a robotic jack of all trades: It can be programmed, in about an hour, to carry out a complex operation, after which it can have its memory erased and be reprogrammed in another hour to do something entirely different. In addition, programming a Unimate requires no special training. The programmer simply uses a teach-cable selector that allows the programmer to move the Unimate arm through the desired operation. This selector consists of a group of pushbutton control boxes, each of which is equipped with buttons in opposed pairs. Each button pair records the motion that will put a Unimate arm through one of five possible motions, in opposite directions. For example, pushing the correct buttons will record a motion in which the robot's arm moves out to one side, aims upward, and angles appropriately to carry out the first portion of its intended job. If the Unimate overshoots, undershoots, or otherwise performs the function incorrectly, the activity can be fine-tuned with the buttons.

Once the desired action has been performed correctly, pressing a "record" button on the robot's main control panel enters the operation into its computer memory. In this fashion, Unimates can be programmed to carry out complex actions that require as many as two hundred commands. Each command tells the Unimate to move its arm or hand in a given way by combining the following five motions: sliding the arm forward, swinging the arm horizontally, tilting the arm up or down, bending the wrist up or down, and swiveling the hand in a half-circle clockwise or counterclockwise.

Before pressing the "record" button on the Unimate's control panel, the operator can also command the hand to grasp an item when in a particular position. Furthermore, the strength of the grasp can be controlled, as can the duration of time between each action. Finally, the Unimate can be instructed to start or stop another routine (such as operating a paint sprayer) at any point. Once the instructor is satisfied with the robot's performance, pressing a "repeat continuous" control starts the Unimate working. The robot will stop repeating its program only when it is turned off.

Inside the base of an original Unimate is a magnetic drum that contains its memory. The drum turns intermittently, moving each of two hundred long strips of metal beneath recording heads. This strip movement brings specific portions of each strip—dictated by particular motions—into position below the heads. When the "record" button is pressed after a motion is completed, the hand position is recorded as a series of numbers that tells the computer the complete hand position in each of the five permissible movement modes.

Once "repeat continuous" is pressed, the computer begins the command series by turning the drum appropriately, carrying out each memorized command in the chosen sequence. When the sequence ends, the computer begins again, and the process repeats until the robot is turned off. If a Unimate user wishes to change the function of such a robot, its drum can be erased and reprogrammed. Users can also remove programmed drums, store them for future use, and replace them with new drums.

Consequences

The first Unimates had a huge impact on industrial manufacturing. In time, different sizes of robots became available so that additional tasks could be performed, and the robots' circuitry was improved. Because they have no eyes and cannot make judgments, Unimates are limited to relatively simple tasks that are coordinated by means of timed operations and simple computer interactions.

Most of the thousands of modern Unimates and their multinational cousins in industry are very similar to the original Unimates in terms of general capabilities, although they can now assemble watches and perform other delicate tasks that the original Unimates could not perform. The crude magnetic drums and computer controls have given way to silicon chips and microcomputers, which have made the robots more accurate and reliable. Some robots can even build other robots, and others can perform tasks such as mowing lawns and walking dogs.

Various improvements have been planned that will ultimately lead to some very interesting and advanced modifications. It is likely that highly sophisticated humanoid robots like those predicted by Karel Čapek will be produced at some future time. One can only hope that these robots will not rebel against their human creators.

See also CAD/CAM; Robot (household); SAINT; Virtual machine.

FURTHER READING

Aleksander, Igor, and Piers Burnett. *Reinventing Man: The Robot Becomes Reality.* London: Kogan Page, 1983.

Asimov, Isaac. *Robots: Machines in Man's Image.* New York: Harmony Books, 1985.

Chakravarty, Subrata N. "Springtime for an Ugly Duckling." *Forbes* 127, no. 9 (April, 1981).

Hartley, J. "Robots Attack the Quiet World of Arc Welding." *Engineer* 246, no. 6376 (June, 1978).

Lamb, W. G. *Unimates at Work.* Edited by C. W. Burckhardt. Basel, Switzerland: Birkhauser Verlag, 1975.

Tuttle, Howard C. "Robots' Contribution: Faster Cycles, Better Quality." *Production* 88, no. 5 (November, 1981).

Rocket

The invention: Liquid-fueled rockets developed by Robert H. Goddard made possible all later developments in modern rocketry, which in turn has made the exploration of space practical.

The person behind the invention:
Robert H. Goddard (1882-1945), an American physics professor

History in a Cabbage Patch

Just as the age of air travel began on an out-of-the-way shoreline at Kitty Hawk, North Carolina, with the Wright brothers' airplane in 1903, so too the seemingly impossible dream of spaceflight began in a cabbage patch in Auburn, Massachusetts, with Robert H. Goddard's launch of a liquid-fueled rocket on March 16, 1926. On that clear, cold day, with snow still on the ground, Goddard launched a three-meter-long rocket using liquid oxygen and gasoline. The flight lasted only about two and one-half seconds, during which the rocket rose 12 meters and landed about 56 meters away.

Although the launch was successful, the rocket's design was clumsy. At first, Goddard had thought that a rocket would be steadier if the motor and nozzles were ahead of the fuel tanks, rather like a horse and buggy. After this first launch, it was clear that the motor needed to be placed at the rear of the rocket. Although Goddard had spent several years working on different pumps to control the flow of fuel to the motor, the first rocket had no pumps or electrical system. Henry Sacks, a Clark University machinist, launched the rocket by turning a valve, placing an alcohol stove beneath the motor, and dashing for safety. Goddard and his coworker Percy Roope watched the launch from behind an iron wall.

Despite its humble setting, this simple event changed the course of history. Many people saw in Goddard's launch the possibilities for high-altitude research, space travel, and modern weaponry. Although Goddard invented and experimented mostly in private, others in the United States, the Soviet Union, and Germany quickly followed in his footsteps. The V-2 rockets used by Nazi Germany in World War II (1939-1945) included many of Goddard's designs and ideas.

A Lifelong Interest

Goddard's success was no accident. He had first become interested in rockets and space travel when he was seventeen, no doubt because of reading books such as H. G. Wells's *The War of the Worlds* (1898) and Garrett P. Serviss's *Edison's Conquest of Mars* (1898). In 1907, he sent to several scientific journals a paper describing his ideas about traveling through a near vacuum. Although the essay was rejected, Goddard began thinking about liquid fuels in 1909. After finishing his doctorate in physics at Clark University and postdoctoral studies at Princeton University, he began to experiment.

One of the things that made Goddard so successful was his ability to combine things he had learned from chemistry, physics, and engineering into rocket design. More than anyone else at the time, Goddard had the ability to combine ideas with practice.

Goddard was convinced that the key for moving about in space was the English physicist and mathematician Sir Isaac Newton's third law of motion (for every action there is an equal and opposite reaction). To prove this, he showed that a gun recoiled when it was fired in a vacuum. During World War I (1914-1918), Goddard moved to the Mount Wilson Observatory in California, where he investigated the use of black powder and smokeless powder as rocket fuel. Goddard's work led to the invention of the bazooka, a weapon that was much used during World War II, as well as bombardment and antiaircraft rockets.

After World War I, Goddard returned to Clark University. By 1920, mostly because of the experiments he had done during the war, he had decided that a liquid-fuel motor, with its smooth thrust, had the best chance of boosting a rocket into space. The most powerful fuel was hydrogen, but it is very difficult to handle. Oxygen had many advantages, but it was hard to find and extremely dangerous, since it boils at −148 degrees Celsius and explodes when it comes in contact with oils, greases, and flames. Other possible fuels were pro-

ROBERT H. GODDARD

In 1920 *The New York Times* made fun of Robert Hutchings Goddard (1882-1945) for claiming that rockets could travel through outer space to the Moon. It was impossible, the newspaper's editorial writer confidently asserted, because in outer space the engine would have no air to push against and so could not move the rocket. A sensitive, quiet man, the Clark University physics professor was stung by the public rebuke, all the more so because it displayed ignorance of basic physics. "Every vision is a joke," Goddard said, somewhat bitterly, "until the first man accomplishes it."

(Library of Congress)

Goddard had already proved that a rocket could move in a vacuum, but he refrained from rebutting the *Times* article. In 1919 he had become the first American to describe mathematically the theory of rocket propulsion in his classic article "A Method of Reaching Extreme Altitude," and during World War I he had acquired experience designing solid-fuel rockets. However, even though he was the world's leading expert on rocketry, he decided to seek privacy for his experiments. His successful launch of a liquid-fuel rocket in 1926, followed by new designs that reached ever higher altitudes, was a source of satisfaction, as were his 214 patents, but real recognition of his achievements did not come his way until World War II. In 1942 he was named director of research at the U.S. Navy's Bureau of Aeronautics, for which he worked on jet-assisted takeoff rockets and variable-thrust liquid-propellant rockets. In 1943 the Curtiss-Wright Corporation hired him as a consulting engineer, and in 1945 he became director of the American Rocket Society.

The New York Times finally apologized to Goddard for its 1920 article on the morning after Apollo 11 took off for the Moon in 1969. However, Goddard, who battled tuberculosis most of his life, had died twenty-four years earlier.

pane, ether, kerosene, or gasoline, but they all had serious disadvantages. Finally, Goddard found a local source of oxygen and was able to begin testing its thrust.

Another problem was designing a fuel pump. Goddard and his assistant Nils Riffolt spent years on this problem before the historic test flight of March, 1926. In the end, because of pressure from the Smithsonian Institution and others who were funding his research, Goddard decided to do without a pump and use an inert gas to push the fuel into the explosion chamber.

Goddard worked without much funding between 1920 and 1925. Riffolt helped him greatly in designing a pump, and Goddard's wife, Esther, photographed some of the tests and helped in other ways. Clark University had granted him some research money in 1923, but by 1925 money was in short supply, and the Smithsonian Institution did not seem willing to grant more. Goddard was convinced that his research would be taken seriously if he could show some serious results, so on March 16, 1926, he launched a rocket even though his design was not yet perfect. The success of that launch not only changed his career but also set the stage for rocketry experiments both in the United States and in Europe.

IMPACT

Goddard was described as being secretive and a loner. He never tried to cash in on his invention but continued his research during the next three years. On July 17, 1929, Goddard launched a rocket carrying a camera and instruments for measuring temperature and air pressure. *The New York Times* published a story about the noisy crash of this rocket and local officials' concerns about public safety. The article also mentioned Goddard's idea that a similar rocket might someday strike the Moon. When American aviation hero Charles A. Lindbergh learned of Goddard's work, Lindbergh helped him to get grants from the Carnegie Institution and the Guggenheim Foundation.

By the middle of 1930, Goddard and a small group of assistants had established a full-time research program near Roswell, New Mexico. Now that money was not so much of a problem, Goddard began to make significant advances in almost every area of astronautics. In 1941, Goddard launched a rocket to a height of 2,700 meters. Flight stability was helped by a gyroscope, and he was finally able to use a fuel pump.

During the 1920's and 1930's, members of the American Rocket Society and the German Society for Space Travel continued their own research. When World War II began, rocket research became a high priority for the American and German governments.

Germany's success with the V-2 rocket was a direct result of Goddard's research and inventions, but the United States did not benefit fully from Goddard's work until after his death. Nevertheless, Goddard remains modern rocketry's foremost pioneer—a scientist with vision, understanding, and practical skill.

See also Airplane; Artificial satellite; Communications satellite; Cruise missile; Hydrogen bomb; Stealth aircraft; Supersonic passenger plane; Turbojet; V-2 rocket; Weather satellite.

FURTHER READING

Alway, Peter. *Retro Rockets: Experimental Rockets, 1926-1941.* Ann Arbor, Mich.: Saturn Press, 1996.

Goddard, Robert Hutchings. *The Autobiography of Robert Hutchings Goddard, Father of the Space Age: Early Years to 1927.* Worcester, Mass.: A. J. St. Onge, 1966.

Lehman, Milton. *Robert H. Goddard: Pioneer of Space Research.* New York: Da Capo Press, 1988.

Rotary dial telephone

THE INVENTION: The first device allowing callers to connect their telephones to other parties without the aid of an operator, the rotary dial telephone preceded the touch-tone phone.

THE PEOPLE BEHIND THE INVENTION:
Alexander Graham Bell (1847-1922), an American inventor
Antoine Barnay (1883-1945), a French engineer
Elisha Gray (1835-1901), an American inventor

ROTARY TELEPHONES DIALS MAKE PHONE LINKUPS AUTOMATIC

The telephone uses electricity to carry sound messages over long distances. When a call is made from a telephone set, the caller speaks into a telephone transmitter and the resultant sound waves are converted into electrical signals. The electrical signals are then transported over a telephone line to the receiver of a second telephone set that was designated when the call was initiated. This receiver reverses the process, converting the electrical signals into the sounds heard by the recipient of the call. The process continues as the parties talk to each other.

The telephone was invented in the 1870's and patented in 1876 by Alexander Graham Bell. Bell's patent application barely preceded an application submitted by his competitor Elisha Gray. After a heated patent battle between Bell and Gray, which Bell won, Bell founded the Bell Telephone Company, which later came to be called the American Telephone and Telegraph Company.

At first, the transmission of phone calls between callers and recipients was carried out manually, by switchboard operators. In 1923, however, automation began with Antoine Barnay's development of the rotary telephone dial. This dial caused the emission of variable electrical impulses that could be decoded automatically and used to link the telephone sets of callers and call recipients. In time, the rotary dial system gave way to push-button dialing and other more modern networking techniques.

Rotary-dial telephone. (Image Club Graphics)

Telephones, Switchboards, and Automation

The carbon transmitter, which is still used in many modern telephone sets, was the key to the development of the telephone by Alexander Graham Bell. This type of transmitter—and its more modern replacements—operates like an electric version of the human ear. When a person talks into the telephone set in a carbon transmitter-equipped telephone, the sound waves that are produced strike an electrically connected metal diaphragm and cause it to vibrate. The speed of vibration of this electric eardrum varies in accordance with the changes in air pressure caused by the changing tones of the speaker's voice.

Behind the diaphragm of a carbon transmitter is a cup filled with powdered carbon. As the vibrations cause the diaphragm to press against the carbon, the electrical signals—electrical currents of varying strength—pass out of the instrument through a telephone wire. Once the electrical signals reach the receiver of the phone being called, they activate electromagnets in the receiver that make a second diaphragm vibrate. This vibration converts the electrical signals into sounds that are very similar to the sounds made by the person who is speaking. Therefore, a telephone receiver may be viewed as an electric mouth.

In modern telephone systems, transportation of the electrical signals between any two phone sets requires the passage of those signals through vast telephone networks consisting of huge numbers of wires, radio systems, and other media. The linkup of any two

ALEXANDER GRAHAM BELL

During the funeral for Alexander Graham Bell in 1922, telephone service throughout the United States stopped for one minute to honor him. To most people he was the inventor of the telephone. In fact, his genius ranged much further.

Bell was born in Edinburgh, Scotland, in 1847. His father, an elocutionist who invented a phonetic alphabet, and his mother, who was deaf, imbued him with deep curiosity, especially about sound. As a boy Bell became an exceptional pianist, and he produced his first invention, for cleaning wheat, at fourteen. After Edinburgh's Royal High School, he attended classes at Edinburgh University and University College, London, but at the age of twenty-three, battling tuberculosis, he left school to move with his parents to Ontario, Canada, to convalesce. Meanwhile, he worked on his idea for a telegraph capable of sending multiple messages at once. From it grew the basic concept for the telephone. He developed it while teaching Visible Speech at the Boston School for Deaf Mutes after 1871. Assisted by Thomas Watson, he succeeded in sending speech over a wire and was issued a patent for his device, among the most valuable ever granted, in 1876. His demonstration of the telephone later that year at Philadelphia's Centennial Exhibition and its subsequent development into a household appliance brought him wealth and fame.

He moved to Nova Scotia, Canada, and continued inventing. He created a photophone, tetrahedron modules for construction, and an airplane, the Silver Dart, which flew in 1909. Even though existing technology made them impracticable, some of his ideas anticipated computers and magnetic sound recording. His last patented invention, tested three years before his death, was a hydrofoil. Capable of reaching seventy-one miles per hour and freighting fourteen thousand pounds, the HD-4 was then the fastest watercraft in the world.

Bell also helped found the National Geographic Society in 1888 and became its president in 1898. He hired Gilbert Grosvenor to edit the society's famous magazine, *National Geographic* and together they planned the format—breathtaking photography and vivid writing—that made it one of the world's best known magazines.

phone sets was originally, however, accomplished manually—on a relatively small scale—by a switchboard operator who made the necessary connections by hand. In such switchboard systems, each telephone set in the network was associated with a jack connector in the switchboard. The operator observed all incoming calls, identified the phone sets for which they were intended, and then used wires to connect the appropriate jacks. At the end of the call, the jacks were disconnected.

This cumbersome methodology limited the size and efficiency of telephone networks and invaded the privacy of callers. The development of automated switching systems soon solved these problems and made switchboard operators obsolete. It was here that Antoine Barnay's rotary dial was used, making possible an exchange that automatically linked the phone sets of callers and call recipients in the following way.

First, a caller lifted a telephone "off the hook," causing a switchhook, like those used in modern phones, to close the circuit that connected the telephone set to the telephone network. Immediately, a dial tone (still familiar to callers) came on to indicate that the automatic switching system could handle the planned call. When the phone dial was used, each number or letter that was dialed produced a fixed number of clicks. Every click indicated that an electrical pulse had been sent to the network's automatic switching system, causing switches to change position slightly. Immediately after a complete telephone number was dialed, the overall operation of the automatic switchers connected the two telephone sets. This connection was carried out much more quickly and accurately than had been possible when telephone operators at manual switchboards made the connection.

IMPACT

The telephone has become the world's most important communication device. Most adults use it between six and eight times per day, for personal and business calls. This widespread use has developed because huge changes have occurred in telephones and telephone networks. For example, automatic switching and the rotary dial system were only the beginning of changes in phone calling.

Touch-tone dialing replaced Barnay's electrical pulses with audio tones outside the frequency of human speech. This much-improved system can be used to send calls over much longer distances than was possible with the rotary dial system, and it also interacts well with both answering machines and computers.

Another advance in modern telephoning is the use of radio transmission techniques in mobile phones, rendering telephone cords obsolete. The mobile phone communicates with base stations arranged in "cells" throughout the service area covered. As the user changes location, the phone link automatically moves from cell to cell in a cellular network.

In addition, the use of microwave, laser, and fiber-optic technologies has helped to lengthen the distance over which phone calls can be transmitted. These technologies have also increased the number of messages that phone networks can handle simultaneously and have made it possible to send radio and television programs (such as cable television), scientific data (via modems), and written messages (via facsimile, or "fax," machines) over phone lines. Many other advances in telephone technology are expected as society's needs change and new technology is developed.

See also Cell phone; Internet; Long-distance telephone; Telephone switching; Touch-tone telephone.

FURTHER READING

Aitken, William. *Who Invented the Telephone?* London: Blackie and Son, 1939.

Coe, Lewis. *The Telephone and Its Several Inventors: A History.* Jefferson, N.C.: McFarland, 1995.

Evenson, A. Edward. *The Telephone Patent Conspiracy of 1876: The Elisha Gray-Alexander Bell Controversy and Its Many Players.* Jefferson, N.C.: McFarland, 2000.

Lisser, Eleena de. "Telecommunications: If You Have a Rotary Phone, Press 1: The Trials of Using the Old Apparatus." *Wall Street Journal* (July 28, 1994).

Mackay, James A. *Alexander Graham Bell: A Life.* New York: J. Wiley, 1997.

SAINT

THE INVENTION: Taking its name from the acronym for symbolic automatic integrator, SAINT is recognized as the first "expert system"—a computer program designed to perform mental tasks requiring human expertise.

THE PERSON BEHIND THE INVENTION:
James R. Slagle (1934-1994), an American computer scientist

THE ADVENT OF ARTIFICIAL INTELLIGENCE

In 1944, the Harvard-IBM Mark I was completed. This was an electromechanical (that is, not fully electronic) digital computer that was operated by means of coding instructions punched into paper tape. The machine took about six seconds to perform a multiplication operation, twelve for a division operation. In the following year, 1945, the world's first fully electronic digital computer, the Electronic Numerical Integrator and Calculator (ENIAC), became operational. This machine, which was constructed at the University of Pennsylvania, was thirty meters long, three meters high, and one meter deep.

At the same time that these machines were being built, a similar machine was being constructed in the United Kingdom: the automated computing engine (ACE). A key figure in the British development was Alan Turing, a mathematician who had used computers to break German codes during World War II. After the war, Turing became interested in the area of "computing machinery and intelligence." He posed the question "Can machines think?" and set the following problem, which is known as the "Turing test." This test involves an interrogator who sits at a computer terminal and asks questions on the terminal about a subject for which he or she seeks intelligent answers. The interrogator does not know, however, whether the system is linked to a human or if the responses are, in fact, generated by a program that is acting intelligently. If the interrogator cannot tell the difference between the human operator and the computer system, then the system is said to have passed the Turing test and has exhibited intelligent behavior.

SAINT: An Expert System

In the attempt to answer Turing's question and create machines that could pass the Turing test, researchers investigated techniques for performing tasks that were considered to require expert levels of knowledge. These tasks included games such as checkers, chess, and poker. These games were chosen because the total possible number of variations in each game was very large. This led the researchers to several interesting questions for study. How do experts make a decision in a particular set of circumstances? How can a problem such as a game of chess be represented in terms of a computer program? Is it possible to know why the system chose a particular solution?

One researcher, James R. Slagle at the Massachusetts Institute of Technology, chose to develop a program that would be able to solve elementary symbolic integration problems (involving the manipulation of integrals in calculus) at the level of a good college freshman. The program that Slagle constructed was known as SAINT, an acronym for symbolic automatic integrator, and it is acknowledged as the first "expert system."

An expert system is a system that performs at the level of a human expert. An expert system has three basic components: a knowledge base, in which domain-specific information is held (for example, rules on how best to perform certain types of integration problems); an inference engine, which decides how to break down a given problem utilizing the rules in the knowledge base; and a human-computer interface that inputs data—in this case, the integral to be solved—and outputs the result of performing the integration. Another feature of expert systems is their ability to explain their reasoning.

The integration problems that could be solved by SAINT were in the form of elementary integral functions. SAINT could perform indefinite integration (also called "antidifferentiation") on these functions. In addition, it was capable of performing definite and indefinite integration on trivial extensions of indefinite integration. SAINT was tested on a set of eighty-six problems, fifty-four of which were drawn from the MIT final examinations in freshman calculus; it succeeded in solving all but two. Slagle added more rules to the knowledge base so that problems of the type it encountered but could not solve could be solved in the future.

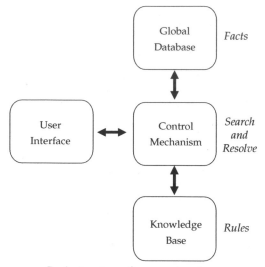

Basic structure of an expert system.

The power of the SAINT system was, in part, based on its ability to perform integration through the adoption of a "heuristic" processing system. A heuristic method is one that helps in discovering a problem's solution by making plausible but feasible guesses about the best strategy to apply next to the current problem situation. A heuristic is a rule of thumb that makes it possible to take short cuts in reaching a solution, rather than having to go through every step in a solution path. These heuristic rules are contained in the knowledge base. SAINT was written in the LISP programming language and ran on an IBM 7090 computer. The program and research were Slagle's doctoral dissertation.

CONSEQUENCES

The SAINT system that Slagle developed was significant for several reasons: First, it was the first serious attempt at producing a program that could come close to passing the Turing test. Second, it brought the idea of representing an expert's knowledge in a computer program together with strategies for solving complex and difficult problems in an area that previously required human expertise. Third, it identified the area of knowledge-based systems and

JAMES R. SLAGLE

James R. Slagle was born in 1934 in Brooklyn, New York, and attended nearby St. John's University. He majored in mathematics and graduated with a bachelor of science degree in 1955, also winning the highest scholastic average award. While earning his master's degree (1957) and doctorate (1961) at the Massachusetts Institute of Technology (MIT), he was a staff mathematician in the university's Lincoln Laboratory.

Slagle taught in MIT's electrical engineering department part-time after completing his dissertation on the first expert computer system and then moved to Lawrence-Livermore National Laboratory near Berkeley, California. While working there he also taught at the University of California. From 1967 until 1974 he was an adjunct member of the computer science faculty of Johns Hopkins University in Baltimore, Maryland, and then was appointed chief of the computer science laboratory at the Naval Research Laboratory (NRL) in Washington, D.C., receiving the Outstanding Handicapped Federal Employee of the Year Award in 1979. In 1984 he was made a special assistant in the Navy Center for Applied Research in Artificial Intelligence at NRL but left in 1984 to become Distinguished Professor of Computer Science at the University of Minnesota.

In these various positions Slagle helped mature the fledgling discipline of artificial intelligence, publishing the influential book *Artificial Intelligence* in 1971. He developed an expert system designed to set up other expert systems—A Generalized Network-based Expert System Shell, or AGNESS. He also worked on parallel expert systems, artificial neural networks, time-based logic, and methods for uncovering causal knowledge in large databases. He died in 1994.

---

showed that computers could feasibly be used for programs that did not relate to business data processing. Fourth, the SAINT system showed how the use of heuristic rules and information could lead to the solution of problems that could not have been solved previously because of the amount of time needed to calculate a solution. SAINT's major impact was in outlining the uses of these techniques, which led to continued research in the subfield of artificial intelligence that became known as expert systems.

See also BASIC programming language; CAD/CAM; COBOL computer language; Differential analyzer; FORTRAN programming language; Robot (industrial).

FURTHER READING

Campbell-Kelly, Martin, and William Aspray. *Computer: A History of the Information Machine.* New York: Basic Books, 1996.
Ceruzzi, Paul E. *A History of Modern Computing.* Cambridge, Mass.: MIT Press, 2000.
Rojas, Paul. *Encyclopedia of Computers and Computer History.* London: Fitzroy Dearborn, 2001.

SALVARSAN

THE INVENTION: The first successful chemotherapeutic for the treatment of syphilis

THE PEOPLE BEHIND THE INVENTION:
Paul Ehrlich (1854-1915), a German research physician and chemist
Wilhelm von Waldeyer (1836-1921), a German anatomist
Friedrich von Frerichs (1819-1885), a German physician and professor
Sahachiro Hata (1872-1938), a Japanese physician and bacteriologist
Fritz Schaudinn (1871-1906), a German zoologist

THE GREAT POX

The ravages of syphilis on humankind are seldom discussed openly. A disease that struck all varieties of people and was transmitted by direct and usually sexual contact, syphilis was both feared and reviled. Many segments of society across all national boundaries were secure in their belief that syphilis was divine punishment of the wicked for their evil ways.

It was not until 1903 that bacteriologists Élie Metchnikoff and Pierre-Paul-Émile Roux demonstrated the transmittal of syphilis to apes, ending the long-held belief that syphilis was exclusively a human disease. The disease destroyed families, careers, and lives, driving its infected victims mad, destroying the brain, or destroying the cardiovascular system. It was methodical and slow, but in every case, it killed with singular precision. There was no hope of a safe and effective cure prior to the discovery of Salvarsan.

Prior to 1910, conventional treatment consisted principally of mercury or, later, potassium iodide. Mercury, however, administered in large doses, led to severe ulcerations of the tongue, jaws, and palate. Swelling of the gums and loosening of the teeth resulted. Dribbling saliva and the attending fetid odor also occurred. These side effects of mercury treatment were so severe that many preferred to suffer the disease to the end rather than undergo the standard cure. About 1906, Metchnikoff and Roux demonstrated that mercurial ointments, applied very early, at the first appearance of the primary lesion, were effective.

Once the spirochete-type bacteria invaded the bloodstream and tissues, the infected person experienced symptoms of varying nature and degree—high fever, intense headaches, and excruciating pain. The patient's skin often erupted in pustular lesions similar in appearance to smallpox. It was the distinguishing feature of these pustular lesions that gave syphilis its other name: the "Great Pox." Death brought the only relief then available.

POISON DYES

Paul Ehrlich became fascinated by the reactions of dyes with biological cells and tissues while a student at the University of Strasbourg under Wilhelm von Waldeyer. It was von Waldeyer who sparked Ehrlich's interest in the chemical viewpoint of medicine. Thus, as a student, Ehrlich spent hours at this laboratory experimenting with different dyes on various tissues. In 1878, he published a book that detailed the discriminate staining of cells and cellular components by various dyes.

Ehrlich joined Friedrich von Frerichs at the Charité Hospital in Berlin, where Frerichs allowed Ehrlich to do as much research as he wanted. Ehrlich began studying atoxyl in 1908, the year he won jointly with Metchnikoff the Nobel Prize in Physiology or Medicine for his work on immunity. Atoxyl was effective against trypanosome—a parasite responsible for a variety of infections, notably sleeping sickness—but also imposed serious side effects upon the patient, not the least of which was blindness. It was Ehrlich's study of atoxyl, and several hundred derivatives sought as alternatives to atoxyl in trypanosome treatment, that led to the development of derivative 606 (Salvarsan). Although compound 606 was the first chemotherapeutic to be used effectively against syphilis, it was discontinued as an atoxyl alternative and shelved as useless for five years.

The discovery and development of compound 606 was enhanced by two critical events. First, the Germans Fritz Schaudinn and Erich

The wonder drug Salvarsan was often called "Ehrlich's silver bullet," after its developer, Paul Ehrlich (left). (Library of Congress)

Hoffmann discovered that syphilis is a bacterially caused disease. The causative microorganism is a spirochete so frail and gossameric in substance that it is nearly impossible to detect by casual microscopic examination; Schaudinn chanced upon it one day in March, 1905. This discovery led, in turn, to German bacteriologist August von Wassermann's development of the now famous test for syphilis: the Wassermann test. Second, a Japanese bacteriologist, Sahachiro Hata, came to Frankfurt in 1909 to study syphilis with Ehrlich. Hata had studied syphilis in rabbits in Japan. Hata's assignment was to test every atoxyl derivative ever developed under Ehrlich for its efficacy in syphilis treatment. After hundreds of tests and clinical trials, Ehrlich and Hata announced Salvarsan as a "magic bullet" that could cure syphilis, at the April, 1910, Congress of Internal Medicine in Wiesbaden, Germany.

The announcement was electrifying. The remedy was immediately and widely sought, but it was not without its problems. A few deaths resulted from its use, and it was not safe for treatment of the gravely ill. Some of the difficulties inherent in Salvarsan were overcome by the development of neosalvarsan in 1912 and sodium salvarsan in 1913. Although Ehrlich achieved much, he fell short of his own assigned goal, a chemotherapeutic that would cure in one injection.

IMPACT

The significance of the development of Salvarsan as an antisyphilitic chemotherapeutic agent cannot be overstated. Syphilis at that time was as frightening and horrifying as leprosy and was a virtual sentence of slow, torturous death. Salvarsan was such a significant development that Ehrlich was recommended for a 1912 and 1913 Nobel Prize for his work in chemotherapy.

It was several decades before any further significant advances in "wonder drugs" occurred, namely, the discovery of prontosil in 1932 and its first clinical use in 1935. On the heels of prontosil—a sulfa drug—came other sulfa drugs. The sulfa drugs would remain supreme in the fight against bacterial infection until the antibiotics, the first being penicillin, were discovered in 1928; however, they were not clinically recognized until World War II (1939-1945). With the discovery of streptomycin in 1943 and Aureomycin in 1944, the assault against bacteria was finally on a sound basis. Medicine possessed an arsenal with which to combat the pathogenic microbes that for centuries before had visited misery and death upon humankind.

See also Abortion pill; Antibacterial drugs; Birth control pill; Penicillin; Reserpine; Syphilis test; Tuberculosis vaccine; Typhus vaccine; Yellow fever vaccine.

FURTHER READING

Bäumler, Ernst. *Paul Ehrlich: Scientist for Life*. New York: Holmes & Meier, 1984.

Leyden, John G. "From Nobel Prize to Courthouse Battle: Paul Ehrlich's 'Wonder Drug' for Syphilis Won Him Acclaim but also Led Critics to Hound Him." *Washington Post* (July 27, 1999).

Quétel, Claude. *History of Syphilis*. Baltimore: Johns Hopkins University Press, 1992.

Scanning tunneling microscope

The invention: A major advance on the field ion microscope, the scanning tunneling microscope has pointed toward new directions in the visualization and control of matter at the atomic level.

The people behind the invention:

Gerd Binnig (1947-   ), a West German physicist who was a cowinner of the 1986 Nobel Prize in Physics

Heinrich Rohrer (1933-   ), a Swiss physicist who was a cowinner of the 1986 Nobel Prize in Physics

Ernst Ruska (1906-1988), a West German engineer who was a cowinner of the 1986 Nobel Prize in Physics

Antoni van Leeuwenhoek (1632-1723), a Dutch naturalist

The Limit of Light

The field of microscopy began at the end of the seventeenth century, when Antoni van Leeuwenhoek developed the first optical microscope. In this type of microscope, a magnified image of a sample is obtained by directing light onto it and then taking the light through a lens system. Van Leeuwenhoek's microscope allowed him to observe the existence of life on a scale that is invisible to the naked eye. Since then, developments in the optical microscope have revealed the existence of single cells, pathogenic agents, and bacteria.

There is a limit, however, to the resolving power of optical microscopes. Known as "Abbe's barrier," after the German physicist and lens maker Ernst Abbe, this limit means that objects smaller than about 400 nanometers (about a millionth of a millimeter) cannot be viewed by conventional microscopes.

In 1925, the physicist Louis de Broglie predicted that electrons would exhibit wave behavior as well as particle behavior. This prediction was confirmed by Clinton J. Davisson and Lester H. Germer of Bell Telephone Laboratories in 1927. It was found that high-energy electrons have shorter wavelengths than low-energy electrons and that electrons with sufficient energies exhibit wavelengths comparable to the diameter of the atom. In 1927, Hans Busch showed in a mathematical analysis that current-carrying coils behave like electron lenses and that they obey the same lens equation that governs optical lenses. Using these findings, Ernst Ruska developed the electron microscope in the early 1930's.

By 1944, the German corporation of Siemens and Halske had manufactured electron microscopes with a resolution of 7 nanometers; modern instruments are capable of resolving objects as small as 0.5 nanometer. This development made it possible to view structures as small as a few atoms across as well as large atoms and large molecules.

The electron beam used in this type of microscope limits the usefulness of the device. First, to avoid the scattering of the electrons, the samples must be put in a vacuum, which limits the applicability of the microscope to samples that can sustain such an environment. Most important, some fragile samples, such as organic molecules, are inevitably destroyed by the high-energy beams required for high resolutions.

VIEWING ATOMS

From 1936 to 1955, Erwin Wilhelm Müller developed the field ion microscope (FIM), which used an extremely sharp needle to hold the sample. This was the first microscope to make possible the direct viewing of atomic structures, but it was limited to samples capable of sustaining the high electric fields necessary for its operation.

In the early 1970's, Russel D. Young and Clayton Teague of the National Bureau of Standards (NBS) developed the "topografiner," a new kind of FIM. In this microscope, the sample is placed at a large distance from the tip of the needle. The tip is scanned across the surface of the sample with a precision of about a nanometer. The precision in the three-dimensional motion of the tip was obtained by using three legs made of piezoelectric crystals. These materials change shape in a reproducible manner when subjected to a voltage. The extent of expansion or contraction of the crystal depends on the amount of voltage that is applied. Thus, the operator can control the motion of the probe by varying the voltage acting on the three legs. The resolution of the topografiner is limited by the size of the probe.

GERD BINNIG AND HEINRICH ROHRER

Both Gerd Binnig and Heinrich Rohrer believe an early and pleasurable introduction to teamwork led to their later success in inventing the scanning tunneling microscope, for which they shared the 1986 Nobel Prize in Physics with Ernst Ruska.

Binnig was born in Frankfurt, Germany, in 1947. He acquired an early interest in physics but was always deeply influenced by classical music, introduced to him by his mother, and the rock music that his younger brother played for him. Binnig played in rock bands as a teenager and learned to enjoy the creative interplay of teamwork. At J. W. Goethe University in Frankfurt he earned a bachelor's degree (1973) and doctorate (1978) in physics and then took a position at International Business Machine's Zurich Research Laboratory. There he recaptured the pleasures of working with a talented team after joining Rohrer in research.

Rohrer had been at the Zurich facility since just after it opened in 1963. He was born in Buch, Switzerland, in 1933, and educated at the Swiss Federal Institute of Technology in Zurich, where he completed his doctorate in 1960. After post-doctoral work at Rutgers University, he joined the IBM research team, a time that he describes as among the most enjoyable passages of his career.

In addition to the Nobel Prize, the pair also received the German Physics Prize, Otto Klung Prize, Hewlett Packard Prize, and King Faisal Prize. Rohrer became an IBM Fellow in 1986 and was selected to manage the physical sciences department at the Zurich Research Laboratory. He retired from IBM in July 1997. Binnig became an IBM Fellow in 1987.

The idea for the scanning tunneling microscope (STM) arose when Heinrich Rohrer of the International Business Machines (IBM) Corporation's Zurich research laboratory met Gerd Binnig in Frankfurt in 1978. The STM is very similar to the topografiner. In the STM, however, the tip is kept at a height of less than a nanometer away from the surface, and the voltage that is applied between the specimen and the probe is low. Under these conditions, the electron cloud of atoms at the end of the tip overlaps with the electron cloud of atoms at the surface of the specimen. This overlapping results in a measurable electrical current flowing through the vacuum or insulating material existing between the tip and the sample. When the probe is moved across the surface and the voltage between the probe and sample is kept constant, the change in the distance between the probe and the surface (caused by surface irregularities) results in a change of the tunneling current.

Two methods are used to translate these changes into an image of the surface. The first method involves changing the height of the probe to keep the tunneling current constant; the voltage used to change the height is translated by a computer into an image of the surface. The second method scans the probe at a constant height away from the sample; the voltage across the probe and sample is changed to keep the tunneling current constant. These changes in voltage are translated into the image of the surface. The main limitation of the technique is that it is applicable only to conducting samples or to samples with some surface treatment.

CONSEQUENCES

In October, 1989, the STM was successfully used in the manipulation of matter at the atomic level. By letting the probe sink into the surface of a metal-oxide crystal, researchers at Rutgers University were able to dig a square hole about 250 atoms across and 10 atoms deep. A more impressive feat was reported in the April 5, 1990, issue of *Nature*; M. Eigler and Erhard K. Schweiser of IBM's Almaden Research Center spelled out their employer's three-letter acronym using thirty-five atoms of xenon. This ability to move and place individual atoms precisely raises several possibilities, which include the creation of custom-made molecules, atomic-scale data storage, and ultrasmall electrical logic circuits.

The success of the STM has led to the development of several new microscopes that are designed to study other features of sample surfaces. Although they all use the scanning probe technique to make measurements, they use different techniques for the actual detection. The most popular of these new devices is the atomic force microscope (AFM). This device measures the tiny electric forces that exist between the electrons of the probe and the electrons of the sample without the need for electron flow, which makes the technique particularly useful in imaging nonconducting surfaces. Other scanned probe microscopes use physical properties such as temperature and magnetism to probe the surfaces.

See also Cyclotron; Electron microscope; Ion field microscope; Mass spectrograph; Neutrino detector; Sonar; Synchrocyclotron; Tevatron accelerator; Ultramicroscope.

FURTHER READING

Morris, Michael D. *Microscopic and Spectroscopic Imaging of the Chemical State*. New York: M. Dekker, 1993.
Wiesendanger, Robert. *Scanning Probe Microscopy: Analytical Methods*. New York: Springer-Verlag, 1998.
_____, and Hans-Joachim Güntherodt. *Scanning Tunneling Microscopy II: Further Applications and Related Scanning Techniques*. 2d ed. New York: Springer, 1995.
_____. *Scanning Tunneling Microscopy III: Theory of STM and Related Scanning Probe Methods*. 2d ed. New York: Springer, 1996.

Silicones

THE INVENTION: Synthetic polymers characterized by lubricity, extreme water repellency, thermal stability, and inertness that are widely used in lubricants, protective coatings, paints, adhesives, electrical insulation, and prosthetic replacements for body parts.

THE PEOPLE BEHIND THE INVENTION:

Eugene G. Rochow (1909-   ), an American research chemist
Frederic Stanley Kipping (1863-1949), a Scottish chemist and professor
James Franklin Hyde (1903-   ), an American organic chemist

SYNTHESIZING SILICONES

Frederic Stanley Kipping, in the first four decades of the twentieth century, made an extensive study of the organic (carbon-based) chemistry of the element silicon. He had a distinguished academic career and summarized his silicon work in a lecture in 1937 ("Organic Derivatives of Silicon"). Since Kipping did not have available any naturally occurring compounds with chemical bonds between carbon and silicon atoms (organosilicon compounds), it was necessary for him to find methods of establishing such bonds. The basic method involved replacing atoms in naturally occurring silicon compounds with carbon atoms from organic compounds.

While Kipping was probably the first to prepare a silicone and was certainly the first to use the term *silicone*, he did not pursue the commercial possibilities of silicones. Yet his careful experimental work was a valuable starting point for all subsequent workers in organosilicon chemistry, including those who later developed the silicone industry.

On May 10, 1940, chemist Eugene G. Rochow of the General Electric (GE) Company's corporate research laboratory in Schenectady, New York, discovered that methyl chloride gas, passed over a heated mixture of elemental silicon and copper, reacted to form compounds with silicon-carbon bonds. Kipping had shown that these silicon compounds react with water to form silicones.

The importance of Rochow's discovery was that it opened the way to a continuous process that did not consume expensive metals, such as magnesium, or flammable ether solvents, such as those used by Kipping and other researchers. The copper acts as a catalyst, and the desired silicon compounds are formed with only minor quantities of by-products. This "direct synthesis," as it came to be called, is now done commercially on a large scale.

SILICONE STRUCTURE

Silicones are examples of what chemists call *polymers*. Basically, a polymer is a large molecule made up of many smaller molecules that are linked together. At the molecular level, silicones consist of long, repeating chains of atoms. In this molecular characteristic, silicones resemble plastics and rubber.

Silicone molecules have a chain composed of alternate silicon and oxygen atoms. Each silicon atom bears two organic groups as substituents, while the oxygen atoms serve to link the silicon atoms into a chain. The silicon-oxygen backbone of the silicones is responsible for their unique and useful properties, such as the ability of a silicone oil to remain liquid over an extremely broad temperature range and to resist oxidative and thermal breakdown at high temperatures.

A fundamental scientific consideration with silicone, as with any polymer, is to obtain the desired physical and chemical properties in a product by closely controlling its chemical structure and molecular weight. Oily silicones with thousands of alternating silicon and oxygen atoms have been prepared. The average length of the molecular chain determines the flow characteristics (viscosity) of the oil. In samples with very long chains, rubber-like elasticity can be achieved by cross-linking the silicone chains in a controlled manner and adding a filler such as silica. High degrees of cross-linking could produce a hard, intractable material instead of rubber.

The action of water on the compounds produced from Rochow's direct synthesis is a rapid method of obtaining silicones, but does not provide much control of the molecular weight. Further development work at GE and at the Dow-Corning company showed that the best procedure for controlled formation of silicone polymers involved treating the crude silicones with acid to produce a mixture

EUGENE G. ROCHOW

Eugene George Rochow was born in 1909 and grew up in the rural New Jersey town of Maplewood. There his father, who worked in the tanning industry, and his big brother maintained a small attic laboratory. They experimented with electricity, radio—Eugene put together his own crystal set in an oatmeal box—and chemistry.

Rochow followed his brother to Cornell University in 1927. The Great Depression began during his junior year, and although he had to take jobs as lecture assistant to get by, he managed to earn his bachelor's degree in chemistry in 1931 and his doctorate in 1935. Luck came his way in the extremely tight job market: General Electric (GE) hired him for his expertise in inorganic chemistry.

In 1938 the automobile industry, among other manufacturers, had a growing need for a high-temperature-resistant insulators. Scientists at Corning were convinced that silicone would have the best properties for the purpose, but they could not find a way to synthesize it cheaply and in large volume. When word about their ideas got back to Rochow at GE, he reasoned that a flexible silicone able to withstand temperatures of 200 to 300 degrees Celsius could be made by bonding silicone to carbon. His research succeeded in producing methyl silicone in 1939, and he devised a way to make it cheaply in 1941. It was the first commercially practical silicone. His process is still used.

After World War II GE asked him to work on an effort to make aircraft carriers nuclear powered. However, Rochow was a Quaker and pacifist, and he refused. Instead, he moved to Harvard University as a chemistry professor in 1948 and remained there until his retirement in 1970. As the founder of a new branch of industrial chemistry, he received most of the discipline's awards and medals, including the Perkin Award, and honorary doctorates.

from which high yields of an intermediate called "D4" could be obtained by distillation. The intermediate D4 could be polymerized in a controlled manner by use of acidic or basic catalysts. Wilton I. Patnode of GE and James F. Hyde of Dow-Corning made important advances in this area. Hyde's discovery of the use of traces of potassium hydroxide as a polymerization catalyst for D4 made possible the manufacture of silicone rubber, which is the most commercially valuable of all the silicones.

IMPACT

Although Kipping's discovery and naming of the silicones occurred from 1901 to 1904, the practical use and impact of silicones started in 1940, with Rochow's discovery of direct synthesis.

Production of silicones in the United States came rapidly enough to permit them to have some influence on military supplies for World War II (1939-1945). In aircraft communication equipment, extensive waterproofing of parts by silicones resulted in greater reliability of the radios under tropical conditions of humidity, where condensing water could be destructive. Silicone rubber, because of its ability to withstand heat, was used in gaskets under high-temperature conditions, in searchlights, and in the engines on B-29 bombers. Silicone grease applied to aircraft engines also helped to protect spark plugs from moisture and promote easier starting.

After World War II, the uses for silicones multiplied. Silicone rubber appeared in many products from caulking compounds to wire insulation to breast implants for cosmetic surgery. Silicone rubber boots were used on the moon walks where ordinary rubber would have failed.

Silicones in their present form owe much to years of patient developmental work in industrial laboratories. Basic research, such as that conducted by Kipping and others, served to point the way and catalyzed the process of commercialization.

See also Buna rubber; Neoprene; Nylon; Plastic; Polystyrene; Teflon.

FURTHER READING

Clarson, Stephen J. *Silicones and Silicone-Modified Materials.* Washington, D.C.: American Chemical Society, 2000.

Koerner, G. *Silicones, Chemistry and Technology.* Boca Raton, Fla.: CRC Press, 1991.

Potter, Michael, and Noel R. Rose. *Immunology of Silicones.* New York: Springer, 1996.

Smith, A. Lee. *The Analytical Chemistry of Silicones.* New York: Wiley, 1991.

Solar Thermal Engine

The invention: The first commercially practical plant for generating electricity from solar energy.

The people behind the invention:
Frank Shuman (1862-1918), an American inventor
John Ericsson (1803-1889), an American engineer
Augustin Mouchout (1825-1911), a French physics professor

Power from the Sun

According to tradition, the Greek scholar Archimedes used reflective mirrors to concentrate the rays of the Sun and set afire the ships of an attacking Roman fleet in 212 b.c.e.. The story illustrates the long tradition of using mirrors to concentrate solar energy from a large area onto a small one, producing very high temperatures.

With the backing of Napoleon III, the Frenchman Augustin Mouchout built, between 1864 and 1872, several steam engines that were powered by the Sun. Mirrors concentrated the sun's rays to a point, producing a temperature that would boil water. The steam drove an engine that operated a water pump. The largest engine had a cone-shaped collector, or "axicon," lined with silver-plated metal. The French government operated the engine for six months but decided it was too expensive to be practical.

John Ericsson, the American famous for designing and building the Civil War ironclad ship *Monitor*, built seven steam-driven solar engines between 1871 and 1878. In Ericsson's design, rays were focused onto a line rather than a point. Long mirrors, curved into a parabolic shape, tracked the Sun. The rays were focused onto a water-filled tube mounted above the reflectors to produce steam. The engineer's largest engine, which used an 11- × 16-foot trough-shaped mirror, delivered nearly 2 horsepower. Because his solar engines were ten times more expensive than conventional steam engines, Ericsson converted them to run on coal to avoid financial loss.

Frank Shuman, a well-known inventor in Philadelphia, Pennsylvania, entered the field of solar energy in 1906. The self-taught engineer believed that curved, movable mirrors were too expensive. His first large solar engine was a hot-box, or flat-plate, collector. It lay flat on the ground and had blackened pipes filled with a liquid that had a low boiling point. The solar-heated vapor ran a 3.5-horsepower engine.

Shuman's wealthy investors formed the Sun Power Company to develop and construct the largest solar plant ever built. The site chosen was in Egypt, but the plant was built near Shuman's home for testing before it was sent to Egypt.

When the inventor added ordinary flat mirrors to reflect more sunlight into each collector, he doubled the heat production of the collectors. The 572 trough-type collectors were assembled in twenty-six rows. Water was piped through the troughs and converted to steam. A condenser converted the steam to water, which reentered the collectors. The engine pumped 3,000 gallons of water per minute and produced 14 horsepower per day; performance was expected to improve 25 percent in the sunny climate of Egypt.

British investors requested that professor C. V. Boys review the solar plant before it was shipped to Egypt. Boys pointed out that the bottom of each collector was not receiving any direct solar energy; in fact, heat was being lost through the bottom. He suggested that each row of flat mirrors be replaced by a single parabolic reflector, and Shuman agreed. Shuman thought Boys's idea was original, but he later realized it was based on Ericsson's design.

The company finally constructed the improved plant in Meadi, Egypt, a farming district on the Nile River. Five solar collectors, spaced 25 feet apart, were built in a north-south line. Each was about 200 feet long and 10 feet wide. Trough-shaped reflectors were made of mirrors held in place by brass springs that expanded and contracted with changing temperatures. The parabolic mirrors shifted automatically so that the rays were always focused on the boiler. Inside the 15-inch boiler that ran down the middle of the collector, water was heated and converted to steam. The engine produced more than 55 horsepower, which was enough to pump 6,000 gallons of water per minute.

The purchase price of Shuman's solar plant was twice as high as

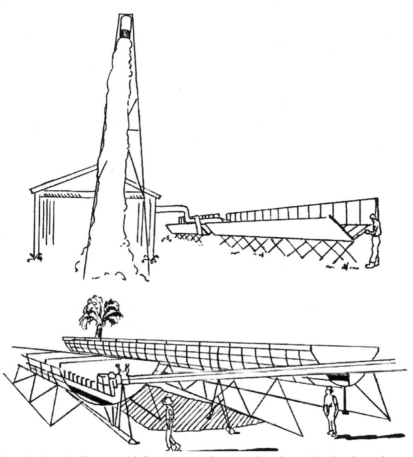

Trough-shaped collectors with flat mirrors (above) produced enough solar thermal energy to pump 3,000 gallons of water per minute. Trough-shaped collectors with parabolic mirrors (below) produced enough solar thermal energy to pump 6,000 gallons of water per minute.

that of a coal-fired plant, but its operating costs were far lower. In Egypt, where coal was expensive, the entire purchase price would be recouped in four years. Afterward, the plant would operate for practically nothing. The first practical solar engine was now in operation, providing enough energy to drive a large-scale irrigation system in the floodplain of the Nile River.

By 1914, Shuman's work was enthusiastically supported, and solar plants were planned for India and Africa. Shuman hoped to build 20,000 reflectors in the Sahara Desert and generate energy equal to all the coal mined in one year, but the outbreak of World

War I ended his dreams of large-scale solar developments. The Meadi project was abandoned in 1915, and Shuman died before the war ended. Powerful nations lost interest in solar power and began to replace coal with oil. Rich oil reserves were discovered in many desert zones that were ideal locations for solar power.

IMPACT

Although World War I ended Frank Shuman's career, his break-through proved to the world that solar power held great promise for the future. His ideas were revived in 1957, when the Soviet Union planned a huge solar project for Siberia. A large boiler was fixed on a platform 140 feet high. Parabolic mirrors, mounted on 1,300 rail-road cars, revolved on circular tracks to focus light on the boiler. The full-scale model was never built, but the design inspired the solar power tower.

In the Mojave desert near Barstow, California, an experimental power tower, Solar One, began operation in 1982. The system collects solar energy to deliver steam to turbines that produce electric power. The 30-story tower is surrounded by more than 1,800 mirrors that adjust continually to track the Sun. Solar One generates about 10 megawatts per day, enough power for 5,000 people.

Solar One was expensive, but future power towers will generate electricity as cheaply as fossil fuels can. If the costs of the air and water pollution caused by coal burning were considered, solar power plants would already be recognized as cost effective. Meanwhile, Frank Shuman's success in establishing and operating a thoroughly practical large-scale solar engine continues to inspire research and development.

See also Compressed-air-accumulating power plant; Fuel cell; Geothermal power; Nuclear power plant; Photoelectric cell; Photovoltaic cell; Tidal power plant.

FURTHER READING

De Kay, James T. *Monitor: The Story of the Legendary Civil War Ironclad and the Man Whose Invention Changed the Course of History*. New York: Ballantine, 1999.

Mancini, Thomas R., James M. Chavez, and Gregory J. Kolb. "Solar Thermal Power Today and Tomorrow." *Mechanical Engineering* 116, no. 8 (August, 1994).

Moore, Cameron M. "Cooking Up Electricity with Sunlight." *The World & I* 12, no. 7 (July, 1997).

Parrish, Michael. "Enron Makes Electrifying Proposal Energy: The Respected Developer Announces a Huge Solar Plant and a Breakthrough Price." *Los Angeles Times* (November 5, 1994).

Sonar

The invention: A device that detects soundwaves transmitted through water, sonar was originally developed to detect enemy submarines but is also used in navigation, fish location, and ocean mapping.

The people behind the invention:
Jacques Curie (1855-1941), a French physicist
Pierre Curie (1859-1906), a French physicist
Paul Langévin (1872-1946), a French physicist

Active Sonar, Submarines, and Piezoelectricity

Sonar, which stands for sound navigation and ranging, is the American name for a device that the British call "asdic." There are two types of sonar. Active sonar, the more widely used of the two types, detects and locates underwater objects when those objects reflect sound pulses sent out by the sonar. Passive sonar merely listens for sounds made by underwater objects. Passive sonar is used mostly when the loud signals produced by active sonar cannot be used (for example, in submarines).

The invention of active sonar was the result of American, British, and French efforts, although it is often credited to Paul Langévin, who built the first working active sonar system by 1917. Langévin's original reason for developing sonar was to locate icebergs, but the horrors of German submarine warfare in World War I led to the new goal of submarine detection. Both Langévin's short-range system and long-range modern sonar depend on the phenomenon of "piezo-electricity," which was discovered by Pierre and Jacques Curie in 1880. (Piezoelectricity is electricity that is produced by certain materials, such as certain crystals, when they are subjected to pressure.) Since its invention, active sonar has been improved and its capabilities have been increased. Active sonar systems are used to detect submarines, to navigate safely, to locate schools of fish, and to map the oceans.

SONAR THEORY, DEVELOPMENT, AND USE

Although active sonar had been developed by 1917, it was not available for military use until World War II. An interesting major use of sonar before that time was measuring the depth of the ocean. That use began when the 1922 German Meteor Oceanographic Expedition was equipped with an active sonar system. The system was to be used to help pay German World War I debts by aiding in the recovery of gold from wrecked vessels. It was not used successfully to recover treasure, but the expedition's use of sonar to determine ocean depth led to the discovery of the Mid-Atlantic Ridge. This development revolutionized underwater geology.

Active sonar operates by sending out sound pulses, often called "pings," that travel through water and are reflected as echoes when they strike large objects. Echoes from these targets are received by the system, amplified, and interpreted. Sound is used instead of light or radar because its absorption by water is much lower. The time that passes between ping transmission and the return of an echo is used to identify the distance of a target from the system by means of a method called "echo ranging." The basis for echo ranging is the normal speed of sound in seawater (5,000 feet per second). The distance of the target from the radar system is calculated by means of a simple equation: range = speed of sound × 0.5 elapsed time. The time is divided in half because it is made up of the time taken to reach the target and the time taken to return.

The ability of active sonar to show detail increases as the energy of transmitted sound pulses is raised by decreasing the sound wavelength. Figuring out active sonar data is complicated by many factors. These include the roughness of the ocean, which scatters sound and causes the strength of echoes to vary, making it hard to estimate the size and identity of a target; the speed of the sound wave, which changes in accordance with variations in water temperature, pressure, and saltiness; and noise caused by waves, sea animals, and ships, which limits the range of active sonar systems.

A simple active pulse sonar system produces a piezoelectric signal of a given frequency and time duration. Then, the signal is amplified and turned into sound, which enters the water. Any echo that is produced returns to the system to be amplified and used to determine the identity and distance of the target.

Most active sonar systems are mounted near surface vessel keels or on submarine hulls in one of three ways. The first and most popular mounting method permits vertical rotation and scanning of a section of the ocean whose center is the system's location. The second method, which is most often used in depth sounders, directs the beam downward in order to measure ocean depth. The third method, called wide scanning, involves the use of two sonar systems, one mounted on each side of the vessel, in such a way that the two beams that are produced scan the whole ocean at right angles to the direction of the vessel's movement.

Active single-beam sonar operation applies an alternating voltage to a piezoelectric crystal, making it part of an underwater loudspeaker (transducer) that creates a sound beam of a particular frequency. When an echo returns, the system becomes an underwater microphone (receiver) that identifies the target and determines its range. The sound frequency that is used is determined by the sonar's

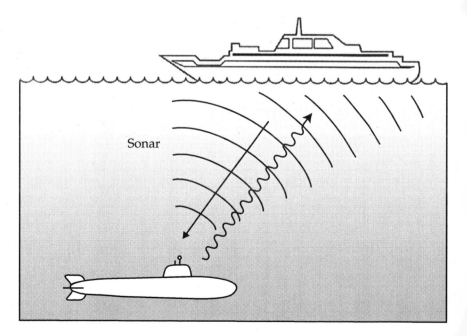

Active sonar detects and locates underwater objects that reflect sound pulses sent out by the sonar.

PAUL LANGÉVIN

If he had not published the Special Theory of Relativity in 1905, Albert Einstein once said, Paul Langévin would have done so not long afterward. Born in Paris in 1872, Langévin was among the foremost physicists of his generation. He studied in the best French schools of science—and with such teachers as Pierre Curie and Jean Perrin—and became a professor of physics at the College de France in 1904. He moved to the Sorbonne in 1909.

Langévin's research was always widely influential. In addition to his invention of active sonar, he was especially noted for his studies of the molecular structure of gases, analysis of secondary X rays from irradiated metals, his theory of magnetism, and work on piezoelectricity and piezoceramics. His suggestion that magnetic properties are linked to the valence electrons of atoms inspired Niels Bohr's classic model of the atom. In his later career, a champion of Einstein's theories of relativity, Langévin worked on the implications of the space-time continuum.

During World War II, Langévin, a pacifist, publicly denounced the Nazis and their occupation of France. They jailed him for it. He escaped to Switzerland in 1944, returning as soon as France was liberated. He died in late 1946.

---

purpose and the fact that the absorption of sound by water increases with frequency. For example, long-range submarine-seeking sonar systems (whose detection range is about ten miles) operate at 3 to 40 kilohertz. In contrast, short-range systems that work at about 500 feet (in mine sweepers, for example) use 150 kilohertz to 2 megahertz.

IMPACT

Modern active sonar has affected military and nonmilitary activities ranging from submarine location to undersea mapping and fish location. In all these uses, two very important goals have been to increase the ability of sonar to identify a target and to increase the effective range of sonar. Much work related to these two goals has involved the development of new piezoelectric materials and the replacement of natural minerals (such as quartz) with synthetic piezoelectric ceramics.

Efforts have also been made to redesign the organization of sonar systems. One very useful development has been changing beam-making transducers from one-beam units to multibeam modules made of many small piezoelectric elements. Systems that incorporate these developments have many advantages, particularly the ability to search simultaneously in many directions. In addition, systems have been redesigned to be able to scan many echo beams simultaneously with electronic scanners that feed into a central receiver.

These changes, along with computer-aided tracking and target classification, have led to the development of greatly improved active sonar systems. It is expected that sonar systems will become even more powerful in the future, finding uses that have not yet been imagined.

See also Aqualung; Bathyscaphe; Bathysphere; Geiger counter; Gyrocompass; Radar; Richter scale; Ultrasound.

FURTHER READING

Curie, Marie. *Pierre Curie*. New York: Dover Publications, 1923.
Hackmann, Willem Dirk. *Seek and Strike: Sonar, Anti-Submarine Warfare, and the Royal Navy, 1914-54*. London: H.M.S.O., 1984.
Segrè, Emilio. *From X-Rays to Quarks: Modern Physicists and Their Discoveries*. San Francisco: W. H. Freeman, 1980.
Senior, John E. *Marie and Pierre Curie*. Gloucestershire: Sutton, 1998.

STEALTH AIRCRAFT

THE INVENTION: The first generation of "radar-invisible" aircraft, stealth planes were designed to elude enemy radar systems.

THE PEOPLE BEHIND THE INVENTION:
Lockhead Corporation, an American research and development firm
Northrop Corporation, an American aerospace firm

RADAR

During World War II, two weapons were developed that radically altered the thinking of the U.S. military-industrial establishment and the composition of U.S. military forces. These weapons were the atomic bombs that were dropped on the Japanese cities of Hiroshima and Nagasaki by U.S. forces and "radio detection and ranging," or radar. Radar saved the English during the Battle of Britain, and it was radar that made it necessary to rethink aircraft design. With radar, attacking aircraft can be detected hundreds of miles from their intended targets, which makes it possible for those aircraft to be intercepted before they can attack. During World War II, radar, using microwaves, was able to relay the number, distance, direction, and speed of German aircraft to British fighter interceptors. This development allowed the fighter pilots of the Royal Air Force, "the few" who were so highly praised by Winston Churchill, to shoot down four times as many planes as they lost.

Because of the development of radar, American airplane design strategy has been to reduce the planes' cross sections, reduce or eliminate the use of metal by replacing it with composite materials, and eliminate the angles that are found on most aircraft control surfaces. These actions help make aircraft less visible—and in some cases, almost invisible—to radar. The Lockheed F-117A Nightrider and the Northrop B-2 Stealth Bomber are the results of these efforts.

AIRBORNE "NINJAS"

Hidden inside Lockheed Corporation is a research and development organization that is unique in the corporate world. This facility has provided the Air Force with the Sidewinder heat-seeking missile; the SR-71, a titanium-skinned aircraft that can fly at four times the speed of sound; and, most recently, the F-117A Nightrider. The Nightrider eluded Iraqi radar so effectively during the 1991 Persian Gulf War that the Iraqis nicknamed it *Shaba*, which is an Arabic word that means ghost. In an unusual move for military projects, the Nightrider was delivered to the Air Force in 1982, before the plane had been perfected. This was done so that Air Force pilots could test fly the plane and provide input that could be used to improve the aircraft before it went into full production.

The Northrop B-2 Stealth Bomber was the result of a design philosophy that was completely different from that of the F-117A Nightrider. The F-117A, for example, has a very angular appearance, but the angles are all greater than 180 degrees. This configuration spreads out radar waves rather than allowing them to be concentrated and sent back to their point of origin. The B-2, however, stays away from angles entirely, opting for a smooth surface that also acts to spread out the radar energy. (The B-2 so closely resembles the YB-49 Flying Wing, which was developed in the late 1940's, that it even has the same wingspan.) The surface of the aircraft is covered with radar-absorbing material and carries its engines and weapons inside to reduce the radar cross section. There are no vertical control surfaces, which has the disadvantage of making the aircraft unstable, so the stabilizing system uses computers to make small adjustments in the control elements on the trailing edges of the wings, thus increasing the craft's stability.

The F-117A Nightrider and the B-2 Stealth Bomber are the "ninjas" of military aviation. Capable of striking powerfully, rapidly, and invisibly, these aircraft added a dimension to the U.S. Air Force that did not exist previously. Before the advent of these aircraft, missions that required radar-avoidance tactics had to be flown below the horizon of ground-based radar, which is 30.5 meters above the ground. Such low-altitude flight is dangerous because of both the increased difficulty of maneuvering and vulnerability to ground fire. Additionally, such flying does not conceal the aircraft from the airborne radar carried by such craft as the American E-3A AWACS and the former Soviet Mainstay. In a major conflict, the only aircraft that could effectively penetrate enemy airspace would be the Nightrider and the B-2.

The purpose of the B-2 was to carry nuclear weapons into hostile airspace undetected. With the demise of the Soviet Union, mainland China seemed the only remaining major nuclear threat. For this reason, many defense experts believed that there was no longer a need for two radar-invisible planes, and cuts in U.S. military expenditures threatened the B-2 program during the early 1990's.

CONSEQUENCES

The development of the Nightrider and the B-2 meant that the former Soviet Union would have had to spend at least $60 billion to upgrade its air defense forces to meet the challenge offered by these aircraft. This fact, combined with the evolution of the Strategic Defense Initiative, commonly called "Star Wars," led to the United States' victory in the arms race. Additionally, stealth technology has found its way onto the conventional battlefield.

As was shown in 1991 during the Desert Storm campaign in Iraq, targets that have strategic importance are often surrounded by a network of anti-air missiles and gun emplacements. During the Desert Storm air war, the F-117A was the only Allied aircraft to be assigned to targets in Baghdad. Nightriders destroyed more than 47 percent of the strategic areas that were targeted, and every pilot and plane returned to base unscathed.

Since the world appears to be moving away from superpower conflicts and toward smaller regional conflicts, stealth aircraft may come to be used more for surveillance than for air attacks. This is particularly true because the SR-71, which previously played the primary role in surveillance, has been retired from service.

See also Airplane; Cruise missile; Hydrogen bomb; Radar; Rocket; Turbojet; V-2 rocket.

FURTHER READING

Chun, Clayton K. S. *The Lockheed F-117A*. Santa Monica, Calif.: Rand, 1991.

Goodall, James C. *America's Stealth Fighters and Bombers*. Osceola, Wis.: Motorbooks, 1992.

Pape, Garry R., and John M. Campbell. *Northrop Flying Wings: A History of Jack Northrop's Visionary Aircraft*. Atglen, Pa.: Schiffer, 1995.

Thornborough, Anthony M. *Stealth*. London: Ian Allen, 1991.

Steelmaking process

The invention: Known as the basic oxygen, or L-D, process, a
method for producing steel that worked about twelve times
faster than earlier methods.

The people behind the invention:
Henry Bessemer (1813-1898), the English inventor of a process
for making steel from iron
Robert Durrer (1890-1978), a Swiss scientist who first proved
the workability of the oxygen process in a laboratory
F. A. Loosley (1891-1966), head of research and development at
Dofasco Steel in Canada
Theodor Suess (1894-1956), works manager at Voest

Ferrous Metal

The modern industrial world is built on ferrous metal. Until
1857, ferrous metal meant cast iron and wrought iron, though a few
specialty uses of steel, especially for cutlery and swords, had existed
for centuries. In 1857, Henry Bessemer developed the first large-
scale method of making steel, the Bessemer converter. By the 1880's,
modification of his concepts (particularly the development of a "basic" process that could handle ores high in phosphor) had made
large-scale production of steel possible.

Bessemer's invention depended on the use of ordinary air, infused
into the molten metal, to burn off excess carbon. Bessemer himself
had recognized that if it had been possible to use pure oxygen instead
of air, oxidation of the carbon would be far more efficient and rapid.
Pure oxygen was not available in Bessemer's day, except at very high
prices, so steel producers settled for what was readily available, ordinary air. In 1929, however, the Linde-Frakl process for separating the
oxygen in air from the other elements was discovered, and for the
first time inexpensive oxygen became available.

Nearly twenty years elapsed before the ready availability of pure
oxygen was applied to refining the method of making steel. The first
experiments were carried out in Switzerland by Robert Durrer. In

1949, he succeeded in making steel expeditiously in a laboratory setting through the use of a blast of pure oxygen. Switzerland, however, had no large-scale metallurgical industry, so the Swiss turned the idea over to the Austrians, who for centuries had exploited the large deposits of iron ore in a mountain in central Austria. Theodor Suess, the works manager of the state-owned Austrian steel complex, Voest, instituted some pilot projects. The results were sufficiently favorable to induce Voest to authorize construction of production converters. In 1952, the first "heat" (as a batch of steel is called) was "blown in," at the Voest works in Linz. The following year, another converter was put into production at the works in Donauwitz. These two initial locations led to the basic oxygen process sometimes being referred to as the L-D process.

THE L-D PROCESS

The basic oxygen, or L-D, process makes use of a converter similar to the Bessemer converter. Unlike the Bessemer, however, the L-D converter blows pure oxygen into the molten metal from above through a water-cooled injector known as a lance. The oxygen burns off the excess carbon rapidly, and the molten metal can then be poured off into ingots, which can later be reheated and formed into the ultimately desired shape. The great advantage of the process is the speed with which a "heat" reaches the desirable metallurgical composition for steel, with a carbon content between 0.1 percent and 2 percent. The basic oxygen process requires about forty minutes. In contrast, the prevailing method of making steel, using an open-hearth furnace (which transferred the technique from the closed Bessemer converter to an open-burning furnace to which the necessary additives could be introduced by hand) requires eight to eleven hours for a "heat" or batch.

The L-D process was not without its drawbacks, however. It was adopted by the Austrians because, by carefully calibrating the timing and amount of oxygen introduced, they could turn their moderately phosphoric ore into steel without further intervention. The process required ore of a standardized metallurgical, or chemical, content, for which the lancing had been calculated. It produced a large amount of iron-oxide dust that polluted the surrounding atmosphere, and it required a lining in the converter of dolomitic brick. The specific chemical content of the brick contributed to the chemical mixture that produced the desired result.

The Austrians quickly realized that the process was an improvement. In May, 1952, the patent specifications for the new process were turned over to a new company, Brassert Oxygen Technik, or BOT, which filed patent applications around the world. BOT embarked on an aggressive marketing campaign, bringing potential customers to Austria to observe the process in action. Despite BOT's efforts, the new process was slow to catch on, even though in 1953 BOT licensed a U.S. firm, Kaiser Engineers, to spread the process in the United States.

Many factors serve to explain the reluctance of steel producers around the world to adopt the new process. One of these was the large investment most major steel producers had in their open-hearth furnaces. Another was uncertainty about the pollution factor. Later, special pollution-control equipment would be developed to deal with this problem. A third concern was whether the necessary refractory liners for the new converters would be available. A fourth was the fact that the new process could handle a load that contained no more than 30 percent scrap, preferably less. In practice, therefore, it would only work where a blast furnace smelting ore was already set up.

One of the earliest firms to show serious interest in the new technology was Dofasco, a Canadian steel producer. Between 1952 and 1954, Dofasco, pushed by its head of research and development, F. A. Loosley, built pilot operations to test the methodology. The results were sufficiently promising that in 1954 Dofasco built the first basic oxygen furnace outside Austria. Dofasco had recently built its own blast furnace, so it had ore available on site. It was able to devise ways of dealing with the pollution problem, and it found refractory liners that would work. It became the first North American producer of basic oxygen steel.

Having bought the licensing rights in 1953, Kaiser Engineers was looking for a U.S. steel producer adventuresome enough to invest in the new technology. It found that producer in McLouth Steel, a small steel plant in Detroit, Michigan. Kaiser Engineers supplied much of the technical advice that enabled McLouth to build the first

HENRY BESSEMER

Henry Bessemer was born in the small village of Charlton, England, in 1813. His father was an early example of a technician, specializing in steam engines, and operated a business making metal type for printing presses. The elder Bessemer wanted his son to attend university, but Henry preferred to study under his father. During his apprenticeship, he learned the properties of alloys. At seventeen he moved to London to open his own business, which fabricated specialty metals.

Three years later the Royal Academy held an exhibition of Bessemer's work. His career, well begun, moved from one invention to another until at his death in 1898 he held 114 patents. Among them were processes for casting type and producing graphite for pencils; methods for manufacturing glass, sugar, bronze powder, and ships; and his best known creation, the Bessemer converter for making steel from iron. Bessemer built his first converter in 1855; fifteen years later Great Britain was producing half of the world's steel.

Bessemer's life and career were models of early Industrial Age industry, prosperity, and longevity. A millionaire from patent royalties, he retired at fifty-nine, lived another twenty-six years, working on yet more inventions and cultivating astronomy as a hobby, and was married for sixty-four years. Among his many awards and honors was a knighthood, bestowed by Queen Victoria.

U.S. basic oxygen steel facility, though McLouth also sent one of its engineers to Europe to observe the Austrian operations. McLouth, which had backing from General Motors, also made use of technical descriptions in the literature.

THE SPECIFICATIONS QUESTION

One factor that held back adoption of basic oxygen steelmaking was the question of specifications. Many major engineering projects came with precise specifications detailing the type of steel to be used and even the method of its manufacture. Until basic oxygen steel was recognized as an acceptable form by the engineering fraternity, so that job specifications included it as appropriate in specific applications, it could not find large-scale markets. It took a number of years for engineers to modify their specifications so that basic oxygen steel could be used.

The next major conversion to the new steelmaking process occurred in Japan. The Japanese had learned of the process early, while Japanese metallurgical engineers were touring Europe in 1951. Some of them stopped off at the Voest works to look at the pilot projects there, and they talked with the Swiss inventor, Robert Durrer. These engineers carried knowledge of the new technique back to Japan. In 1957 and 1958, Yawata Steel and Nippon Kokan, the largest and third-largest steel producers in Japan, decided to implement the basic oxygen process. An important contributor to this decision was the Ministry of International Trade and Industry, which brokered a licensing arrangement through Nippon Kokan, which in turn had signed a one-time payment arrangement with BOT. The licensing arrangement allowed other producers besides Nippon Kokan to use the technique in Japan.

The Japanese made two important technical improvements in the basic oxygen technology. They developed a multiholed lance for blowing in oxygen, thus dispersing it more effectively in the molten metal and prolonging the life of the refractory lining of the converter vessel. They also pioneered the OG process for recovering some of the gases produced in the converter. This procedure reduced the pollution generated by the basic oxygen converter.

The first large American steel producer to adopt the basic oxygen process was Jones and Laughlin, which decided to implement the new process for several reasons. It had some of the oldest equipment in the American steel industry, ripe for replacement. It also had experienced significant technical difficulties at its Aliquippa plant, difficulties it was unable to solve by modifying its open-hearth procedures. It therefore signed an agreement with Kaiser Engineers to build some of the new converters for Aliquippa. These converters were constructed on license from Kaiser Engineers by Pennsylvania Engineering, with the exception of the lances, which were imported from Voest in Austria. Subsequent lances, however, were built in the United States. Some of Jones and Laughlin's production managers were sent to Dofasco for training, and technical advisers were brought to the Aliquippa plant both from Kaiser Engineers and from Austria.

Other European countries were somewhat slower to adopt the new process. A major cause for the delay was the necessary modification of the process to fit the high phosphoric ores available in Germany and France. Europeans also experimented with modifications of the basic oxygen technique by developing converters that revolved. These converters, known as Kaldo in Sweden and Rotor in Germany, proved in the end to have sufficient technical difficulties that they were abandoned in favor of the standard basic oxygen converter. The problems they had been designed to solve could be better dealt with through modifications of the lance and through adjustments in additives.

By the mid-1980's, the basic oxygen process had spread throughout the world. Neither Japan nor the European Community was producing any steel by the older, open-hearth method. In conjunction with the electric arc furnace, fed largely on scrap metal, the basic oxygen process had transformed the steel industry of the world.

IMPACT

The basic oxygen process has significant advantages over older procedures. It does not require additional heat, whereas the open-hearth technique calls for the infusion of nine to twelve gallons of fuel oil to raise the temperature of the metal to the level necessary to burn off all the excess carbon. The investment cost of the converter is about half that of an open-hearth furnace. Fewer refractories are required, less than half those needed in an open-hearth furnace. Most important of all, however, a "heat" requires less than an hour, as compared with the eight or more hours needed for a "heat" in an open-hearth furnace.

There were some disadvantages to the basic oxygen process. Perhaps the most important was the limited amount of scrap that could be included in a "heat," a maximum of 30 percent. Because the process required at least 70 percent new ore, it could be operated most effectively only in conjunction with a blast furnace. Counterbalancing this last factor was the rapid development of the electric arc furnace, which could operate with 100 percent scrap. A firm with its own blast furnace could, with both an oxygen converter and an electric arc furnace, handle the available raw material.

The advantages of the basic oxygen process overrode the disadvantages. Some other new technologies combined to produce this effect. The most important of these was continuous casting. Because of the short time required for a "heat," it was possible, if a plant had two or three converters, to synchronize output with the fill needs of a continuous caster, thus largely canceling out some of the economic drawbacks of the batch process. Continuous production, always more economical, was now possible in the basic steel industry, particularly after development of computer-controlled rolling mills.

These new technologies forced major changes in the world's steel industry. Labor requirements for the basic oxygen converter were about half those for the open-hearth furnace. The high speed of the new technology required far less manual labor but much more technical expertise. Labor requirements were significantly reduced, producing major social dislocations in steel-producing regions. This effect was magnified by the fact that demand for steel dropped sharply in the 1970's, further reducing the need for steelworkers.

The U.S. steel industry was slower than either the Japanese or the European to convert to the basic oxygen technique. The U.S. industry generally operated with larger quantities, and it took a number of years before the basic oxygen technique was adapted to converters with an output equivalent to that of the open-hearth furnace. By the time that had happened, world steel demand had begun to drop. U.S. companies were less profitable, failing to generate internally the capital needed for the major investment involved in abandoning open-hearth furnaces for oxygen converters. Although union contracts enabled companies to change work assignments when new technologies were introduced, there was stiff resistance to reducing employment of steelworkers, most of whom had lived all their lives in one-industry towns. Finally, engineers at the steel firms were wedded to the old methods and reluctant to change, as were the large bureaucracies of the big U.S. steel firms.

The basic oxygen technology in steel is part of a spate of new technical developments that have revolutionized industrial production, drastically reducing the role of manual labor and dramatically increasing the need for highly skilled individuals with technical expertise. Because capital costs are significantly lower than for alternative processes, it has allowed a number of developing countries to enter a heavy industry and compete successfully with the old industrial giants. It has thus changed the face of the steel industry.

See also Assembly line; Buna rubber; Disposable razor; Laminated glass; Memory metal; Neoprene; Oil-well drill bit; Pyrex glass.

FURTHER READING

Bain, Trevor. *Banking the Furnace: Restructuring of the Steel Industry in Eight Countries.* Kalamazoo, Mich.: W. E. Upjohn Institute for Employment Research, 1992.

Gold, Bela, Gerhard Rosegger, and Myles G. Boylan, Jr. *Evaluating Technological Innovations: Methods, Expectations, and Findings.* Lexington, Mass.: Lexington Books, 1980.

Hall, Christopher. *Steel Phoenix: The Fall and Rise of the U.S. Steel Industry.* New York: St. Martin's Press, 1997.

Hoerr, John P. *And the Wolf Finally Came: The Decline of the American Steel Industry.* Pittsburgh, Pa.: University of Pittsburgh Press, 1988.

Lynn, Leonard H. *How Japan Innovates: A Comparison with the United States in the Case of Oxygen Steelmaking.* Boulder, Colo.: Westview Press, 1982.

Seely, Burce Edsall. *Iron and Steel in the Twentieth Century.* New York: Facts on File, 1994.

Supercomputer

The invention: A computer that had the greatest computational power that then existed.

The person behind the invention:
Seymour R. Cray (1928-1996), American computer architect and designer

The Need for Computing Power

Although modern computers have roots in concepts first proposed in the early nineteenth century, it was only around 1950 that they became practical. Early computers enabled their users to calculate equations quickly and precisely, but it soon became clear that even more powerful computers—machines capable of receiving, computing, and sending out data with great precision and at the highest speeds—would enable researchers to use computer "models," which are programs that simulate the conditions of complex experiments.

Few computer manufacturers gave much thought to building the fastest machine possible, because such an undertaking is expensive and because the business use of computers rarely demands the greatest processing power. The first company to build computers specifically to meet scientific and governmental research needs was Control Data Corporation (CDC). The company had been founded in 1957 by William Norris, and its young vice president for engineering was the highly respected computer engineer Seymour R. Cray. When CDC decided to limit high-performance computer design, Cray struck out on his own, starting Cray Research in 1972. His goal was to design the most powerful computer possible. To that end, he needed to choose the principles by which his machine would operate; that is, he needed to determine its architecture.

The Fastest Computer

All computers rely upon certain basic elements to process data. Chief among these elements are the central processing unit, or CPU (which handles data), memory (where data are stored temporarily before and after processing), and the bus (the interconnection between memory and the processor, and the means by which data are transmitted to or from other devices, such as a disk drive or a monitor). The structure of early computers was based on ideas developed by the mathematician John von Neumann, who, in the 1940's, conceived a computer architecture in which the CPU controls all events in a sequence: It fetches data from memory, performs calculations on those data, and then stores the results in memory. Because it functions in sequential fashion, the speed of this "scalar processor" is limited by the rate at which the processor is able to complete each cycle of tasks.

Before Cray produced his first supercomputer, other designers tried different approaches. One alternative was to link a vector processor to a scalar unit. A vector processor achieves its speed by performing computations on a large series of numbers (called a vector) at one time rather than in sequential fashion, though specialized and complex programs were necessary to make use of this feature. In fact, vector processing computers spent most of their time operating as traditional scalar processors and were not always efficient at switching back and forth between the two processing types.

Another option chosen by Cray's competitors was the notion of "pipelining" the processor's tasks. A scalar processor often must wait while data are retrieved or stored in memory. Pipelining techniques allow the processor to make use of idle time for calculations in other parts of the program being run, thus increasing the effective speed. A variation on this technique is "parallel processing," in which multiple processors are linked. If each of, for example, eight central processors is given a portion of a computing task to perform, the task will be completed more quickly than the traditional von Neumann architecture, with its single processor, would allow.

Ever the pragmatist, however, Cray decided to employ proved technology rather than use advanced techniques in his first supercomputer, the Cray 1, which was introduced in 1976. Although the Cray 1 did incorporate vector processing, Cray used a simple form of vector calculation that made the technique practical and easy to use. Most striking about this computer was its shape, which was far more modern than its internal design. The Cray 1 was shaped like a

SEYMOUR R. CRAY

Seymour R. Cray was born in 1928 in Chippewa Falls, Wisconsin. The son of a civil engineer, he became interested in radio and electronics as a boy. After graduating from high school in 1943, he joined the U.S. Army, was posted to Europe in an infantry communications platoon, and fought in the Battle of the Bulge. Back from the war, he pursued his interest in electronics in college while majoring in mathematics at the University of Minnesota. Upon graduation in 1950, he took a job at Engineering Research Associates. It was there that he first learned about computers. In fact, he helped design the first digital computer, UNIVAC.

Cray co-founded Control Data Corporation in 1957. Based on his ideas, the company built large-scale, high-speed computers. In 1972 he founded his own company, Cray Research Incorporated, with the intention of employing new processing methods and simplifying architecture and software to build the world's fastest computers. He succeeded, and the series of computers that the company marketed made possible computer modeling as a central part of scientific research in areas as diverse as meteorology, oil exploration, and nuclear weapons design. Through the 1970's and 1980's Cray Research was at the forefront of supercomputer technology, which became one of the symbols of American technological leadership.

In 1989 Cray left Cray Research to form still another company, Cray Computer Corporation. He planned to build the next generation supercomputer, the Cray 5, but advances in microprocessor technology undercut the demand for supercomputers. Cray Computer entered bankruptcy in 1995. A year later he died from injuries sustained in an automobile accident near Colorado Springs, Colorado.

cylinder with a small section missing and a hollow center, with what appeared to be a bench surrounding it. The shape of the machine was designed to minimize the length of the interconnecting wires that ran between circuit boards to allow electricity to move the shortest possible distance. The bench concealed an important part of the cooling system that kept the system at an appropriate operating temperature.

The measurements that describe the performance of supercomputers are called MIPS (millions of instructions per second) for scalar processors and megaflops (millions of floating-point operations per second) for vector processors. (Floating-point numbers are numbers expressed in scientific notation; for example, 10^{27}.) Whereas the fastest computer before the Cray 1 was capable of some 35 MIPS, the Cray 1 was capable of 80 MIPS. Moreover, the Cray 1 was theoretically capable of vector processing at the rate of 160 megaflops, a rate unheard of at the time.

CONSEQUENCES

Seymour Cray first estimated that there would be few buyers for a machine as advanced as the Cray 1, but his estimate turned out to be incorrect. There were many scientists who wanted to perform computer modeling (in which scientific ideas are expressed in such a way that computer-based experiments can be conducted) and who needed raw processing power.

When dealing with natural phenomena such as the weather or geological structures, or in rocket design, researchers need to make calculations involving large amounts of data. Before computers, advanced experimental modeling was simply not possible, since even the modest calculations for the development of atomic energy, for example, consumed days and weeks of scientists' time. With the advent of supercomputers, however, large-scale computation of vast amounts of information became possible. Weather researchers can design a detailed program that allows them to analyze complex and seemingly unpredictable weather events such as hurricanes; geologists searching for oil fields can gather data about successful finds to help identify new ones; and spacecraft designers can "describe" in computer terms experimental ideas that are too costly or too dangerous to carry out. As supercomputer performance evolves, there is little doubt that scientists will make ever greater use of its power.

See also Apple II computer; BINAC computer; Colossus computer; ENIAC computer; IBM Model 1401 computer; Personal computer; UNIVAC computer.

FURTHER READING

Edwards, Owen. "Seymour Cray." *Forbes* 154, no. 5 (August 29, 1994).

Lloyd, Therese, and Stanley N. Wellborn. "Computers' Next Frontiers." *U.S. News & World Report* 99 (August 26, 1985).

Slater, Robert. *Portraits in Silicon.* Cambridge, Mass.: MIT Press, 1987.

Zipper, Stuart. "Chief Exec. Leaves Cray Computer." *Electronic News* 38, no. 1908 (April, 1992).

Supersonic passenger plane

The invention: The first commercial airliner that flies passengers at speeds in excess of the speed of sound.

The people behind the invention:
Sir Archibald Russell (1904-), a designer with the British Aircraft Corporation
Pierre Satre (1909-), technical director at Sud-Aviation
Julian Amery (1919-), British minister of aviation, 1962-1964
Geoffroy de Cource (1912-), French minister of aviation, 1962
William T. Coleman, Jr. (1920-), U.S. secretary of transportation, 1975-1977

Birth of Supersonic Transportations

On January 21, 1976, the Anglo-French Concorde became the world's first supersonic airliner to carry passengers on scheduled commercial flights. British Airways flew a Concorde from London's Heathrow Airport to the Persian Gulf emirate of Bahrain in three hours and thirty-eight minutes. At about the same time, Air France flew a Concorde from Paris's Charles de Gaulle Airport to Rio de Janeiro, Brazil, in seven hours and twenty-five minutes. The Concordes' cruising speeds were about twice the speed of sound, or 1,350 miles per hour. On May 24, 1976, the United States and Europe became linked for the first time with commercial supersonic air transportation. British Airways inaugurated flights between Dulles International Airport in Washington, D.C., and Heathrow Airport. Likewise, Air France inaugurated flights between Dulles International Airport and Charles de Gaulle Airport. The London-Washington, D.C., flight was flown in an unprecedented time of three hours and forty minutes. The Paris-Washington, D.C., flight was flown in a time of three hours and fifty-five minutes.

THE DECISION TO BUILD THE SST

Events leading to the development and production of the Anglo-French Concorde went back almost twenty years and included approximately $3 billion in investment costs. Issues surrounding the development and final production of the supersonic transport (SST) were extremely complex and at times highly emotional. The concept of developing an SST brought with it environmental concerns and questions, safety issues both in the air and on the ground, political intrigue of international proportions, and enormous economic problems from costs of operations, research, and development.

In England, the decision to begin the SST project was made in October, 1956. Under the promotion of Morien Morgan with the Royal Aircraft Establishment in Farnborough, England, it was decided at the Aviation Ministry headquarters in London to begin development of a supersonic aircraft. This decision was based on the intense competition from the American Boeing 707 and Douglas DC-8 subsonic jets going into commercial service. There was little point in developing another subsonic plane; the alternative was to go above the speed of sound. In November, 1956, at Farnborough, the first meeting of the Supersonic Transport Aircraft Committee, known as STAC, was held.

Members of the STAC proposed that development costs would be in the range of $165 million to $260 million, depending on the range, speed, and payload of the chosen SST. The committee also projected that by 1970, there would be a world market for at least 150 to 500 supersonic planes. Estimates were that the supersonic plane would recover its entire research and development cost through thirty sales. The British, in order to continue development of an SST, needed a European partner as a way of sharing the costs and preempting objections to proposed funding by England's Treasury.

In 1960, the British government gave the newly organized British Aircraft Corporation (BAC) $1 million for an SST feasibility study. Sir Archibald Russell, BAC's chief supersonic designer, visited Pierre Satre, the technical director at the French firm of Sud-Aviation. Satre's suggestion was to evolve an SST from Sud-Aviation's highly successful subsonic Caravelle transport. By September, 1962, an agreement was reached by Sud and BAC design teams on a new SST, the Super Caravelle.

There was a bitter battle over the choice of engines with two British engine firms, Bristol-Siddeley and Rolls-Royce, as contenders. Sir Arnold Hall, the managing director of Bristol-Siddeley, in collaboration with the French aero-engine company SNECMA, was eventually awarded the contract for the engines. The engine chosen was a "civilianized" version of the Olympus, which Bristol had been developing for the multirole TRS-2 combat plane.

THE CONCORDE CONSORTIUM

On November 29, 1962, the Concorde Consortium was created by an agreement between England and the French Republic, signed by Ministers of Aviation Julian Amery and Geoffroy de Cource (1912-   ). The first Concorde, Model 001, rolled out from Sud-Aviation's St. Martin-du-Touch assembly plant on December 11, 1968. The second, Model 002, was completed at the British Aircraft Corporation a few months later. Eight years later, on January 21, 1976, the Concorde became the world's first supersonic airliner to carry passengers on scheduled commercial flights.

Development of the SST did not come easily for the Anglo-French consortium. The nature of supersonic flight created numerous problems and uncertainties not present for subsonic flight. The SST traveled faster than the speed of sound. Sound travels at 760 miles per hour at sea level at a temperature of 59 degrees Fahrenheit. This speed drops to about 660 miles per hour at sixty-five thousand feet, cruising altitude for the SST, where the air temperature drops to 70 degrees below zero.

The Concorde was designed to fly at a maximum of 1,450 miles per hour. The European designers could use an aluminum alloy construction and stay below the critical skin-friction temperatures that required other airframe alloys, such as titanium. The Concorde was designed with a slender curved wing surface. The design incorporated widely separated engine nacelles, each housing two Olympus 593 jet engines. The Concorde was also designed with a "droop snoot," providing three positions: the supersonic configuration, a heat-visor retracted position for subsonic flight, and a nose-lowered position for landing patterns.

IMPACT

Early SST designers were faced with questions such as the intensity and ionization effect of cosmic rays at flight altitudes of sixty to seventy thousand feet. The "cascade effect" concerned the intensification of cosmic radiation when particles from outer space struck a metallic cover. Scientists looked for ways to shield passengers from this hazard inside the aluminum or titanium shell of an SST flying high above the protective blanket of the troposphere. Experts questioned whether the risk of being struck by meteorites was any greater for the SST than for subsonic jets and looked for evidence on wind shear of jet streams in the stratosphere.

Other questions concerned the strength and frequency of clear air turbulence above forty-five thousand feet, whether the higher ozone content of the air at SST cruise altitude would affect the materials of the aircraft, whether SST flights would upset or destroy the protective nature of the earth's ozone layer, the effect of aerodynamic heating on material strength, and the tolerable strength of sonic booms over populated areas. These and other questions consumed the designers and researchers involved in developing the Concorde.

Through design research and flight tests, many of the questions were resolved or realized to be less significant than anticipated. Several issues did develop into environmental, economic, and international issues. In late 1975, the British and French governments requested permission to use the Concorde at New York's John F. Kennedy International Airport and at Dulles International Airport for scheduled flights between the United States and Europe. In December, 1975, as a result of strong opposition from anti-Concorde environmental groups, the U.S. House of Representatives approved a six-month ban on SST's coming into the United States so that the impact of flights could be studied. Secretary of Transportation William T. Coleman, Jr., held hearings to prepare for a decision by February 5, 1976, as to whether to allow the Concorde into U.S. airspace. The British and French, if denied landing rights, threatened to take the United States to an international court, claiming that treaties had been violated.

The treaties in question were the Chicago Convention and Bermuda agreements of February 11, 1946, and March 27, 1946. These treaties prohibited the United States from banning aircraft that both France and Great Britain had certified to be safe. The Environmental Defense Fund contended that the United States had the right to ban SST aircraft on environmental grounds.

Under pressure from both sides, Coleman decided to allow limited Concorde service at Dulles and John F. Kennedy airports for a sixteen-month trial period. Service into John F. Kennedy Airport, however, was delayed by a ban by the Port Authority of New York and New Jersey until a pending suit was pursued by the airlines. During the test period, detailed records were to be kept on the Concorde's noise levels, vibration, and engine emission levels. Other provisions included that the plane would not fly at supersonic speeds over the continental United States; that all flights could be cancelled by the United States with four months notice, or immediately if they proved harmful to the health and safety of Americans; and that at the end of a year, four months of study would begin to determine if the trial period should be extended.

The Concorde's noise was one of the primary issues in determining whether the plane should be allowed into U.S. airports. The Federal Aviation Administration measured the effective perceived noise in decibels. After three months of monitoring the Concorde's departure noise at 3.5 nautical miles was found to vary from 105 to 130 decibels. The Concorde's approach noise at one nautical mile from threshold varied from 115 to 130 decibels. These readings were approximately equal to noise levels of other four-engine jets, such as the Boeing 747, on landing but were twice as loud on takeoff.

THE ECONOMICS OF OPERATION

Another issue of significance was the economics of Concorde's operation and its tremendous investment costs. In 1956, early predictions of Great Britain's STAC were for a world market of 150 to 500 supersonic planes. In November, 1976, Great Britain's Gerald Kaufman and France's Marcel Cavaille said that production of the Concorde would not continue beyond the sixteen vehicles then contracted for with BAC and Sud-Aviation. There was no demand by U.S. airline corporations for the plane. Given that the planes could not fly at supersonic speeds over populated areas because of the sonic boom phenomenon, markets for the SST had to be separated by at least three thousand miles, with flight paths over mostly water or desert. Studies indicated that there were only twelve to fifteen routes in the world for which the Concorde was suitable. The planes were expensive, at a price of approximately $74 million each and had a limited seating capacity of one hundred passengers. The plane's range was about four thousand miles.

These statistics compared to a Boeing 747 with a cost of $35 million, seating capacity of 360, and a range of six thousand miles. In addition, the International Air Transport Association negotiated that the fares for the Concorde flights should be equivalent to current first-class fares plus 20 percent. The marketing promotion for the Anglo-French Concorde was thus limited to the elite business traveler who considered speed over cost of transportation. Given these factors, the recovery of research and development costs for Great Britain and France would never occur.

See also Airplane; Bullet train; Dirigible; Rocket; Stealth aircraft; Turbojet; V-2 rocket.

FURTHER READING

Ellingsworth, Rosalind K. "Concorde Stresses Time, Service." *Aviation Week and Space Technology* 105 (August 16, 1976).
Kozicharow, Eugene. "Concorde Legal Questions Raised." *Aviation Week and Space Technology* 104 (January 12, 1976).
Ropelewski, Robert. "Air France Poised for Concorde Service." *Aviation Week and Space Technology* 104 (January 19, 1976).
Sparaco, Pierre. "Official Optimism Grows for Concorde's Return." *Aviation Week and Space Technology* 154, no. 8 (February 19, 2001).
Trubshaw, Brian. *Concorde: The Inside Story.* Thrupp, Stroud: Sutton, 2000.

Synchrocyclotron

The invention: A powerful particle accelerator that performed better than its predecessor, the cyclotron.

The people behind the invention:
Edwin Mattison McMillan (1907-1991), an American physicist who won the Nobel Prize in Chemistry in 1951
Vladimir Iosifovich Veksler (1907-1966), a Soviet physicist
Ernest Orlando Lawrence (1901-1958), an American physicist
Hans Albrecht Bethe (1906-   ), a German American physicist

The First Cyclotron

The synchrocyclotron is a large electromagnetic apparatus designed to accelerate atomic and subatomic particles at high energies. Therefore, it falls under the broad class of scientific devices known as "particle accelerators." By the early 1920's, the experimental work of physicists such as Ernest Rutherford and George Gamow demanded that an artificial means be developed to generate streams of atomic and subatomic particles at energies much greater than those occurring naturally. This requirement led Ernest Orlando Lawrence to develop the cyclotron, the prototype for most modern accelerators. The synchrocyclotron was developed in response to the limitations of the early cyclotron.

In September, 1930, Lawrence announced the basic principles behind the cyclotron. Ionized—that is, electrically charged—particles are admitted into the central section of a circular metal drum. Once inside the drum, the particles are exposed to an electric field alternating within a constant magnetic field. The combined action of the electric and magnetic fields accelerates the particles into a circular path, or orbit. This increases the particles' energy and orbital radii. This process continues until the particles reach the desired energy and velocity and are extracted from the machine for use in experiments ranging from particle-to-particle collisions to the synthesis of radioactive elements.

Although Lawrence was interested in the practical applications of his invention in medicine and biology, the cyclotron also was applied to a variety of experiments in a subfield of physics called "high-energy physics." Among the earliest applications were studies of the subatomic, or nuclear, structure of matter. The energetic particles generated by the cyclotron made possible the very type of experiment that Rutherford and Gamow had attempted earlier. These experiments, which bombarded lithium targets with streams of highly energetic accelerated protons, attempted to probe the inner structure of matter.

Although funding for scientific research on a large scale was scarce before World War II (1939-1945), Lawrence nevertheless conceived of a 467-centimeter cyclotron that would generate particles with energies approaching 100 million electronvolts. By the end of the war, increases in the public and private funding of scientific research and a demand for higher-energy particles created a situation in which this plan looked as if it would become reality, were it not for an inherent limit in the physics of cyclotron operation.

OVERCOMING THE PROBLEM OF MASS

In 1937, Hans Albrecht Bethe discovered a severe theoretical limitation to the energies that could be produced in a cyclotron. Physicist Albert Einstein's special theory of relativity had demonstrated that as any mass particle gains velocity relative to the speed of light, its mass increases. Bethe showed that this increase in mass would eventually slow the rotation of each particle. Therefore, as the rotation of each particle slows and the frequency of the alternating electric field remains constant, particle velocity will decrease eventually. This factor set an upper limit on the energies that any cyclotron could produce.

Edwin Mattison McMillan, a colleague of Lawrence at Berkeley, proposed a solution to Bethe's problem in 1945. Simultaneously and independently, Vladimir Iosifovich Veksler of the Soviet Union proposed the same solution. They suggested that the frequency of the alternating electric field be slowed to meet the decreasing rotational frequencies of the accelerating particles—in essence, "synchronizing" the electric field with the moving particles. The result was the synchrocyclotron.

Prior to World War II, Lawrence and his colleagues had obtained the massive electromagnet for the new 100-million-electronvolt cyclotron. This 467-centimeter magnet would become the heart of the new Berkeley synchrocyclotron. After initial tests proved successful, the Berkeley team decided that it would be reasonable to convert the cyclotron magnet for use in a new synchrocyclotron. The apparatus was operational in November of 1946.

These high energies combined with economic factors to make the synchrocyclotron a major achievement for the Berkeley Radiation Laboratory. The synchrocyclotron required less voltage to produce higher energies than the cyclotron because the obstacles cited by Bethe were virtually nonexistent. In essence, the energies produced by synchrocyclotrons are limited only by the economics of building them. These factors led to the planning and construction of other synchrocyclotrons in the United States and Europe. In 1957, the Berkeley apparatus was redesigned in order to achieve energies of 720 million electronvolts, at that time the record for cyclotrons of any kind.

Impact

Previously, scientists had had to rely on natural sources for highly energetic subatomic and atomic particles with which to experiment. In the mid-1920's, the American physicist Robert Andrews Millikan began his experimental work in cosmic rays, which are one natural source of energetic particles called "mesons." Mesons are charged particles that have a mass more than two hundred times that of the electron and are therefore of great benefit in high-energy physics experiments. In February of 1949, McMillan announced the first synthetically produced mesons using the synchrocyclotron.

McMillan's theoretical development led not only to the development of the synchrocyclotron but also to the development of the electron synchrotron, the proton synchrotron, the microtron, and the linear accelerator. Both proton and electron synchrotrons have been used successfully to produce precise beams of muons and pi-mesons, or pions (a type of meson).

The increased use of accelerator apparatus ushered in a new era of physics research, which has become dominated increasingly by large accelerators and, subsequently, larger teams of scientists and engineers required to run individual experiments. More sophisticated machines have generated energies in excess of 2 trillion electronvolts at the United States' Fermi National Accelerator Laboratory, or Fermilab, in Illinois. Part of the huge Tevatron apparatus at Fermilab, which generates these particles, is a proton synchrotron, a direct descendant of McMillan and Lawrence's early efforts.

See also Atomic bomb; Cyclotron; Electron microscope; Field ion microscope; Geiger counter; Hydrogen bomb; Mass spectrograph; Neutrino detector; Scanning tunneling microscope; Tevatron accelerator.

FURTHER READING

Bernstein, Jeremy. *Hans Bethe: Prophet of Energy.* New York: Basic Books, 1980.
McMillan, Edwin. "The Synchrotron: A Proposed High-Energy Particle Accelerator." *Physical Review* 68 (September, 1945).
_____. "Vladimir Iosifovich Veksler." *Physics Today* (November, 1966).
"Witness to a Century." *Discover* 20 (December, 1999).

SYNTHETIC AMINO ACID

THE INVENTION: A method for synthesizing amino acids by combining water, hydrogen, methane, and ammonia and exposing the mixture to an electric spark.

THE PEOPLE BEHIND THE INVENTION:
Stanley Lloyd Miller (1930-   ), an American professor of chemistry
Harold Clayton Urey (1893-1981), an American chemist who won the 1934 Nobel Prize in Chemistry
Aleksandr Ivanovich Oparin (1894-1980), a Russian biochemist
John Burdon Sanderson Haldane (1892-1964), a British scientist

PREBIOLOGICAL EVOLUTION

The origin of life on Earth has long been a tough problem for scientists to solve. While most scientists can envision the development of life through geologic time from simple single-cell bacteria to complex mammals by the processes of mutation and natural selection, they have found it difficult to develop a theory to define how organic materials were first formed and organized into lifeforms. This stage in the development of life before biologic systems arose, which is called "chemical evolution," occurred between 4.5 and 3.5 billion years ago. Although great advances in genetics and biochemistry have shown the intricate workings of the cell, relatively little light has been shed on the origins of this intricate machinery of the cell. Some experiments, however, have provided important data from which to build a scientific theory of the origin of life. The first of these experiments was the classic work of Stanley Lloyd Miller.

Miller worked with Harold Clayton Urey, a Nobel laureate, on the environments of the early earth. John Burdon Sanderson Haldane, a British biochemist, had suggested in 1929 that the earth's early atmosphere was a reducing one—that it contained no free oxygen. In 1952, Urey published a seminal work in planetology, *The Planets*, in which he elaborated on Haldane's suggestion, and he postulated that the earth had formed from a cold stellar dust cloud. Urey thought that the earth's primordial atmosphere probably contained elements in the approximate relative abundances found in the solar system and the universe.

It had been discovered in 1929 that the Sun is approximately 87 percent hydrogen, and by 1935 it was known that hydrogen encompassed the vast majority (92.8 percent) of atoms in the universe. Urey reasoned that the earth's early atmosphere contained mostly hydrogen, with the oxygen, nitrogen, and carbon atoms chemically bonded to hydrogen to form water, ammonia, and methane. Most important, free oxygen could not exist in the presence of such an abundance of hydrogen.

As early as the mid-1920's, Aleksandr Ivanovich Oparin, a Russian biochemist, had argued that the organic compounds necessary for life had been built up on the early earth by chemical combinations in a reducing atmosphere. The energy from the Sun would have been sufficient to drive the reactions to produce life. Haldane later proposed that the organic compounds would accumulate in the oceans to produce a "dilute organic soup" and that life might have arisen by some unknown process from that mixture of organic compounds.

PRIMORDIAL SOUP IN A BOTTLE

Miller combined the ideas of Oparin and Urey and designed a simple, but elegant, experiment. He decided to mix the gases presumed to exist in the early atmosphere (water vapor, hydrogen, ammonia, and methane) and expose them to an electrical spark to determine which, if any, organic compounds were formed. To do this, he constructed a relatively simple system, essentially consisting of two Pyrex flasks connected by tubing in a roughly circular pattern. The water and gases in the smaller flask were boiled and the resulting gas forced through the tubing into a larger flask that contained tungsten electrodes. As the gases passed the electrodes, an electrical spark was generated, and from this larger flask the gases and any other compounds were condensed. The gases were recycled through the system, whereas the organic compounds were trapped in the bottom of the system.

Miller was trying to simulate conditions that had prevailed on the early earth. During the one week of operation, Miller extracted and analyzed the residue of compounds at the bottom of the system. The results were truly astounding. He found that numerous organic compounds had, indeed, been formed in only that one week. As much as 15 percent of the carbon (originally in the gas methane) had been combined into organic compounds, and at least 5 percent of the carbon was incorporated into biochemically important compounds. The most important compounds produced were some of the twenty amino acids essential to life on Earth.

The formation of amino acids is significant because they are the building blocks of proteins. Proteins consist of a specific sequence of amino acids assembled into a well-defined pattern. Proteins are necessary for life for two reasons. First, they are important structural

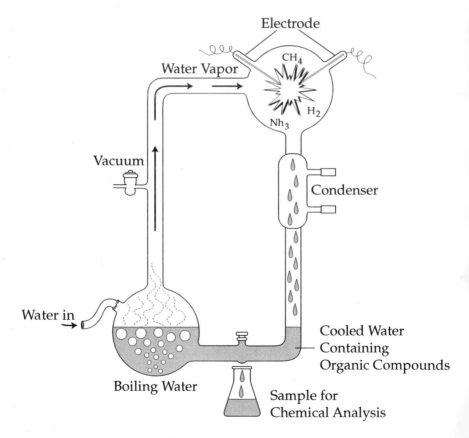

The Miller-Urey experiment.

materials used to build the cells of the body. Second, the enzymes that increase the rate of the multitude of biochemical reactions of life are also proteins. Miller not only had produced proteins in the laboratory but also had shown clearly that the precursors of proteins—the amino acids—were easily formed in a reducing environment with the appropriate energy.

Perhaps the most important aspect of the experiment was the ease with which the amino acids were formed. Of all the thousands of organic compounds that are known to chemists, amino acids were among those that were formed by this simple experiment. This strongly implied that one of the first steps in chemical evolution was not only possible but also highly probable. All that was necessary for the synthesis of amino acids were the common gases of the solar system, a reducing environment, and an appropriate energy source, all of which were present on early Earth.

CONSEQUENCES

Miller opened an entirely new field of research with his pioneering experiments. His results showed that much about chemical evolution could be learned by experimentation in the laboratory. As a result, Miller and many others soon tried variations on his original experiment by altering the combination of gases, using other gases, and trying other types of energy sources. Almost all the essential amino acids have been produced in these laboratory experiments.

Miller's work was based on the presumed composition of the primordial atmosphere of Earth. The composition of this atmosphere was calculated on the basis of the abundance of elements in the universe. If this reasoning is correct, then it is highly likely that there are many other bodies in the universe that have similar atmospheres and are near energy sources similar to the Sun. Moreover, Miller's experiment strongly suggests that amino acids, and perhaps life as well, should have formed on other planets.

See also Artificial hormone; Artificial kidney; Synthetic DNA; Synthetic RNA.

FURTHER READING

Dronamraju, Krishna R., and J. B. S. Haldane. *Haldane's Daedalus Revisited*. New York: Oxford University Press, 1995.

Lipkin, Richard. "Early Earth May Have Had Two Key RNA Bases." *Science News* 148, no. 1 (July 1, 1995).

Miller, Stanley L., and Leslie E. Orgel. *The Origins of Life on the Earth*. Englewood Cliffs, N.J.: Prentice-Hall, 1974.

Nelson, Kevin E., Matthew Levy, and Stanley L. Miller. "Peptide Nucleic Acids Rather than RNA May Have Been the First Genetic Molecule." *Proceedings of the National Academy of Sciences of the United States of America* 97, no. 8 (April 11, 2000).

Yockey, Hubert P. "Walther Lob, Stanley L. Miller, and Prebiotic 'Building Blocks' in the Silent Electrical Discharge." *Perspectives in Biology and Medicine* 41, no. 1 (Autumn, 1997).

Synthetic DNA

THE INVENTION: A method for replicating viral deoxyribonucleic acid (DNA) in a test tube that paved the way for genetic engineering.

THE PEOPLE BEHIND THE INVENTION:
Arthur Kornberg (1918-   ), an American physician and biochemist
Robert L. Sinsheimer (1920-   ), an American biophysicist
Mehran Goulian (1929-   ), a physician and biochemist

THE ROLE OF DNA

Until the mid-1940's, it was believed that proteins were the carriers of genetic information, the source of heredity. Proteins appeared to be the only biological molecules that had the complexity necessary to encode the enormous amount of genetic information required to reproduce even the simplest organism. Nevertheless, proteins could not be shown to have genetic properties, and by 1944, it was demonstrated conclusively that deoxyribonucleic acid (DNA) was the material that transmitted hereditary information. It was discovered that DNA isolated from a strain of infective bacteria that can cause pneumonia was able to transform a strain of noninfective bacteria into an infective strain; in addition, the infectivity trait was transmitted to future generations. Subsequently, it was established that DNA is the genetic material in virtually all forms of life.

Once DNA was known to be the transmitter of genetic information, scientists sought to discover how it performs its role. DNA is a polymeric molecule composed of four different units, called "deoxynucleotides." The units consist of a sugar, a phosphate group, and a base; they differ only in the nature of the base, which is always one of four related compounds: adenine, guanine, cytosine, or thymine. The way in which such a polymer could transmit genetic information, however, was difficult to discern. In 1953, biophysicists James D. Watson and Francis Crick brilliantly determined the three-dimensional structure of DNA by analyzing X-ray diffraction photographs of DNA fibers. From their analysis of the structure of DNA, Watson and Crick inferred DNA's mechanism of replication. Their work led to an understanding of gene function in molecular terms.

Watson and Crick showed that DNA has a very long double-stranded (duplex) helical structure. DNA has a duplex structure because each base forms a link to a specific base on the opposite strand. The discovery of this complementary pairing of bases provided a model to explain the two essential functions of a hereditary molecule: It must preserve the genetic code from one generation to the next, and it must direct the development of the cell.

Watson and Crick also proposed that DNA is able to serve as a mold (or template) for its own reproduction because the two strands of DNA polymer can separate. Upon separation, each strand acts as a template for the formation of a new complementary strand. An adenine base in the existing strand gives rise to cytosine, and so on. In this manner, a new double-stranded DNA is generated that is identical to the parent DNA.

DNA IN A TEST TUBE

Watson and Crick's theory provided a valuable model for the reproduction of DNA, but it did not explain the biological mechanism by which the process occurs. The biochemical pathway of DNA reproduction and the role of the enzymes required for catalyzing the reproduction process were discovered by Arthur Kornberg and his coworkers. For his success in achieving DNA synthesis in a test tube and for discovering and isolating an enzyme—DNA polymerase—that catalyzed DNA synthesis, Kornberg won the 1959 Nobel Prize in Physiology or Medicine.

To achieve DNA replication in a test tube, Kornberg found that a small amount of preformed DNA must be present, in addition to DNA polymerase enzyme and all four of the deoxynucleotides that occur in DNA. Kornberg discovered that the base composition of the newly made DNA was determined solely by the base composition of the preformed DNA, which had been used as a template in the test-tube synthesis. This result showed that DNA polymerase obeys instructions dictated by the template DNA. It is thus said to be "template-directed." DNA polymerase was the first template-directed enzyme to be discovered.

Although test-tube synthesis was a most significant achievement, important questions about the precise character of the newly made DNA were still unanswered. Methods of analyzing the order, or sequence, of the bases in DNA were not available, and hence it could not be shown directly whether DNA made in the test tube was an exact copy of the template of DNA or merely an approximate copy. In addition, some DNAs prepared by DNA polymerase appeared to be branched structures. Since chromosomes in living cells contain long, linear, unbranched strands of DNA, this branching might have indicated that DNA synthesized in a test tube was not equivalent to DNA synthesized in the living cell.

Kornberg realized that the best way to demonstrate that newly synthesized DNA is an exact copy of the original was to test the new DNA for biological activity in a suitable system. Kornberg reasoned that a demonstration of infectivity in viral DNA produced in a test tube would prove that polymerase-catalyzed synthesis was virtually error-free and equivalent to natural, biological synthesis. The experiment, carried out by Kornberg, Mehran Goulian at Stanford University, and Robert L. Sinsheimer at the California Institute of Technology, was a complete success. The viral DNAs produced in a test tube by the DNA polymerase enzyme, using a viral DNA template, were fully infective. This synthesis showed that DNA polymerase could copy not merely a single gene but also an entire chromosome of a small virus without error.

CONSEQUENCES

The purification of DNA polymerase and the preparation of biologically active DNA were major achievements that influenced biological research on DNA for decades. Kornberg's methodology proved to be invaluable in the discovery of other enzymes that synthesize DNA. These enzymes have been isolated from *Escherichia coli* bacteria and from other bacteria, viruses, and higher organisms.

The test-tube preparation of viral DNA also had significance in the studies of genes and chromosomes. In the mid-1960's, it had not been established that a chromosome contains a continuous strand of

DNA. Kornberg and Sinsheimer's synthesis of a viral chromosome proved that it was, indeed, a very long strand of uninterrupted DNA.

Kornberg and Sinsheimer's work laid the foundation for subsequent recombinant DNA research and for genetic engineering technology. This technology promises to revolutionize both medicine and agriculture. The enhancement of food production and the generation of new drugs and therapies are only a few of the subsequent benefits that may be expected.

See also Artificial chromosome; Artificial hormone; Cloning; Genetic "fingerprinting"; Genetically engineered insulin; In vitro plant culture; Synthetic amino acid; Synthetic RNA.

FURTHER READING

Baker, Tania A., and Arthur Kornberg. *DNA Replication.* 2d ed. New York: W. H. Freeman, 1991.

Kornberg, Arthur. *The Golden Helix: Inside Biotech Ventures.* Sausalito, Calif.: University Science Books, 1995.

_____. *For the Love of Enzymes: The Odyssey of a Biochemist.* Harvard University Press, 1991.

Sinsheimer, Robert. *The Strands of a Life: The Science of DNA and the Art of Education.* Berkeley: University of California Press, 1994.

Synthetic RNA

THE INVENTION: A method for synthesizing the biological molecule RNA established that this process can occur outside the living cell.

THE PEOPLE BEHIND THE INVENTION:
Severo Ochoa (1905-1993), a Spanish biochemist who shared the 1959 Nobel Prize in Physiology or Medicine
Marianne Grunberg-Manago (1921-   ), a French biochemist
Marshall W. Nirenberg (1927-   ), an American biochemist who won the 1968 Nobel Prize in Physiology or Medicine
Peter Lengyel (1929-   ), a Hungarian American biochemist

RNA OUTSIDE THE CELLS

In the early decades of the twentieth century, genetics had not been experimentally united with biochemistry. This merging soon occurred, however, with work involving the mold *Neurospora crassa*. This Nobel award-winning work by biochemist Edward Lawrie Tatum and geneticist George Wells Beadle showed that genes control production of proteins, which are major functional molecules in cells. Yet no one knew the chemical composition of genes and chromosomes, or, rather, the molecules of heredity.

The American bacteriologist Oswald T. Avery and his colleagues at New York's Rockefeller Institute determined experimentally that the molecular basis of heredity was a large polymer known as deoxyribonucleic acid (DNA). Avery's discovery triggered a furious worldwide search for the particular structural characteristics of DNA, which allow for the known biological characteristics of genes.

One of the most famous studies in the history of science solved this problem in 1953. Scientists James D. Watson, Francis Crick, and Maurice H. F. Wilkins postulated that DNA exists as a double helix. That is, two long strands twist about each other in a predictable pattern, with each single strand held to the other by weak, reversible linkages known as "hydrogen bonds." About this time, researchers recognized also that a molecule closely related to DNA, ribonucleic acid (RNA), plays an important role in transcribing the genetic information as well as in other biological functions.

Severo Ochoa was born in Spain as the science of genetics was developing. In 1942, he moved to New York University, where he studied the bacterium *Azobacter vinelandii*. Specifically, Ochoa was focusing on the question of how cells process energy in the form of organic molecules such as the sugar glucose to provide usable biological energy in the form of adenosine triphosphate (ATP). With postdoctoral fellow Marianne Grunberg-Manago, he studied enzymatic reactions capable of incorporating inorganic phosphate (a compound consisting of one atom of phosphorus and four atoms of oxygen) into adenosine diphosphate (ADP) to form ATP.

One particularly interesting reaction was followed by monitoring the amount of radioactive phosphate reacting with ADP. Following separation of the reaction products, it was discovered that the main product was not ATP, but a much larger molecule. Chemical characterization demonstrated that this product was a polymer of adenosine monophosphate. When other nucleocide diphosphates, such as inosine diphosphate, were used in the reaction, the corresponding polymer of inosine monophosphate was formed. Thus, in each case, a polymer (a long string of building-block units) was formed. The polymers formed were synthetic RNAs, and the enzyme responsible for the conversion became known as "polynucleotide phosphorylase." This finding, once the early skepticism was resolved, was received by biochemists with great enthusiasm because no technique outside the cell had ever been discovered previously in which a nucleic acid similar to RNA could be synthesized.

Learning the Language

Ochoa, Peter Lengyel, and Marshall W. Nirenberg at the National Institute of Health took advantage of this breakthrough to synthesize different RNAs useful in cracking the genetic code. Crick had postulated that the flow of information in biological systems is from DNA to RNA to protein. In other words, genetic information contained in the DNA structure is transcribed into complementary RNA structures, which, in turn, are translated into the protein. Protein synthesis, an extremely complex process, involves bringing a type of RNA, known as messenger RNA, together with amino acids and huge cellular organelles known as ribosomes.

Yet investigators did not know the nature of the nucleic acid alphabet—for example, how many single units of the RNA polymer code were needed for each amino acid, and the order that the units must be in to stand for a "word" in the nucleic acid language. In 1961, Nirenberg demonstrated that the polymer of synthetic RNA with multiple units of uracil (poly U) would "code" only for a protein containing the amino acid phenylalanine. Each three units (U's) gave one phenylalanine. Therefore, genetic words each contain three letters. UUU translates into phenylalanine. Poly A, the first polymer discovered with polynucleotide phosphorylase, was coded for a protein containing multiple lysines. That is, AAA translates into the amino acid lysine.

The words, containing combinations of letters, such as AUG, were not as easily studied, but Nirenberg, Ochoa, and Gobind Khorana of the University of Wisconsin eventually uncovered the exact translation for each amino acid. In RNA, there are four possible letters (A, U, G, and C) and three letters in each word. Accordingly, there are sixty-four possible words. With only twenty amino acids, it became clear that more than one RNA word can translate into a given amino acid. Yet, no given word stands for any more than one amino acid. A few words do not translate into any amino acid; they are stop signals, telling the ribosome to cease translating RNA.

The question of which direction an RNA is translated is critical. For example, CAA codes for the amino acid glutamine, but the reverse, AAC, translates to the amino acid asparagine. Such a difference is critical because the exact sequence of a protein determines its activity—that is, what it will do in the body and therefore what genetic trait it will express.

Consequences

Synthetic RNAs provided the key to understanding the genetic code. The genetic code is universal; it operates in all organisms, simple or complex. It is used by viruses, which are nearly life but are not alive. Spelling out the genetic code was one of the top discoveries of the twentieth century. Nearly all work in molecular biology depends on this knowledge.

The availability of synthetic RNAs has provided hybridization tools for molecular geneticists. Hybridization is a technique in which an RNA is allowed to bind in a complementary fashion to DNA under investigation. The greater the similarity between RNA and DNA, the greater the amount of binding. The differential binding allows for seeking, finding, and ultimately isolating a target DNA from a large, diverse pool of DNA—in short, finding a needle in a haystack. Hybridization has become an indispensable aid in experimental molecular genetics as well as in applied sciences, such as forensics.

See also Artificial chromosome; Artificial hormone; Cloning; Genetic "fingerprinting"; Genetically engineered insulin; In vitro plant culture; Synthetic amino acid; Synthetic DNA.

FURTHER READING

"Biochemist Severo Ochoa Dies: Won Nobel Prize." *Washington Post* (November 3, 1993).

Santesmases, Maria Jesus. "Severo Ochoa and the Biomedical Sciences in Spain Under Franco, 1959-1975." *Isis* 91, no. 4 (December, 2000).

"Severo Ochoa, 1905-1993." *Nature* 366, no. 6454 (December, 1993).

SYPHILIS TEST

THE INVENTION: The first simple test for detecting the presence of the venereal disease syphilis led to better syphilis control and other advances in immunology.

THE PEOPLE BEHIND THE INVENTION:
Reuben Leon Kahn (1887-1974), a Soviet-born American serologist and immunologist
August von Wassermann (1866-1925), a German physician and bacteriologist

COLUMBUS'S DISCOVERIES

Syphilis is one of the chief venereal diseases, a group of diseases whose name derives from Venus, the Roman goddess of love. The term "venereal" arose from the idea that the diseases were transmitted solely by sexual contact with an infected individual. Although syphilis is almost always passed from one person to another in this way, it occasionally arises after contact with objects used by infected people in highly unclean surroundings, particularly in the underdeveloped countries of the world.

It is believed by many that syphilis was introduced to Europe by the members of Spanish explorer Christopher Columbus's crew— supposedly after they were infected by sexual contact with West Indian women—during their voyages of exploration. Columbus is reported to have died of heart and brain problems very similar to symptoms produced by advanced syphilis. At that time, according to many historians, syphilis spread rapidly over sixteenth century Europe. The name "syphilis" was coined by the Italian physician Girolamo Fracastoro in 1530 in an epic poem he wrote.

Modern syphilis is much milder than the original disease and relatively uncommon. Yet, if it is not identified and treated appropriately, syphilis can be devastating and even fatal. It can also be passed from pregnant mothers to their unborn children. In these cases, the afflicted children will develop serious health problems that can include paralysis, insanity, and heart disease. Therefore, the understanding, detection, and cure of syphilis are important worldwide.

Syphilis is caused by a spiral-shaped germ called a "spirochete." Spirochetes enter the body through breaks in the skin or through the mucous membranes, regardless of how they are transmitted. Once spirochetes enter the body, they spread rapidly. During the first four to six weeks after infection, syphilis—said to be in its primary phase—is very contagious. During this time, it is identified by the appearance of a sore, or chancre, at the entry site of the infecting spirochetes.

The chancre disappears quickly, and within six to twenty-four weeks, the disease shows itself as a skin rash, feelings of malaise, and other flulike symptoms (secondary-phase syphilis). These problems also disappear quickly in most cases, and here is where the real trouble—latent syphilis—begins. In latent syphilis, now totally without symptoms, spirochetes that have spread through the body may lodge in the brain or the heart. When this happens, paralysis, mental incapacitation, and death may follow.

TESTING BEFORE MARRIAGE

Because of the danger to unborn children, Americans wishing to marry must be certified as being free of the disease before a marriage license is issued. The cure for syphilis is easily accomplished through the use of penicillin or other types of antibiotics, though no vaccine is yet available to prevent the disease. It is for this reason that syphilis detection is particularly important.

The first viable test for syphilis was originated by August von Wassermann in 1906. In this test, blood samples are taken and treated in a medical laboratory. The treatment of the samples is based on the fact that the blood of infected persons has formed antibodies to fight the syphilis spirochete, and that these antibodies will react with certain body chemicals to cause the blood sample to clot. This indicates the person has the disease. After the syphilis has been cured, the antibodies disappear, as does the clotting.

Although the Wassermann test was effective in 95 percent of all infected persons, it was very time-consuming (requiring a two-day incubation period) and complex. In 1923, Reuben Leon Kahn developed a modified syphilis test, "the standard Kahn test," that was simpler and faster: The test was complete after only a few minutes. By 1925, Kahn's test had become the standard syphilis test of the United States Navy and later became a worldwide test for the detection of the disease.

Kahn soon realized that his test was not perfect and that in some cases, the results were incorrect. This led him to a broader study of the immune reactions at the center of the Kahn test. He investigated the role of various tissues in immunity, as compared to the role of white blood antibodies and white blood cells. Kahn showed, for example, that different tissues of immunized or nonimmunized animals possessed differing immunologic capabilities. Furthermore, the immunologic capabilities of test animals varied with their age, being very limited in newborns and increasing as they matured.

This effort led, by 1951, to Kahn's "universal serological reaction," a precipitation reaction in which blood serum was tested against a reagent composed of tissue lipids. Kahn viewed it as a potentially helpful chemical indicator of how healthy or ill an individual was. This effort is viewed as an important landmark in the development of the science of immunology.

IMPACT

At the time that Kahn developed his standard Kahn test for syphilis, the Wassermann test was used all over the world for the diagnosis of syphilis. As has been noted, one of the great advantages of the standard Kahn test was its speed, minutes versus days. For example, in October, 1923, Kahn is reported to have tested forty serum samples in fifteen minutes.

Kahn's efforts have been important to immunology and to medicine. Among the consequences of his endeavors was the stimulation of other developments in the field, including the VDRL test (originated by the Venereal Disease Research Laboratory), which has replaced the Kahn test as one of the most often used screening tests for syphilis. Even more specific syphilis tests developed later include a fluorescent antibody test to detect the presence of the antibody to the syphilis spirochete.

See also Abortion pill; Amniocentesis; Antibacterial drugs; Birth control pill; Mammography; Pap test; Penicillin; Ultrasound.

FURTHER READING

Cates, William, Jr., Richard B. Rothenberg, and Joseph H. Blount. "Syphilis Control." *Sexually Transmitted Diseases* 23, no. 1 (January, 1996).
Cobb, W. Montague. "Reuben Leon Kahn." *Journal of the National Medical Association* 63 (September, 1971).
Quétel, Claude. *History of Syphilis*. Baltimore: Johns Hopkins University Press, 1992.
St. Louis, Michael E., and Judith N. Wasserheit. "Elimination of Syphilis in the United States." *Science* 281, no. 5375 (July, 1998).

Talking Motion Pictures

The invention: The first practical system for linking sound with moving pictures.

The people behind the invention:
Harry Warner (1881-1958), the brother who used sound to fashion a major filmmaking company
Albert Warner (1884-1967), the brother who persuaded theater owners to show Warner films
Samuel Warner (1887-1927), the brother who adapted sound-recording technology to filmmaking
Jack Warner (1892-1978), the brother who supervised the making of Warner films

Taking the Lead

The silent films of the early twentieth century had live sound accompaniment featuring music and sound effects. Neighborhood theaters made do with a piano and violin; larger "picture palaces" in major cities maintained resident orchestras of more than seventy members. During the late 1920's, Warner Bros. led the American film industry in producing motion pictures with their own soundtracks, which were first recorded on synchronized records and later added on to the film beside the images.

The ideas that led to the addition of sound to film came from corporate-sponsored research by American Telephone and Telegraph Company (AT&T) and the Radio Corporation of America (RCA). Both companies worked to improve sound recording and playback, AT&T to help in the design of long-distance telephone equipment and RCA as part of the creation of better radio sets. Yet neither company could, or would, enter filmmaking. AT&T was willing to contract its equipment out to Paramount or one of the other major Hollywood studios of the day; such studios, however, did not want to risk their sizable profit positions by junking silent films. The giants of the film industry were doing fine with what they had and did not want to switch to something that had not been proved.

In 1924, Warner Bros. was a prosperous, though small, corporation that produced films with the help of outside financial backing. That year, Harry Warner approached the important Wall Street investment banking house of Goldman, Sachs and secured the help he needed.

As part of this initial wave of expansion, Warner Bros. acquired a Los Angeles radio station in order to publicize its films. Through this deal, the four Warner brothers learned of the new technology that the radio and telephone industries had developed to record sound, and they succeeded in securing the necessary equipment from AT&T. During the spring of 1925, the brothers devised a plan by which they could record the most popular musical artists on film and then offer these "shorts" as added attractions to theaters that booked its features. As a bonus, Warner Bros. could add recorded orchestral music to its feature films and offer this music to theaters that relied on small musical ensembles.

"Vitaphone"

On August 6, 1926, Warner Bros. premiered its new "Vitaphone" technology. The first package consisted of a traditional silent film (*Don Juan*) with a recorded musical accompaniment, plus six recordings of musical talent highlighted by a performance from Giovanni Martineli, the most famous opera tenor of the day.

The first Vitaphone feature was *The Jazz Singer*, which premiered in October, 1927. The film was silent during much of the movie, but as soon as Al Jolson, the star, broke into song, the new technology would be implemented. The film was an immediate hit. *The Jazz Singer* package, which included accompanying shorts with sound, forced theaters in cities that rarely held films over for more than a single week to ask to have the package stay for two, three, and sometimes four straight weeks.

The Jazz Singer did well at the box office, but skeptics questioned the staying power of talkies. If sound was so important, they wondered, why hadn't *The Jazz Singer* moved to the top of the all-time box-office list? Such success, though, would come a year later with *The Singing Fool*, also starring Jolson. From its opening day (September 20, 1928), it was the financial success of its time; produced for an estimated $200,000, it took in $5 million. In New York City, *The*

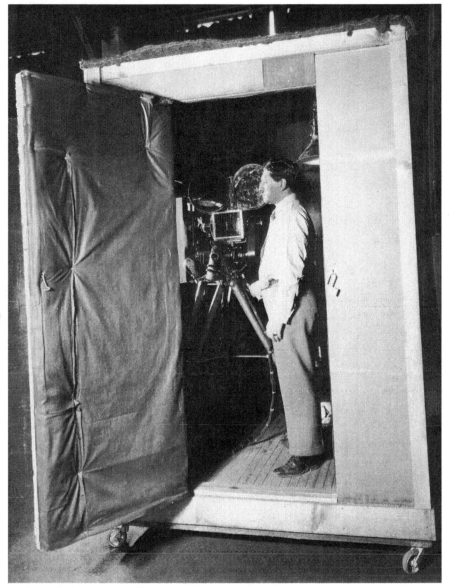

In the early days of sound films, cameras had to be soundproofed so their operating noises would not be picked up by the primitive sound-recording equipment. (Library of Congress)

Singing Fool registered the heaviest business in Broadway history, with an advance sale that exceeded more than $100,000 (equivalent to more than half a million dollars in 1990's currency).

Impact

The coming of sound transformed filmmaking, ushering in what became known as the golden age of Hollywood. By 1930, there were more reporters stationed in the filmmaking capital of the world than in any capital of Europe or Asia.

The Warner Brothers

Businessmen rather than inventors, the four Warner brothers were hustlers who knew a good thing when they saw it. They started out running theaters in 1903, evolved into film distributors, and began making their own films in 1909, in defiance of the Patents Company, a trust established by Thomas A. Edison to eliminate competition from independent filmmakers. Harry Warner was the president of the company, Sam and Jack were vice presidents in charge of production, and Abe (or Albert) was the treasurer.

Theirs was a small concern. Their silent films and serials attracted few audiences, and during World War I they made training films for the government. In fact, their film about syphilis, *Open Your Eyes*, was their first real success. In 1918, however, they released *My Four Years in Germany*, a dramatized documentary, and it was their first blockbuster. Although considered gauche upstarts, they were suddenly taken seriously by the movie industry.

When Sam first heard an actor talk on screen in an experimental film at the Bell lab in New York in 1925, he recognized a revolutionary opportunity. He soon convinced Jack that talking movies would be a gold mine. However, Harry and Abe were against the idea because of its costs—and because earlier attempts at "talkies" had been dismal failures. Sam and Jack tricked Harry into a seeing a experimental film of an orchestra, however, and he grew enthusiastic despite his misgivings. Within a year, the brothers released the all-music *Don Juan*. The rave notices from critics astounded Harry and Abe.

Still, they thought sound in movies was simply a novelty. When Sam pointed out that they could make movies in which the actors talked, as on stage, Harry, who detested actors, snorted, "Who the hell wants to hear actors talk?" Sam and Jack pressed for dramatic talkies, nonetheless, and prevailed upon Harry to finance them. The silver screen has seldom been silent since.

As a result of its foresight, Warner Bros. was the sole small competitor of the early 1920's to succeed in the Hollywood elite, producing successful films for consumption throughout the world.

After Warner Bros.' innovation, the soundtrack became one of the features that filmmakers controlled when making a film. Indeed, sound became a vital part of the filmmaker's art; music, in particular, could make or break a film.

Finally, the coming of sound helped make films a dominant medium of mass culture, both in the United States and throughout the world. Innumerable fashions, expressions, and designs were soon created or popularized by filmmakers. Many observers had not viewed the silent cinema as especially significant; with the coming of the talkies, however, there was no longer any question about the social and cultural importance of films. As one clear consequence of the new power of the movie industry, within a few years of the coming of sound, the notorious Hays Code mandating prior restraint of film content went into effect. The pairing of images and sound caused talking films to be deemed simply too powerful for uncensored presentation to audiences; although the Hays Code was gradually weakened and eventually abandoned, less onerous "rating systems" would continue to be imposed on filmmakers by various regulatory bodies.

See also Autochrome plate; Dolby noise reduction; Electronic synthesizer; Television.

FURTHER READING

Brayer, Elizabeth. *George Eastman: A Biography.* Baltimore: Johns Hopkins University Press, 1996.

Crafton, Donald. *The Talkies: American Cinema's Transition to Sound, 1926-1931.* Berkeley: University of California Press, 1999.

Geduld, Harry M. *The Birth of the Talkies: From Edison to Jolson.* Bloomington: Indiana University Press, 1975.

Neale, Stephen. *Cinema and Technology: Image, Sound, Colour.* London: Macmillan Education, 1985.

Wagner, A. F. *Recollections of Thomas A. Edison: A Personal History of the Early Days of the Phonograph, the Silent and Sound Film, and Film Censorship.* 2d ed. London: City of London Phonograph & Gramophone Society, 1996.

Teflon

THE INVENTION: A fluorocarbon polymer whose chemical inertness and physical properties have made it useful for many applications, from nonstick cookware coatings to suits for astronauts.

THE PERSON BEHIND THE INVENTION:
Roy J. Plunkett (1910-1994), an American chemist

Nontoxic Refrigerant Sought

As the use of mechanical refrigeration increased in the late 1930's, manufacturers recognized the need for a material to replace sulfur dioxide and ammonia, which, although they were the commonly used refrigerants of the time, were less than ideal for the purpose. The material sought had to be nontoxic, odorless, colorless, and not flammable. Thomas Midgley, Jr., and Albert Henne of General Motors Corporation's Frigidaire Division concluded, from studying published reports listing properties of a wide variety of chemicals, that hydrocarbon-like materials with hydrogen atoms replaced by chlorine and fluorine atoms would be appropriate.

Their conclusion led to the formation of a joint effort between the General Motors Corporation's Frigidaire Division and E. I. Du Pont de Nemours to research and develop the chemistry of fluorocarbons. In this research effort, a number of scientists began making and studying the large number of individual chemicals in the general class of compounds being investigated. It fell to Roy J. Plunkett to do a detailed study of tetrafluoroethylene, a compound consisting of two carbon atoms, each of which is attached to the other as well as to two fluorine atoms.

The "Empty" Tank

Tetrafluoroethylene, at normal room temperature and pressure, is a gas that is supplied to users in small pressurized cylinders. On the morning of the day of the discovery, Plunkett attached such a tank to his experimental apparatus and opened the tank's valve. To his great surprise, no gas flowed from the tank. Plunkett's subsequent actions transformed this event from an experiment gone wrong into a historically significant discovery. Rather than replacing the tank with another and going on with the work planned for the day, Plunkett, who wanted to know what had happened, examined the "empty" tank. When he weighed the tank, he discovered that it was not empty; it did contain the chemical that was listed on the label. Opening the valve and running a wire through the opening proved that what had happened had not been caused by a malfunctioning valve. Finally, Plunkett sawed the cylinder in half and discovered what had happened. The chemical in the tank was no longer a gas; instead, it was a waxy white powder.

Plunkett immediately recognized the meaning of the presence of the solid. The six-atom molecules of the tetrafluoroethylene gas had somehow linked with one another to form much larger molecules. The gas had *polymerized*, becoming polytetrafluoroethylene, a solid with a high molecular weight. Capitalizing on this occurrence, Plunkett, along with other Du Pont chemists, performed a series of experiments and soon learned to control the polymerization reaction so that the product could be produced, its properties could be studied, and applications for it could be developed.

The properties of the substance were remarkable indeed. It was unaffected by strong acids and bases, withstood high temperatures without reacting or melting, and was not dissolved by any solvent that the scientists tried. In addition to this highly unusual behavior, the polymer had surface properties that made it very slick. It was so slippery that other materials placed on its surface slid off in much the same way that beads of water slide off the surface of a newly waxed automobile.

Although these properties were remarkable, no applications were suggested immediately for the new material. The polymer might have remained a laboratory curiosity if a conversation had not taken place between Leslie R. Groves, the head of the Manhattan Project (which engineered the construction of the first atomic bombs), and a Du Pont chemist who described the polymer to him. The Manhattan Project research team was hunting for an inert material to use for gaskets to seal pumps and piping. The gaskets had to be able to withstand the highly corrosive uranium hexafluoride with

Roy J. Plunkett

Roy J. Plunkett was born in 1910 in New Carlisle, Ohio. In 1932 he received a bachelor's degree in chemistry from Manchester College and transferred to Ohio State University for graduate school, earning a master's degree in 1933 and a doctorate in 1936. The same year he went to work for E. I. Du Pont de Nemours and Company as a research chemist at the Jackson Laboratory in Deepwater, New Jersey. Less then two years later, when he was only twenty-seven years old, he found the strange polymer tetrafluoroethylene, whose trade name became Teflon. It would turn out to be among Du Pont's most famous products.

In 1938 Du Pont appointed Plunkett the chemical supervisor at its largest plant, the Chamber Works in Deepwater, which produced tetraethyl lead. He held the position until 1952 and afterward directed the company's Freon Products Division. He retired in 1975. In 1985 he was inducted into the Inventor's Hall of Fame, and after his death in 1994, Du Pont created the Plunkett Award, presented to inventors who find new uses for Teflon and Tefzel, a related fluoropolymer, in aerospace, automotive, chemical, or electrical applications.

which the team was working. This uranium compound is fundamental to the process of upgrading uranium for use in explosive devices and power reactors. Polytetrafluoroethylene proved to be just the material that they needed, and Du Pont proceeded, throughout World War II and after, to manufacture gaskets for use in uranium enrichment plants.

The high level of secrecy of the Manhattan Project in particular and atomic energy in general delayed the commercial introduction of the polymer, which was called Teflon, until the late 1950's. At that time, the first Teflon-coated cooking utensils were introduced.

IMPACT

Plunkett's thoroughness in following up a chance observation gave the world a material that has found a wide variety of uses, ranging from home kitchens to outer space. Some applications make use

An important space application for Teflon is its use on the outer skins of suits worn by astronauts. (PhotoDisc)

of Teflon's slipperiness, others make use of its inertness, and others take advantage of both properties.

The best-known application of Teflon is as a non-stick coating for cookware. Teflon's very slippery surface initially was troublesome, when it proved to be difficult to attach to other materials. Early versions of Teflon-coated cookware shed their surface coatings easily, even when care was taken to avoid scraping it off. A suitable bonding process was soon developed, however, and the present coated surfaces are very rugged and provide a noncontaminating coating that can be cleaned easily.

Teflon has proved to be a useful material in making devices that are implanted in the human body. It is easily formed into various shapes and is one of the few materials that the human body does not reject. Teflon has been used to make heart valves, pacemakers, bone and tendon substitutes, artificial corneas, and dentures.

Teflon's space applications have included its use as the outer skin of the suits worn by astronauts, as insulating coating on wires and cables in spacecraft that must resist high-energy cosmic radiation, and as heat-resistant nose cones and heat shields on spacecraft.

See also Buna rubber; Neoprene; Nylon; Plastic; Polystyrene; Pyrex glass; Tupperware.

FURTHER READING

Friedel, Robert. "The Accidental Inventor." *Discover* 17, no. 10 (October, 1996).
"Happy Birthday, Teflon." *Design News* 44, no. 8 (April, 1988).
"Teflon." *Newsweek* 130, 24a (Winter, 1997/1998).

Telephone switching

The invention: The first completely automatic electronic system for switching telephone calls.

The people behind the invention:
Almon B. Strowger (1839-1902), an American inventor
Charles Wilson Hoover, Jr. (1925-   ), supervisor of memory system development
Wallace Andrew Depp (1914-   ), director of Electronic Switching
Merton Brown Purvis (1923-   ), designer of switching matrices

Electromechanical Switching Systems

The introduction of electronic switching technology into the telephone network was motivated by the desire to improve the quality of the telephone system, add new features, and reduce the cost of switching technology. Telephone switching systems have three features: signaling, control, and switching functions. There were several generations of telephone switching equipment before the first fully electronic switching "office" (device) was designed.

The first automatic electromechanical (partly electronic and partly mechanical) switching office was the Strowger step-by-step switch. Strowger switches relied upon the dial pulses generated by rotary dial telephones to move their switching elements to the proper positions to connect one telephone with another. In the step-by-step process, the first digit dialed moved the first mechanical switch into position, the second digit moved the second mechanical switch into position, and so forth, until the proper telephone connection was established. These Strowger switching offices were quite large, and they lacked flexibility and calling features.

The second generation of automatic electromechanical telephone switching offices was of the "crossbar" type. Initially, crossbar switches relied upon a specialized electromechanical controller called a "marker" to establish call connections. Electromechanical telephone switching offices had difficulty implementing additional features and were unable to handle large numbers of incoming calls.

Electronic Switching Systems

In the early 1940's, research into the programmed control of switching offices began at the American Telephone and Telegraph Company's Bell Labs. This early research resulted in a trial office being put into service in Morris, Illinois, in 1960. The Morris switch used a unique memory called the "flying spot store." It used a photographic plate as a program memory, and the memory was accessed optically. In order to change the memory, one had to scratch out or cover parts of the photographic plate.

Before the development of the Morris switch, gas tubes had been used to establish voice connections. This was accomplished by applying a voltage difference across the end points of the conversation. When this voltage difference was applied, the gas tubes would conduct electricity, thus establishing the voice connection. The Morris trial showed that gas tubes could not support the voltages that the new technology required to make telephones ring or to operate pay telephones.

The knowledge gained from the Morris trial led to the development of the first full-scale, commercial, computer-controlled electronic switch, the electronic switching system 1 (ESS-1). The first ESS-1 went into service in New Jersey in 1965. In the ESS-1, electromechanical switching elements, or relays, were controlled by computer software. A centralized computer handled call processing. Because the telephone service of an entire community depends on the reliability of the telephone switching office, the ESS-1 had two central processors, so that one would be available if the other broke down. The switching system of the ESS-1 was composed of electromechanical relays; the control of the switching system was electronic, but the switching itself remained mechanical.

Bell Labs developed models to demonstrate the concept of integrating digital transmission and switching systems. Unfortunately, the solid state electronics necessary for such an undertaking had not developed sufficiently at that time, so the commercial development

ALMON B. STROWGER

Some people thought Almon B. Strowger was strange, perhaps even demented. Certainly, he was hot-tempered, restless, and argumentative. One thing he was not, however, was unimaginative.

Born near Rochester, New York, in 1839, Strowger was old enough to fight for the Union at the second battle of Manassas during the American Civil War. The bloody battle apparently shattered and embittered him. He wandered slowly west after the war, taught himself undertaking, and opened a funeral home in Topeka, Kansas, in 1882. There began his running war with telephone operators, which continued when he moved his business to Kansas City.

With the help of technicians (whom he later cheated) he built the first "collar box," an automatic switching device, in 1887. The round contraption held a pencil that could be revolved to different pins arrange around it in order to change phone connections. Two years later he produced a more sophisticated device that was operated by push-button, and despite initial misgivings brought out a rotary dial device in 1896. That same year he sold the rights to his patents to business partners for $1,800 and his share in Strowger Automatic Dial Telephone Exchange for $10,000 in 1898. He moved to St. Petersburg, Florida, and opened a small hotel, dying there in 1902. It surely would have done his temper no good to learn that fourteen years later the Bell system bought his patents for $2.5 million.

of digital switching was not pursued. New versions of the ESS continued to employ electromechanical technology, although mechanical switching elements can cause impulse noise in voice signals and are larger and more difficult to maintain than electronic switching elements. Ten years later, however, Bell Labs began to develop a digital toll switch, the ESS-4, in which both switching and control functions were electronic.

Although the ESS-1 was the first electronically controlled switching system, it did not switch voices electronically. The ESS-1 used computer control to move mechanical contacts in order to establish a conversation. In a fully electronic switching system, the voices are digitized before switching is performed. This technique, which is called "digital switching," is still used.

The advent of electronically controlled switching systems made possible features such as call forwarding, call waiting, and detailed billing for long-distance calls. Changing these services became a matter of simply changing tables in computer programs. Telephone maintenance personnel could communicate with the central processor of the ESS-1 by using a teletype, and they could change numbers simply by typing commands on the teletype. In electromechanically controlled telephone switching systems, however, changing numbers required rewiring.

CONSEQUENCES

Electronic switching has greatly decreased the size of switching offices. Digitization of the voice prior to transmission improves voice quality. When telephone switches were electromechanical, a large area was needed to house the many mechanical switches that were required. In the era of electronic switching, voices are switched digitally by computer. In this method, voice samples are read into a computer memory and then read out of the memory when it is time to connect a caller with a desired number. Basically, electronic telephone systems are specialized computer systems that move digitized voice samples between customers.

Telephone networks are moving toward complete digitization. Digitization was first applied to the transmission of voice signals. This made it possible for a single pair of copper wires to be shared by a number of telephone users. Currently, voices are digitized upon their arrival at the switching office. If the final destination of the telephone call is not connected to the particular switching office, the voice is sent to the remote office by means of digital circuits.

Currently, voice signals are sent between the switching office and homes or businesses. In the future, digitization of the voice signal will occur in the telephone sets themselves. Digital voice signals will be sent directly from one telephone to another. This will provide homes with direct digital communication. A network that provides such services is called the "integrated services digital network" (ISDN).

See also Cell phone; Long-distance telephone; Rotary dial telephone; Touch-tone telephone.

FURTHER READING

Briley, Bruce E. *Introduction to Telephone Switching*. Reading, Mass.: Addison-Wesley, 1983.

Talley, David. *Basic Electronic Switching for Telephone Systems*. 2d ed. Rochelle Park, N.J.: Hayden, 1982.

Thompson, Richard A. *Telephone Switching Systems*. Boston: Artech House, 2000.

Television

THE INVENTION: System that converts moving pictures and sounds into electronic signals that can be broadcast at great distances.

THE PEOPLE BEHIND THE INVENTION:

Vladimir Zworykin (1889-1982), a Soviet electronic engineer and recipient of the National Medal of Science in 1967

Paul Gottlieb Nipkow (1860-1940), a German engineer and inventor

Alan A. Campbell Swinton (1863-1930), a Scottish engineer and Fellow of the Royal Society

Charles F. Jenkins (1867-1934), an American physicist, engineer, and inventor

The Persistence of Vision

In 1894, an American inventor, Charles F. Jenkins, described a scheme for electrically transmitting moving pictures. Jenkins's idea, however, was only one in an already long tradition of theoretical television systems. In 1842, for example, the English physicist Alexander Bain had invented an automatic copying telegraph for sending still pictures. Bain's system scanned images line by line. Similarly, the wide recognition of the persistence of vision—the mind's ability to retain a visual image for a short period of time after the image has been removed—led to experiments with systems in which the image to be projected was repeatedly scanned line by line. Rapid scanning of images became the underlying principle of all television systems, both electromechanical and all-electronic.

In 1884, a German inventor, Paul Gottlieb Nipkow, patented a complete television system that utilized a mechanical sequential scanning system and a photoelectric cell sensitized with selenium for transmission. The selenium photoelectric cell converted the light values of the image being scanned into electrical impulses to be transmitted to a receiver where the process would be reversed. The electrical impulses led to light of varying brightnesses being produced and projected on to a rotating disk that was scanned to repro-

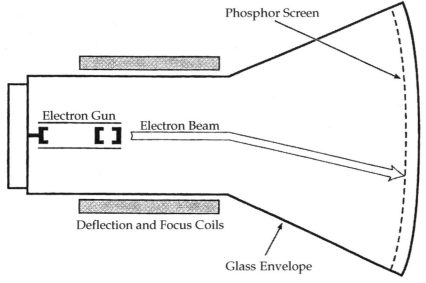

Schematic of a television picture tube.

duce the original image. If the system—that is, the transmitter and the receiver—were in perfect synchronization and if the disk ro tated quickly enough, persistence of vision enabled the viewer to see a complete image rather than a series of moving points of light.

For a television image to be projected onto a screen of reasonable size and retain good quality and high resolution, any system employing only thirty to one hundred lines (as early mechanical systems did) is inadequate. A few systems were developed that utilized two hundred or more lines, but the difficulties these presented made the possibility of an all-electronic system increasingly attractive. These difficulties were not generally recognized until the early 1930's, when television began to move out of the laboratory and into commercial production.

Interest in all-electronic television paralleled interest in mechanical systems, but solutions to technical problems proved harder to achieve. In 1908, a Scottish engineer, Alan A. Campbell Swinton, proposed what was essentially an all-electronic television system. Swinton theorized that the use of magnetically deflected cathode-ray tubes for both the transmitter and receiver in a system was possible. In 1911, Swinton formally presented his idea to the Röntgen

---

Vladimir Zworykin

Born in 1889, Vladimir Kosma Zworykin grew up in Murom, a small town two hundred miles east of Moscow. His father ran a riverboat service, and Zworykin sometimes helped him, but his mind was on electricity, which he studied on his own while aboard his father's boats. In 1906, he entered the St. Petersburg Institute of Technology, and there he became acquainted with the idea of television through the work of Professor Boris von Rosing.

Zworykin assisted Rosing in his attempts to transmit pictures with a cathode-ray tube. He served with the Russian Signal Corps during World War I, but then fled to the United States after the Bolshevist Revolution. In 1920 he got a job at Westinghouse's research laboratory in Pittsburgh, helping develop radio tubes and photoelectric cells. He became an American citizen in 1924 and completed a doctorate at the University of Pittsburgh in 1926. By then he had already demonstrated his iconoscope and applied for a patent. Unable to interest Westinghouse in his invention, he moved to the Radio Corporation of America (RCA) in 1929, and later became director of its electronics research laboratory. RCA's president, David Sarnoff, also a Russian immigrant, had faith in Zworykin and his ideas. Before Zworykin retired in 1954, RCA had invested $50 million in television.

Among the many awards Zworykin received for his culture-changing invention was the National Medal of Science, presented by President Lyndon Johnson in 1966. Zworykin died on his birthday in 1982.

Society in London, but the technology available did not allow for practical experiments.

Zworykin's Picture Tube

In 1923, Vladimir Zworykin, a Soviet electronic engineer working for the Westinghouse Electric Corporation, filed a patent application for the "iconoscope," or television transmission tube. On March 17, 1924, Zworykin applied for a patent for a two-way system. The first cathode-ray tube receiver had a cathode, a modulating grid, an anode, and a fluorescent screen.

Early console model television. (PhotoDisc)

Zworykin later admitted that the results were very poor and the system, as shown, was still far removed from a practical television system. Zworykin's employers were so unimpressed that they admonished him to forget television and work on something more useful. Zworykin's interest in television was thereafter confined to his nonworking hours, as he spent the next year working on photographic sound recording.

It was not until the late 1920's that he was able to devote his full attention to television. Ironically, Westinghouse had by then resumed research in television, but Zworykin was not part of the team. After he returned from a trip to France, where in 1928 he had witnessed an exciting demonstration of an electrostatic tube, Westinghouse indicated that it was not interested. This lack of corporate support in Pittsburgh led Zworykin to approach the Radio Corporation of America (RCA). According to reports, Zworykin demonstrated his system to the Institute of Radio Engineers at Rochester, New York, on November 18, 1929, claiming to have developed a working picture tube, a tube that would revolutionize television development. Finally, RCA recognized the potential.

IMPACT

The picture tube, or "kinescope," developed by Zworykin changed the history of television. Within a few years, mechanical systems disappeared and television technology began to utilize systems similar to Zworykin's by use of cathode-ray tubes at both ends of the system. At the transmitter, the image is focused upon a mosaic screen composed of light-sensitive cells. A stream of electrons sweeps the image, and each cell sends off an electric current pulse as it is hit by the electrons, the light and shade of the focused image regulating the amount of current.

This string of electrical impulses, after amplification and modification into ultrahigh frequency wavelengths, is broadcast by antenna to be picked up by any attuned receiver, where it is retransformed into a moving picture in the cathode-ray tube receiver. The cathode-ray tubes contain no moving parts, as the electron stream is guided entirely by electric attraction.

Although both the iconoscope and the kinescope were far from perfect when Zworykin initially demonstrated them, they set the stage for all future television development.

See also Color television; Community antenna television; Communications satellite; Fiber-optics; FM radio; Holography; Internet; Radio; Talking motion pictures.

FURTHER READING

Abramson, Albert. *Zworykin: Pioneer of Television*. Urbana: University of Illinois Press, 1995.
Sconce, Jeffrey. *Haunted Media: Electronic Presence from Telegraphy to Television*. Durham, N.C.: Duke University Press, 2000.
Zworykin, Vladimir Kosma, and George Ashmun Morton. *Television: The Electronics of Image Transmission in Color and Monochrome*. 2d ed. New York: J. Wiley, 1954.

Tevatron accelerator

THE INVENTION: A particle accelerator that generated collisions between beams of protons and antiprotons at the highest energies ever recorded.

THE PEOPLE BEHIND THE INVENTION:
Robert Rathbun Wilson (1914-   ), an American physicist and director of Fermilab from 1967 to 1978
John Peoples (1933-   ), an American physicist and deputy director of Fermilab from 1987

PUTTING SUPERMAGNETS TO USE

The Tevatron is a particle accelerator, a large electromagnetic device used by high-energy physicists to generate subatomic particles at sufficiently high energies to explore the basic structure of matter. The Tevatron is a circular, tubelike track 6.4 kilometers in circumference that employs a series of superconducting magnets to accelerate beams of protons, which carry a positive charge in the atom, and antiprotons, the proton's negatively charged equivalent, at energies up to 1 trillion electronvolts (equal to 1 teraelectronvolt, or 1 TeV; hence the name Tevatron). An electronvolt is the unit of energy that an electron gains through an electrical potential of 1 volt.

The Tevatron is located at the Fermi National Accelerator Laboratory, which is also known as Fermilab. The laboratory was one of several built in the United States during the 1960's.

The heart of the original Fermilab was the 6.4-kilometer main accelerator ring. This main ring was capable of accelerating protons to energies approaching 500 billion electronvolts, or 0.5 teraelectronvolt. The idea to build the Tevatron grew out of a concern for the millions of dollars spent annually on electricity to power the main ring, the need for higher energies to explore the inner depths of the atom and the consequences of new theories of both matter and energy, and the growth of superconductor technology. Planning for a second accelerator ring, the Tevatron, to be installed beneath the main ring began in 1972.

Robert Rathbun Wilson, the director of Fermilab at that time, realized that the only way the laboratory could achieve the higher energies needed for future experiments without incurring intolerable electricity costs was to design a second accelerator ring that employed magnets made of superconducting material. Extremely powerful magnets are the heart of any particle accelerator; charged particles such as protons are given a "push" as they pass through an electromagnetic field. Each successive push along the path of the circular accelerator track gives the particle more and more energy. The enormous magnetic fields required to accelerate massive particles such as protons to energies approaching 1 trillion electronvolts would require electricity expenditures far beyond Fermilab's operating budget. Wilson estimated that using superconducting materials, however, which have virtually no resistance to electrical current, would make it possible for the Tevatron to achieve double the main ring's magnetic field strength, doubling energy output without significantly increasing energy costs.

TEVATRON TO THE RESCUE

The Tevatron was conceived in three phases. Most important, however, were Tevatron I and Tevatron II, where the highest energies were to be generated and where it was hoped new experimental findings would emerge. Tevatron II experiments were designed to be very similar to other proton beam experiments, except that in this case, the protons would be accelerated to an energy of 1 trillion electronvolts. More important still are the proton-antiproton colliding beam experiments of Tevatron I. In this phase, beams of protons and antiprotons rotating in opposite directions are caused to collide in the Tevatron, producing a combined, or center-of-mass, energy approaching 2 trillion electronvolts, nearly three times the energy achievable at the largest accelerator at Centre Européen de Recherche Nucléaire (the European Center for Nuclear Research, or CERN).

John Peoples was faced with the problem of generating a beam of antiprotons of sufficient intensity to collide efficiently with a beam of protons. Knowing that he had the use of a large proton accelerator—the old main ring—Peoples employed the two-ring mode in which 120 billion electronvolt protons from the main ring are aimed at a fixed tungsten target, generating antiprotons, which scatter from the target. These particles were extracted and accumulated in a smaller storage ring. These particles could be accelerated to relatively low energies. After sufficient numbers of antiprotons were collected, they were injected into the Tevatron, along with a beam of protons for the colliding beam experiments. On October 13, 1985, Fermilab scientists reported a proton-antiproton collision with a center-of-mass energy measured at 1.6 trillion electronvolts, the highest energy ever recorded.

CONSEQUENCES

The Tevatron's success at generating high-energy proton-antiproton collisions affected future plans for accelerator development in the United States and offered the potential for important discoveries in high-energy physics at energy levels that no other accelerator could achieve.

Physics recognized four forces in nature: the electromagnetic force, the gravitational force, the strong nuclear force, and the weak nuclear force. A major goal of the physics community is to formulate a theory that will explain all these forces: the so-called grand unification theory. In 1967, one of the first of the so-called gauge theories was developed that unified the weak nuclear force and the electromagnetic force. One consequence of this theory was that the weak force was carried by massive particles known as "bosons." The search for three of these particles—the intermediate vector bosons W^+, W^-, and Z^0—led to the rush to conduct colliding beam experiments to the early 1970's. Because the Tevatron was in the planning phase at this time, these particles were discovered by a team of international scientists based in Europe. In 1989, Tevatron physicists reported the most accurate measure to date of the Z^0 mass.

The Tevatron is thought to be the only particle accelerator in the world with sufficient power to conduct further searches for the elusive Higgs boson, a particle attributed to weak interactions by University of Edinburgh physicist Peter Higgs in order to account for the large masses of the intermediate vector bosons. In addition, the Tevatron has the ability to search for the so-called top quark. Quarks are believed to be the constituent particles of protons and neutrons.

Evidence has been gathered of five of the six quarks believed to exist. Physicists have yet to detect evidence of the most massive quark, the top quark.

See also Atomic bomb; Cyclotron; Electron microscope; Field ion microscope; Geiger counter; Hydrogen bomb; Mass spectrograph; Neutrino detector; Scanning tunneling microscope; Synchrocyclotron.

FURTHER READING

Hilts, Philip J. *Scientific Temperaments: Three Lives in Contemporary Science.* New York: Simon and Schuster, 1984.
Ladbury, Ray. "Fermilab Tevatron Collider Group Goes over the Top—Cautiously." *Physics Today* 47, no. 6 (June, 1994).
Lederman, Leon M. "The Tevatron." *Scientific American* 264, no. 3 (March, 1991).
Wilson, Robert R., and Raphael Littauer. *Accelerators: Machines of Nuclear Physics.* London: Heinemann, 1962.

THERMAL CRACKING PROCESS

THE INVENTION: Process that increased the yield of refined gasoline extracted from raw petroleum by using heat to convert complex hydrocarbons into simpler gasoline hydrocarbons, thereby making possible the development of the modern petroleum industry.

THE PEOPLE BEHIND THE INVENTION:
William M. Burton (1865-1954), an American chemist
Robert E. Humphreys (1942-   ), an American chemist

GASOLINE, MOTOR VEHICLES, AND THERMAL CRACKING

Gasoline is a liquid mixture of hydrocarbons (chemicals made up of only hydrogen and carbon) that is used primarily as a fuel for internal combustion engines. It is produced by petroleum refineries that obtain it by processing petroleum (crude oil), a naturally occurring mixture of thousands of hydrocarbons, the molecules of which can contain from one to sixty carbon atoms.

Gasoline production begins with the "fractional distillation" of crude oil in a fractionation tower, where it is heated to about 400 degrees Celsius at the tower's base. This heating vaporizes most of the hydrocarbons that are present, and the vapor rises in the tower, cooling as it does so. At various levels of the tower, various portions (fractions) of the vapor containing simple hydrocarbon mixtures become liquid again, are collected, and are piped out as "petroleum fractions." Gasoline, the petroleum fraction that boils between 30 and 190 degrees Celsius, is mostly a mixture of hydrocarbons that contain five to twelve carbon atoms.

Only about 25 percent of petroleum will become gasoline via fractional distillation. This amount of "straight run" gasoline is not sufficient to meet the world's needs. Therefore, numerous methods have been developed to produce the needed amounts of gasoline. The first such method, "thermal cracking," was developed in 1913 by William M. Burton of Standard Oil of Indiana. Burton's cracking process used heat to convert complex hydrocarbons (whose molecules contain many carbon atoms) into simpler gasoline hydrocarbons (whose molecules contain fewer carbon atoms), thereby increasing the yield of gasoline from petroleum. Later advances in petroleum technology, including both an improved Burton method and other methods, increased the gasoline yield still further.

More Gasoline!

Starting in about 1900, gasoline became important as a fuel for the internal combustion engines of the new vehicles called automobiles. By 1910, half a million automobiles traveled American roads. Soon, the great demand for gasoline—which was destined to grow and grow—required both the discovery of new crude oil fields around the world and improved methods for refining the petroleum mined from these new sources. Efforts were made to increase the yield of gasoline—at that time, about 15 percent—from petroleum. The Burton method was the first such method.

At the time that the cracking process was developed, Burton was the general superintendent of the Whiting refinery, owned by Standard Oil of Indiana. The Burton process was developed in collaboration with Robert E. Humphreys and F. M. Rogers. This three-person research group began work knowing that heating petroleum fractions that contained hydrocarbons more complex than those present in gasoline—a process called "coking"—produced kerosene, coke (a form of carbon), and a small amount of gasoline. The process needed to be improved substantially, however, before it could be used commercially.

Initially, Burton and his coworkers used the "heavy fuel" fraction of petroleum (the 66 percent of petroleum that boils at a temperature higher than the boiling temperature of kerosene). Soon, they found that it was better to use only the part of the material that contained its smaller hydrocarbons (those containing fewer carbon atoms), all of which were still much larger than those present in gasoline. The cracking procedure attempted first involved passing the starting material through a hot tube. This hot-tube treatment vaporized the material and broke down 20 to 30 percent of the larger hydrocarbons into the hydrocarbons found in gasoline. Various tarry products were also produced, however, that reduced the quality of the gasoline that was obtained in this way.

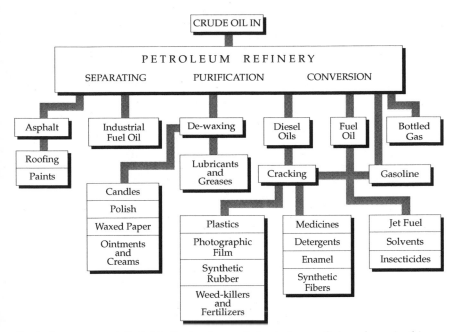

Burton's process contributed to the development of petroleum refining, shown in this diagram.

Next, the investigators attempted to work at a higher temperature by bubbling the starting material through molten lead. More gasoline was made in this way, but it was so contaminated with gummy material that it could not be used. Continued investigation showed, however, that moderate temperatures (between those used in the hot-tube experiments and that of molten lead) produced the best yield of useful gasoline.

The Burton group then had the idea of using high pressure to "keep starting materials still." Although the theoretical basis for the use of high pressure was later shown to be incorrect, the new method worked quite well. In 1913, the Burton method was patented and put into use. The first cracked gasoline, called Motor Spirit, was not very popular, because it was yellowish and had a somewhat unpleasant odor. The addition of some minor refining procedures, however, soon made cracked gasoline indistinguishable from straight run gasoline. Standard Oil of Indiana made huge profits from cracked gasoline over the next ten years. Ultimately, thermal cracking subjected the petroleum fractions that were utilized to temperatures between 550 and 750 degrees Celsius, under pressures between 250 and 750 pounds per square inch.

IMPACT

In addition to using thermal cracking to make gasoline for sale, Standard Oil of Indiana also profited by licensing the process for use by other gasoline producers. Soon, the method was used throughout the oil industry. By 1920, it had been perfected as much as it could be, and the gasoline yield from petroleum had been significantly increased. The disadvantages of thermal cracking include a relatively low yield of gasoline (compared to those of other methods), the waste of hydrocarbons in fractions converted to tar and coke, and the relatively high cost of the process.

A partial solution to these problems was found in "catalytic cracking"—the next logical step from the Burton method—in which petroleum fractions to be cracked are mixed with a catalyst (a substance that causes a chemical reaction to proceed more quickly, without reacting itself). The most common catalysts used in such cracking were minerals called "zeolites." The wide use of catalytic cracking soon enabled gasoline producers to work at lower temperatures (450 to 550 degrees Celsius) and pressures (10 to 50 pounds per square inch). This use decreased manufacturing costs because catalytic cracking required relatively little energy, produced only small quantities of undesirable side products, and produced high-quality gasoline.

Various other methods of producing gasoline have been developed—among them catalytic reforming, hydrocracking, alkylation, and catalytic isomerization—and now about 60 percent of the petroleum starting material can be turned into gasoline. These methods, and others still to come, are expected to ensure that the world's needs for gasoline will continue to be satisfied—as long as petroleum remains available.

See also Fuel cell; Gas-electric car; Geothermal power; Internal combustion engine; Oil-well drill bit; Solar thermal engine.

FURTHER READING

Gorman, Hugh S. *Redefining Efficiency: Pollution Concerns, Regulatory Mechanisms, and Technological Change in the U.S. Petroleum Industry.* Akron, Ohio: University of Akron Press, 2001.
Sung, Hsun-chang, Robert Roy White, and George Granger Brown. *Thermal Cracking of Petroleum.* Ann Arbor: University of Michigan, 1945.
William Meriam Burton: A Pioneer in Modern Petroleum Technology. Cambridge, Mass.: University Press, 1952.

TIDAL POWER PLANT

THE INVENTION: Plant that converts the natural ocean tidal forces into electrical power.

THE PEOPLE BEHIND THE INVENTION:
Mariano di Jacopo detto Taccola (Mariano of Siena, 1381-1453), an Italian notary, artist, and engineer
Bernard Forest de Bélidor (1697 or 1698-1761), a French engineer
Franklin D. Roosevelt (1882-1945), president of the United States

TIDAL ENERSGY

Ocean tides have long been harnessed to perform useful work. Ancient Greeks, Romans, and medieval Europeans all left records and ruins of tidal mills, and Mariano di Jacopo included tidal power in his treatise *De Ingeneis* (1433; on engines). Some mills consisted of water wheels suspended in tidal currents, others lifted weights that powered machinery as they fell, and still others trapped the high tide to run a mill.

Bernard Forest de Bélidor's *Architecture hydraulique* (1737; hydraulic architecture) is often cited as initiating the modern era of tidal power exploitation. Bélidor was an instructor in the French École d'Artillerie et du Génie (School of Artillery and Engineering).

Industrial expansion between 1700 and 1800 led to the construction of many tidal mills. In these mills, waterwheels or simple turbines rotated shafts that drove machinery by means of gears or belts. They powered small enterprises located on the seashore. Steam engines, however, soon began to replace tidal mills. Steam could be generated wherever it was needed, and steam mills were not dependent upon the tides or limited in their production capacity by the amount of tidal flow. Thus, tidal mills gradually were abandoned, although a few still operate in New England, Great Britain, France, and elsewhere.

ELECTRIC POWER FROM TIDES

Modern society requires tremendous amounts of electric energy generated by large power stations. This need was first met by using coal and by damming rivers. Later, oil and nuclear power became important. Although small mechanical tidal mills are inadequate for modern needs, tidal power itself remains an attractive source of energy. Periodic alarms about coal or oil supplies and concern about the negative effects on the environment of using coal, oil, or nuclear energy continue to stimulate efforts to develop renewable energy sources with fewer negative effects. Every crisis—for example, the perceived European coal shortages in the early 1900's, oil shortages in the 1920's and 1970's, and growing anxiety about nuclear power—revives interest in tidal power.

In 1912, a tidal power plant was proposed at Busum, Germany. The English, in 1918 and more recently, promoted elaborate schemes for the Severn Estuary. In 1928, the French planned a plant at Aber-Wrach in Brittany. In 1935, under the leadership of Franklin Delano Roosevelt, the United States began construction of a tidal power plant at Passamaquoddy, Maine. These plants, however, were never built. All of them had to be located at sites where tides were extremely high, and such sites are often far from power users. So much electricity was lost in transmission that profitable quantities of power could not be sent where they were needed. Also, large tidal power stations were too expensive to compete with existing steam plants and river dams. In addition, turbines and generators capable of using the large volumes of slow-moving tidal water that reversed flow had not been invented. Finally, large tidal plants inevitably hampered navigation, fisheries, recreation, and other uses of the sea and shore.

French engineers, especially Robert Gibrat, the father of the La Rance project, have made the most progress in solving the problems of tidal power plants. France, a highly industrialized country, is short of coal and petroleum, which has brought about an intense search by the French for alternative energy supplies.

La Rance, which was completed in December, 1967, is the first full-scale tidal electric power plant in the world. The Chinese, however, have built more than a hundred small tidal electric stations about the size of the old mechanical tidal mills, and the Canadians and the Russians have both operated plants of pilot-plant size.

La Rance, which was selected from more than twenty competing localities in France, is one of a few places in the world where the tides are extremely high. It also has a large reservoir that is located above a narrow constriction in the estuary. Finally, interference with navigation, fisheries, and recreational activities is minimal at La Rance.

Submersible "bulbs" containing generators and mounting propeller turbines were specially designed for the La Rance project. These turbines operate using both incoming and outgoing tides, and they can pump water either into or out of the reservoir. These features allow daily and seasonal changes in power generation to be "smoothed out." These turbines also deliver electricity most economically. Many engineering problems had to be solved, however, before the dam could be built in the tidal estuary.

The La Rance plant produces 240 megawatts of electricity. Its twenty-four highly reliable turbine generator sets operate about 95 percent of the time. Output is coordinated with twenty-four other hydroelectric plants by means of a computer program. In this system, pump-storage stations use excess La Rance power during periods of low demand to pump water into elevated reservoirs. Later, during peak demand, this water is fed through a power plant, thus "saving" the excess generated at La Rance when it was not immediately needed. In this way, tidal energy, which must be used or lost as the tides continue to flow, can be saved.

CONSEQUENCES

The operation of La Rance proved the practicality of tide-generated electricity. The equipment, engineering practices, and operating procedures invented for La Rance have been widely applied. Submersible, low-head, high-flow reversible generators of the La Rance type are now used in Austria, Switzerland, Sweden, Russia, Canada, the United States, and elsewhere.

Economic problems have prevented the building of more large tidal power plants. With technological advances, the inexorable depletion of oil and coal resources, and the increasing cost of nuclear power, tidal power may be used more widely in the future. Construction costs may be significantly lowered by using preconstructed power units and dam segments that are floated into place and submerged, thus making unnecessary expensive dams and reducing pumping costs.

See also Compressed-air-accumulating power plant; Geothermal power; Nuclear power plant; Nuclear reactor; Solar thermal engine; Thermal cracking process.

FURTHER READING

Bernshtein, L. B. *Tidal Power Plants*. Seoul, Korea: Korea Ocean Research and Development Institute, 1996.

Boyle, Godfrey. *Renewable Energy: Power for a Sustainable Future*. Oxford: Oxford University Press, 1998.

Ross, David. *Power from the Waves*. New York: Oxford University Press, 1995.

Seymour, Richard J. *Ocean Energy Recovery: The State of the Art*. New York: American Society of Civil Engineers, 1992.

Touch-tone telephone

THE INVENTION: A push-button dialing system for telephones that replaced the earlier rotary-dial phone.

THE PERSON BEHIND THE INVENTION:
Bell Labs, the research and development arm of the American Telephone and Telegraph Company

Dialing Systems

A person who wishes to make a telephone call must inform the telephone switching office which number he or she wishes to reach. A telephone call begins with the customer picking up the receiver and listening for a dial tone. The action of picking up the telephone causes a switch in the telephone to close, allowing electric current to flow between the telephone and the switching office. This signals the telephone office that the user is preparing to dial a number. To acknowledge its readiness to receive the digits of the desired number, the telephone office sends a dial tone to the user. Two methods have been used to send telephone numbers to the telephone office: dial pulsing and touch-tone dialing.

"Dial pulsing" is the method used by telephones that have rotary dials. In this method, the dial is turned until it stops, after which it is released and allowed to return to its resting position. When the dial is returning to its resting position, the telephone breaks the current between the telephone and the switching office. The switching office counts the number of times that current flow is interrupted, which indicates the number that had been dialed.

Introduction of Touch-tone Dialing

The dial-pulsing technique was particularly appropriate for use in the first electromechanical telephone switching offices, because the dial pulses actually moved mechanical switches in the switching office to set up the telephone connection. The introduction of touch-tone dialing into electromechanical systems was made possible by a special device that converted the touch-tones into rotary dial pulses that controlled the switches. At the American Telephone and Telegraph Company's Bell Labs, experimental studies were pursued that explored the use of "multifrequency key pulsing" (in other words, using keys that emitted tones of various frequencies) by both operators and customers. Initially, plucked tuned reeds were proposed. These were, however, replaced with "electronic transistor oscillators," which produced the required signals electronically.

The introduction of "crossbar switching" made dial pulse signaling of the desired number obsolete. The dial pulses of the telephone were no longer needed to control the mechanical switching process at the switching office. When electronic control was introduced into switching offices, telephone numbers could be assigned by computer rather than set up mechanically. This meant that a single touch-tone receiver at the switching office could be shared by a large number of telephone customers.

Before 1963, telephone switching offices relied upon rotary dial pulses to move electromechanical switching elements. Touch-tone dialing was difficult to use in systems that were not computer controlled, such as the electromechanical step-by-step method. In about 1963, however, it became economically feasible to implement centralized computer control and touch-tone dialing in switching offices. Computerized switching offices use a central touch-tone receiver to detect dialed numbers, after which the receiver sends the number to a call processor so that a voice connection can be established.

Touch-tone dialing transmits two tones simultaneously to represent a digit. The tones that are transmitted are divided into two groups: a high-band group and a low-band group. For each digit that is dialed, one tone from the low-frequency (low-band) group and one tone from the high-frequency (high-band) group are transmitted. The two frequencies of a tone are selected so that they are not too closely related harmonically. In addition, touch-tone receivers must be designed so that false digits cannot be generated when people are speaking into the telephone.

For a call to be completed, the first digit dialed must be detected in the presence of a dial tone, and the receiver must not interpret background noise or speech as valid digits. In order to avoid such misinterpretation, the touch-tone receiver uses both the relative and the absolute strength of the two simultaneous tones of the first digit dialed to determine what that digit is.

A system similar to the touch-tone system is used to send telephone numbers between telephone switching offices. This system, which is called "multifrequency signaling," also uses two tones to indicate a single digit, but the frequencies used are not the same frequencies that are used in the touch-tone system. Multifrequency signaling is currently being phased out; new computer-based systems are being introduced to replace it.

IMPACT

Touch-tone dialing has made new caller features available. The touch-tone system can be used not only to signal the desired number to the switching office but also to interact with voice-response systems. This means that touch-tone dialing can be used in conjunction with such devices as bank teller machines. A customer can also dial many more digits per second with a touch-tone telephone than with a rotary dial telephone.

Touch-tone dialing has not been implemented in Europe, and one reason may be that the economics of touch-tone dialing change as a function of technology. In the most modern electronic switching offices, rotary signaling can be performed at no additional cost, whereas the addition of touch-tone dialing requires a centralized touch-tone receiver at the switching office. Touch-tone signaling was developed in an era of analog telephone switching offices, and since that time, switching offices have become overwhelmingly digital. When the switching network becomes entirely digital, as will be the case when the integrated services digital network (ISDN) is implemented, touch-tone dialing will become unnecessary. In the future, ISDN telephone lines will use digital signaling methods exclusively.

See also Cell phone; Rotary dial telephone; Telephone switching.

FURTHER READING

Coe, Lewis. *The Telephone and Its Several Inventors: A History.* Jefferson, N.C.: McFarland, 1995.
Young, Peter. *Person to Person: The International Impact of the Telephone.* Cambridge: Granta Editions, 1991.

TRANSISTOR

THE INVENTION: A miniature electronic device, comprising a tiny semiconductor and multiple electrical contacts, used in circuits as an amplifier, detector, or switch, that revolutionized electronics in the mid-twentieth century.

THE PEOPLE BEHIND THE INVENTION:
  William B. Shockley (1910-1989), an American physicist who led the Bell Laboratories team that produced the first transistors
  Akio Morita (1921-1999), a Japanese physicist and engineer who was the cofounder of the Sony electronics company
  Masaru Ibuka (1908-1997), a Japanese electrical engineer and businessman who cofounded Sony with Morita

THE BIRTH OF SONY

In 1952, a Japanese engineer visiting the United States learned that the Western Electric company was granting licenses to use its transistor technology. He was aware of the development of this device and thought that it might have some commercial applications. Masaru Ibuka told his business partner in Japan about the opportunity, and they decided to raise the $25,000 required to obtain a license. The following year, his partner, Akio Morita, traveled to New York City and concluded negotiations with Western Electric. This was a turning point in the history of the Sony company and in the electronics industry, for transistor technology was to open profitable new fields in home entertainment.

The origins of the Sony corporation were in the ruins of postwar Japan. The Tokyo Telecommunications Company was incorporated in 1946 and manufactured a wide range of electrical equipment based on the existing vacuum tube technology. Morita and Ibuka were involved in research and development of this technology during the war and intended to put it to use in the peacetime economy. In the United States and Europe, electrical engineers who had done the same sort of research founded companies to build advanced audio products such as high-performance amplifiers, but Morita and Ibuka did not have the resources to make such sophisticated products and concentrated on simple items such as electric water heaters and small electric motors for record players.

In addition to their experience as electrical engineers, both men were avid music lovers, as a result of their exposure to American-built phonographs and gramophones exported to Japan in the early twentieth century. They decided to combine their twin interests by devising innovative audio products and looked to the new field of magnetic recording as a likely area for exploitation. They had learned about tape recorders from technical journals and had seen them in use by the American occupation force.

They developed a reel-to-reel tape recorder and introduced it in 1950. It was a large machine with vacuum tube amplifiers, so heavy that they transported it by truck. Although it worked well, they had a hard job selling it. Ibuka went to the United States in 1952 partly on a fact-finding mission and partly to get some ideas about marketing the tape recorder to schools and businesses. It was not seen as a consumer product.

Ibuka and Morita had read about the invention of the transistor in Western Electric's laboratories shortly after the war. John Bardeen and Walter H. Brattain had discovered that a semiconducting material could be used to amplify or control electric current. Their point contact transistor of 1948 was a crude laboratory apparatus that served as the basis for further research. The project was taken over by William B. Shockley, who had suggested the theory of the transistor effect. A new generation of transistors was devised; they were simpler and more efficient than the original. The junction transistors were the first to go into production.

Ongoing Research

Bell Laboratories had begun transistor research because Western Electric, one of its parent companies along with American Telephone and Telegraph, was interested in electronic amplification. This was seen as a means to increase the strength of telephone signals traveling over long distances, a job carried out by vacuum tubes. The junction transistor was developed as an amplifier. Western Electric thought that the hearing aid was the only consumer product that could be based on it and saw the transistor solely as a telecommunications technology. The Japanese purchased the license with only the slightest understanding of the workings of semiconductors and despite the belief that transistors could not be used at the high frequencies associated with radio.

The first task of Ibuka and Morita was to develop a high-frequency transistor. Once this was accomplished, in 1954, a method had to be found to manufacture it cheaply. Transistors were made from crystals, which had to be grown and doped with impurities to form different layers of conductivity. This was not an exact science, and Sony engineers found that the failure rate for high-frequency transistors was very high. This increased costs and put the entire project into doubt, because the adoption of transistors was based on simplicity, reliability, and low cost.

The introduction of the first Sony transistor radio, the TR-55, in 1955 was the result of basic research combined with extensive industrial engineering. Morita admitted that its sound was poor, but because it was the only transistor radio in Japan, it sold well. These were not cheap products, nor were they particularly compact. The selling point was that they consumed much less battery power than the old portable radios.

The TR-55 carried the brand name Sony, a relative of the Soni magnetic tape made by the company and a name influenced by the founders' interest in sound. Morita and Ibuka had already decided that the future of their company would be in international trade and wanted its name to be recognized all over the world. In 1957, they changed the company's name from Tokyo Telecomunications Engineering to Sony.

The first product intended for the export market was a small transistor radio. Ibuka was disappointed at the large size of the TR-55 because one of the advantages of the transistor over the vacuum tube was supposed to be smaller size. He saw a miniature radio as a promising consumer product and gave his engineers the task of designing one small enough to fit into his shirt pocket.

All elements of the radio had to be reduced in size: amplifier, transformer, capacitor, and loudspeaker. Like many other Japanese manufacturers, Sony bought many of the component parts of its products from small manufacturers, all of which had to be cajoled into decreasing the size of their parts. Morita and Ibuka stated that the hardest task in developing this new product was negotiating with the subcontractors. Finally, the Type 63 pocket transistor radio—the "Transistor Six"—was introduced in 1957.

IMPACT

When the transistor radio was introduced, the market for radios was considered to be saturated. People had rushed to buy them when they were introduced in the 1920's, and by the time of the Great Depression, the majority of American households had one. Improvements had been made to the receiver and more attractive radio/phonograph console sets had been introduced, but these developments did not add many new customers. The most manufacturers could hope for was the replacement market with a few sales as children moved out of their parents' homes and established new households.

The pocket radio created a new market. It could be taken anywhere and used at any time. Its portability was its major asset, and it became an indispensable part of youth-oriented popular culture of the 1950's and 1960's. It provided an outlet for the crowded airwaves of commercial AM radio and was the means to bring the new music of rock and roll to a mass audience.

As soon as Sony introduced the Transistor Six, it began to redesign it to reduce manufacturing cost. Subsequent transistor radios were smaller and cheaper. Sony sold them by the millions, and millions more were made by other companies under brand names such as "Somy" and "Sonny." By 1960, more than twelve million transistor radios had been sold.

The transistor radio was the product that established Sony as an international audio concern. Morita had resisted the temptation to make radios for other companies to sell under their names. Exports of Sony radios increased name recognition and established a bridgehead in the United States, the biggest market for electronic consumer products. Morita planned to follow the radio with other transistorized products.

The television had challenged radio's position as the mechanical entertainer in the home. Like the radio, it stood in nearly every

WILLIAM SHOCKLEY

William Shockley's reputation contains extremes. He helped invent one of the basic devices supporting modern technological society, the transistor. He also tried to revive one of the most infamous social theories, eugenics.

His parents, mining engineer William Hillman Shockley, and surveyor May Bradford Shockley, were on assignment in England in 1910 when he was born. The family returned to Northern California when the younger William was three, and they schooled him at home until he was eight. He acquired an early interest in physics from a neighbor who taught at Stanford University. Shockley pursed that interest at the California Institute of Technology and the Massachusetts Institute of Technology, which awarded him a doctorate in 1936.

Shockley went to work for Bell Telephone Laboratories in the same year. While trying to design a vacuum tube that could amplify current, it occurred to him that solid state components might work better than the fragile tubes. He experimented with the semiconductors germanium and silicon, but the materials available were too impure for his purpose. World War II interrupted the experiments, and he worked instead to improve radar and anti-submarine devices for the military. Back at Bell Labs in 1945, Shockley teamed with theorist John Bardeen and experimentalist Walter Brattain. Two years later they succeeded in making the first amplifier out of semiconductor materials and called it a transistor (short for *trans*fer res*istor*). Its effect on the electronics industry was revolutionary, and the three shared the 1956 Nobel Prize in Physics for their achievement.

In the mid-1950's Shockley left Bell Labs to start Shockley Transistor, then switched to academia in 1963, becoming Stanford University's Alexander M. Poniatoff Professor of Engineering and Applied Science. He grew interested in the relation between race and intellectual ability. Teaching himself psychology and genetics, he conceived the theory that Caucasians were inherently more intelligent than other races because of their genetic make-up. When he lectured on his brand of eugenics, he was denounced by the public as a racist and by scientists for shoddy thinking. Shockley retired in 1975 and died in 1989.

American living room and used the same vacuum tube amplification unit. The transistorized portable television set did for images what the transistor radio did for sound. Sony was the first to develop an all-transistor television, in 1959. At a time when the trend in television receivers was toward larger screens, Sony produced extremely small models with eight-inch screens. Ignoring the marketing experts who said that Americans would never buy such a product, Sony introduced these models into the United States in 1960 and found that there was a huge demand for them.

As in radio, the number of television stations on the air and broadcasts for the viewer to choose from grew. A personal television or radio gave the audience more choices. Instead of one machine in the family room, there were now several around the house. The transistorization of mechanical entertainers allowed each family member to choose his or her own entertainment. Sony learned several important lessons from the success of the transistor radio and television. The first was that small size and low price could create new markets for electronic consumer products. The second was that constant innovation and cost reduction were essential to keep ahead of the numerous companies that produced cheaper copies of original Sony products.

In 1962, Sony introduced a tiny television receiver with a five-inch screen. In the 1970's and 1980's, it produced even smaller models, until it had a TV set that could sit in the palm of the hand—the Video Walkman. Sony's scientists had developed an entirely new television screen that worked on a new principle and gave better color resolution; the company was again able to blend the fruits of basic scientific research with innovative industrial engineering.

The transistorized amplifier unit used in radio and television sets was applied to other products, including amplifiers for record players and tape recorders. Japanese manufacturers were slow to take part in the boom in high-fidelity audio equipment that began in the United States in the 1950's. The leading manufacturers of high-quality audio components were small American companies based on the talents of one engineer, such as Avery Fisher or Henry Koss. They sold expensive amplifiers and loudspeakers to audiophiles. The transistor reduced the size, complexity, and price of these components. The Japanese took the lead devising complete audio units based on transistorized integrated circuits, thus developing the basic home stereo.

In the 1960's, companies such as Sony and Matsushita dominated the market for inexpensive home stereos. These were the basic radio/phonograph combination, with two detached speakers. The finely crafted wooden consoles that had been the standard for the home phonograph were replaced by small plastic boxes. The Japanese were also quick to exploit the opportunities of the tape cassette. The Philips compact cassette was enthusiastically adopted by Japanese manufacturers and incorporated into portable tape recorders. This was another product with its ancestry in the transistor radio. As more of them were sold, the price dropped, encouraging more consumers to buy. The cassette player became as commonplace in American society in the 1970's as the transistor radio had been in the 1960's.

THE WALKMAN

The transistor took another step in miniaturization in the Sony Walkman, a personal stereo sound system consisting of a cassette player and headphones. It was based on the same principles as the transistor radio and television. Sony again confounded marketing experts by creating a new market for a personal electronic entertainer. In the ten years following the introduction of the Walkman in 1979, Sony sold fifty million units worldwide, half of those in the United States. Millions of imitation products were sold by other companies.

Sony's acquisition of the Western Electric transistor technology was a turning point in the fortunes of that company and of Japanese manufacturers in general. Less than ten years after suffering defeat in a disastrous war, Japanese industry served notice that it had lost none of its engineering capabilities and innovative skills. The production of the transistor radio was a testament to the excellence of Japanese research and development. Subsequent products proved that the Japanese had an uncanny sense of the potential market for consumer products based on transistor technology. The ability to incorporate solid-state electronics into innovative home entertainment products allowed Japanese manufacturers to dominate the world market for electronic consumer products and to eliminate most of their American competitors.

The little transistor radio was the vanguard of an invasion of new products unparalleled in economic history. Japanese companies such as Sony and Panasonic later established themselves at the leading edge of digital technology, the basis of a new generation of entertainment products. Instead of Japanese engineers scraping together the money to buy a license for an American technology, the great American companies went to Japan to license compact disc and other digital technologies.

See also Cassette recording; Color television; FM radio; Radio; Television; Transistor radio; Videocassette recorder; Walkman cassette player.

FURTHER READING

Lyons, Nick. *The Sony Vision*. New York: Crown Publishers, 1976.

Marshall, David V. *Akio Morita and Sony*. Watford: Exley, 1995.

Morita, Akio, with Edwin M. Reingold, and Mitsuko Shimomura. *Made in Japan: Akio Morita and Sony*. London: HarperCollins, 1994.

Reid, T. R. *The Chip: How Two Americans Invented the Microchip and Launched a Revolution*. New York: Simon and Schuster, 1984.

Riordan, Michael. *Crystal Fire: The Invention of the Transistor and the Birth of the Information Age*. New York: Norton, 1998.

Scott, Otto. *The Creative Ordeal: The Story of Raytheon*. New York: Atheneum, 1974.

TRANSISTOR RADIO

THE INVENTION: Miniature portable radio that used transistors and created a new mass market for electronic products.

THE PEOPLE BEHIND THE INVENTION:
John Bardeen (1908-1991), an American physicist
Walter H. Brattain (1902-1987), an American physicist
William Shockley (1910-1989), an American physicist
Akio Morita (1921-1999), a Japanese physicist and engineer
Masaru Ibuka (1907-1997), a Japanese electrical engineer and industrialist

A Replacement for Vacuum Tubes

The invention of the first transistor by William Shockley, John Bardeen, and Walter H. Brattain of Bell Labs in 1947 was a scientific event of great importance. Its commercial importance at the time, however, was negligible. The commercial potential of the transistor lay in the possibility of using semiconductor materials to carry out the functions performed by vacuum tubes, the fragile and expensive tubes that were the electronic hearts of radios, sound amplifiers, and telephone systems. Transistors were smaller, more rugged, and less power-hungry than vacuum tubes. They did not suffer from overheating. They offered an alternative to the unreliability and short life of vacuum tubes.

Bell Labs had begun the semiconductor research project in an effort to find a better means of electronic amplification. This was needed to increase the strength of telephone signals over long distances. Therefore, the first commercial use of the transistor was sought in speech amplification, and the small size of the device made it a perfect component for hearing aids. Engineers from the Raytheon Company, the leading manufacturer of hearing aids, were invited to Bell Labs to view the new transistor and to help assess the commercial potential of the technology. The first transistorized consumer product, the hearing aid, was soon on the market. The early models built by Raytheon used three junction-type transistors and cost more than two hundred dollars. They were small enough to go directly into the ear or to be incorporated into eyeglasses.

The commercial application of semiconductors was aimed largely at replacing the control and amplification functions carried out by vacuum tubes. The perfect vehicle for this substitution was the radio set. Vacuum tubes were the most expensive part of a radio set and the most prone to break down. The early junction transistors operated best at low frequencies, and subsequently more research was needed to produce a commercial high-frequency transistor. Several of the licensees embarked on this quest, including the Radio Corporation of America (RCA), Texas Instruments, and the Tokyo Telecommunications Engineering Company of Japan.

PERFECTING THE TRANSISTOR

The Tokyo Telecommunications Engineering Company of Japan, formed in 1946, had produced a line of instruments and consumer products based on vacuum-tube technology. Its most successful product was a magnetic tape recorder. In 1952, one of the founders of the company, Masaru Ibuka, visited the United States to learn more about the use of tape recorders in schools and found out that Western Electric was preparing to license the transistor patent. With only the slightest understanding of the workings of semiconductors, Tokyo Telecommunications purchased a license in 1954 with the intention of using transistors in a radio set.

The first task facing the Japanese was to increase the frequency response of the transistor to make it suitable for radio use. Then a method of manufacturing transistors cheaply had to be found. At the time, junction transistors were made from slices of germanium crystal. Growing the crystal was not an exact science, nor was the process of "doping" it with impurities to form the different layers of conductivity that made semiconductors useful. The Japanese engineers found that the failure rate for high-frequency transistors was extremely high. The yield of good transistors from one batch ran as low as 5 percent, which made them extremely expensive and put the whole project in doubt. The effort to replace vacuum tubes with components made of semiconductors was motivated by cost rather than performance; if transistors proved to be more expensive, then it was not worth using them.

Engineers from Tokyo Telecommunications again came to the United States to search for information about the production of transistors. In 1954, the first high-frequency transistor was produced in Japan. The success of Texas Instruments in producing the components for the first transistorized radio (introduced by the Regency Company in 1954) spurred the Japanese to greater efforts. Much of their engineering and research work was directed at the manufacture and quality control of transistors. In 1955, they introduced their transistor radio, the TR-55, which carried the brand name "Sony." The name was chosen because the executives of the company believed that the product would have an international appeal and therefore needed a brand name that could be recognized easily and remembered in many languages. In 1957, the name of the entire company was changed to Sony.

IMPACT

Although Sony's transistor radios were successful in the marketplace, they were still relatively large and cumbersome. Ibuka saw a consumer market for a miniature radio and gave his engineers the task of designing a radio small enough to fit into a shirt pocket. The realization of this design—"Transistor Six"—was introduced in 1957. It was an immediate success. Sony sold the radios by the millions, and numerous imitations were also marketed under brand names such as "Somy" and "Sonny." The product became an indispensable part of popular culture of the late 1950's and 1960's; its low cost enabled the masses to enjoy radio wherever there were broadcasts.

The pocket-sized radio was the first of a line of electronic consumer products that brought technology into personal contact with the user. Sony was convinced that miniaturization did more than make products more portable; it established a one-on-one relationship between people and machines. Sony produced the first all-transistor television in 1960. Two years later, it began to market a miniature television in the United States. The continual reduction in the size of Sony's tape recorders reached a climax with the portable tape player introduced in the 1980's. The Sony Walkman was a marketing triumph and a further reminder that Japanese companies led the way in the design and marketing of electronic products.

John Bardeen

The transistor reduced the size of electronic circuits and at the same time the amount of energy lost from them as heat. Superconduction gave rise to electronic circuits with practically no loss of energy at all. John Bardeen helped unlock the secrets of both.

Bardeen was born in 1908 in Madison, Wisconsin, where his mother was an artist and his father was a professor of anatomy at the University of Wisconsin. Bardeen attended the university, earning a bachelor's degree in electrical engineering in 1928 and a master's degree in geophysics in 1929. After working as a geophysicist, he entered Princeton University, studying with Eugene Wigner, the leading authority on solid-state physics, and received a doctorate in mathematics and physics in 1936.

Bardeen taught at Harvard University and the University of Minnesota until World War II, when he moved to the Naval Ordnance Laboratory. Finding academic salaries too low to support his family after the war, he accepted a position at Bell Telephone Laboratories. There, with Walter Brattain, he turned William Shockley's theory of semiconductors into a practical device the transfer resistor, or transistor.

He returned to academia as a professor at the University of Illinois and began to investigate a long-standing mystery in physics, superconductivity, with a postdoctoral associate, Leon Cooper, and a graduate student, J. Robert Schrieffer. In 1956 Cooper made a key discovery—superconducting electrons travel in pairs. And while Bardeen was in Stockholm, Sweden, collecting a share of the 1956 Nobel Prize in Physics for his work on transistors, Schrieffer worked out a mathematical analysis of the phenomenon. The theory that the three men published since became known as BCS theory from the first letters of their last names, and as well as explain superconductors, it pointed toward a great deal of technology and additional basic research.

The team won the 1972 Nobel Prize in Physics for BCS theory, making Bardeen the only person to ever win two Nobel Prizes for physics. He retired in 1975 and died sixteen years later.

See also Compact disc; FM radio; Radio; Radio crystal sets; Television; Transistor; Walkman cassette player.

FURTHER READING

Handy, Roger, Maureen Erbe, and Aileen Antonier. *Made in Japan: Transistor Radios of the 1950s and 1960s.* San Francisco: Chronicle Books, 1993.

Marshall, David V. *Akio Morita and Sony.* Watford: Exley, 1995.

Morita, Akio, with Edwin M. Reingold, and Mitsuko Shimomura. *Made in Japan: Akio Morita and Sony.* London: HarperCollins, 1994.

Nathan, John. *Sony: The Private Life.* London: HarperCollins-Business, 2001.

TUBERCULOSIS VACCINE

THE INVENTION: Vaccine that uses an avirulent (nondisease) strain of bovine tuberculosis bacilli that is safer than earlier vaccines.

THE PEOPLE BEHIND THE INVENTION:
Albert Calmette (1863-1933), a French microbiologist
Camille Guérin (1872-1961), a French veterinarian and microbiologist
Robert Koch (1843-1910), a German physician and microbiologist

ISOLATING BACTERIA

Tuberculosis, once called "consumption," is a deadly, contagious disease caused by the bacterium *Mycobacterium tuberculosis*, first identified by the eminent German physician Robert Koch in 1882. The bacterium can be transmitted from person to person by physical contact or droplet infection (for example, sneezing). The condition eventually inflames and damages the lungs, causing difficulty in breathing and failure of the body to deliver sufficient oxygen to various tissues. It can spread to other body tissues, where further complications develop. Without treatment, the disease progresses, disabling and eventually killing the victim. Tuberculosis normally is treated with a combination of antibiotics and other drugs.

Koch developed his approach for identifying bacterial pathogens (disease producers) with simple equipment, primarily microscopy. Having taken blood samples from diseased animals, he would identify and isolate the bacteria he found in the blood. Each strain of bacteria would be injected into a healthy animal. The latter would then develop the disease caused by the particular strain.

In 1890, he discovered that a chemical released from tubercular bacteria elicits a hypersensitive (allergic) reaction in individuals previously exposed to or suffering from tuberculosis. This chemical, called "tuberculin," was isolated from culture extracts in which tubercular bacteria were being grown.

When small amounts of tuberculin are injected into a person subcutaneously (beneath the skin), a reddened, inflamed patch approximately the size of a quarter develops if the person has been exposed to or is suffering from tuberculosis. Injection of tuberculin into an uninfected person yields a negative response (that is, no inflammation). Tuberculin does not harm those being tested.

TUBERCULOSIS'S WEAKER GRANDCHILDREN

The first vaccine to prevent tuberculosis was developed in 1921 by two French microbiologists, Albert Calmette and Camille Guérin. Calmette was a student of the eminent French microbiologist Louis Pasteur at Pasteur's Institute in Paris. Guérin was a veterinarian who joined Calmette's laboratory in 1897. At Lille, Calmette and Guérin focused their research upon the microbiology of infectious diseases, especially tuberculosis.

In 1906, they discovered that individuals who had been exposed to tuberculosis or who had mild infections were developing resistance to the disease. They found that resistance to tuberculosis was initiated by the body's immune system. They also discovered that tubercular bacteria grown in culture over many generations become progressively weaker and avirulent, losing their ability to cause disease.

From 1906 through 1921, Calmette and Guérin cultured tubercle bacilli from cattle. With proper nutrients and temperature, bacteria can reproduce by fission (that is, one bacterium splits into two bacteria) in as little time as thirty minutes. Calmette and Guérin cultivated these bacteria in a bile-derived food medium for thousands of generations over fifteen years, periodically testing the bacteria for virulence by injecting them into cattle. After many generations, the bacteria lost their virulence, their ability to cause disease. Nevertheless, these weaker, or "avirulent" bacteria still stimulated the animals' immune systems to produce antibodies. Calmette and Guérin had successfully bred a strain of avirulent bacteria that could not cause tuberculosis in cows but could also stimulate immunity against the disease.

There was considerable concern over whether the avirulent strain was harmless to humans. Calmette and Guérin continued cultivating weaker versions of the avirulent strain that retained antibody-stimulating capacity. By 1921, they had isolated an avirulent anti-body-stimulating strain that was harmless to humans, a strain they called "Bacillus Calmette-Guérin" (BCG).

In 1922, they began BCG-vaccinating newborn children against tuberculosis at the Charité Hospital in Paris. The immunized children exhibited no ill effects from the BCG vaccination. Calmette and Guérin's vaccine was so successful in controlling the spread of tuberculosis in France that it attained widespread use in Europe and Asia beginning in the 1930's.

IMPACT

Most bacterial vaccines involve the use of antitoxin or heat- or chemical-treated bacteria. BCG is one of the few vaccines that use specially bred live bacteria. Its use sparked some controversy in the United States and England, where the medical community questioned its effectiveness and postponed BCG immunization until the late 1950's. Extensive testing of the vaccine was performed at the University of Illinois before it was adopted in the United States. Its effectiveness is questioned by some physicians to this day.

Some of the controversy stems from the fact that the avirulent, antibody-stimulating BCG vaccine conflicts with the tuberculin skin test. The tuberculin skin test is designed to identify people suffering from tuberculosis so that they can be treated. A BCG-vaccinated person will have a positive tuberculin skin test similar to that of a tuberculosis sufferer. If a physician does not know that a patient has had a BCG vaccination, it will be presumed (incorrectly) that the patient has tuberculosis. Nevertheless, the BCG vaccine has been invaluable in curbing the worldwide spread of tuberculosis, although it has not eradicated the disease.

See also Antibacterial drugs; Birth control pill; Penicillin; Polio vaccine (Sabin); Polio vaccine (Salk); Salvarsan; Typhus vaccine; Yellow fever vaccine.

FURTHER READING

Daniel, Thomas M. *Pioneers of Medicine and their Impact on Tuberculosis*. Rochester, N.Y.: University of Rochester Press, 2000.
DeJauregui, Ruth. *100 Medical Milestones That Shaped World History*. San Mateo, Calif.: Bluewood Books, 1998.
Fry, William F. "Prince Hamlet and Professor Koch." *Perspectives in Biology and Medicine* 40, no. 3 (Spring, 1997).
Lutwick, Larry I. *New Vaccines and New Vaccine Technology*. Philadelphia: Saunders, 1999.

Tungsten filament

THE INVENTION: Metal filament used in the incandescent light bulbs that have long provided most of the world's electrical lighting.

THE PEOPLE BEHIND THE INVENTION:
William David Coolidge (1873-1975), an American electrical engineer
Thomas Alva Edison (1847-1931), an American inventor

THE INCANDESCENT LIGHT BULB

The electric lamp developed along with an understanding of electricity in the latter half of the nineteenth century. In 1841, the first patent for an incandescent lamp was granted in Great Britain. A patent is a legal claim that protects the patent holder for a period of time from others who might try to copy the invention and make a profit from it. Although others tried to improve upon the incandescent lamp, it was not until 1877, when Thomas Alva Edison, the famous inventor, became interested in developing a successful electric lamp, that real progress was made. The Edison Electric Light Company was founded in 1878, and in 1892, it merged with other companies to form the General Electric Company.

Early electric lamps used platinum wire as a filament. Because platinum is expensive, alternative filament materials were sought. After testing many substances, Edison finally decided to use carbon as a filament material. Although carbon is fragile, making it difficult to manufacture filaments, it was the best choice available at the time.

THE MANUFACTURE OF DUCTILE TUNGSTEN

Edison and others had tested tungsten as a possible material for lamp filaments but discarded it as unsuitable. Tungsten is a hard, brittle metal that is difficult to shape and easy to break, but it possesses properties that are needed for lamp filaments. It has the highest melting point (3,410 degrees Celsius) of any known metal; therefore, it can be heated to a very high temperature, giving off a relatively large amount of radiation without melting (as platinum does) or decomposing (as carbon does). The radiation it emits when heated is primarily visible light. Its resistance to the passage of electricity is relatively high, so it requires little electric current to reach its operating voltage. It also has a high boiling point (about 5,900 degrees Celsius) and therefore does not tend to boil away, or vaporize, when heated. In addition, it is mechanically strong, resisting breaking caused by mechanical shock.

William David Coolidge, an electrical engineer with the General Electric Company, was assigned in 1906 the task of transforming tungsten from its natural state into a form suitable for lamp filaments. The accepted procedure for producing fine metal wires was (and still is) to force a wire rod through successively smaller holes in a hard metal block until a wire of the proper diameter is achieved. The property that allows a metal to be drawn into a fine wire by means of this procedure is called "ductility." Tungsten is not naturally ductile, and it was Coolidge's assignment to make it into a ductile form. Over a period of five years, and after many failures, Coolidge and his workers achieved their goal. By 1911, General Electric was selling lamps that contained tungsten filaments.

Originally, Coolidge attempted to mix powdered tungsten with a suitable substance, form a paste, and squirt that paste through a die to form the wire. The paste-wire was then sintered (heated at a temperature slightly below its melting point) in an effort to fuse the powder into a solid mass. Because of its higher boiling point, the tungsten would remain after all the other components in the paste boiled away. At about 300 degrees Celsius, tungsten softens sufficiently to be hammered into an elongated form. Upon cooling, however, tungsten again becomes brittle, which prevents it from being shaped further into filaments. It was suggested that impurities in the tungsten caused the brittleness, but specially purified tungsten worked no better than the unpurified form.

Many metals can be reduced from rods to wires if the rods are passed through a series of rollers that are successively closer together. Some success was achieved with this method when the rollers were heated along with the metal, but it was still not possible to produce sufficiently fine wire. Next, Coolidge tried a procedure called "swaging," in which a thick wire is repeatedly and rapidly struck by a series of rotating hammers as the wire is drawn past them. After numerous failures, a fine wire was successfully produced using this procedure. It was still too thick for lamp filaments, but it was ductile at room temperature.

Microscopic examination of the wire revealed a change in the crystalline structure of tungsten as a result of the various treatments. The individual crystals had elongated, taking on a fiberlike appearance. Now the wire could be drawn through a die to achieve the appropriate thickness. Again, the wire had to be heated, and if the temperature was too high, the tungsten reverted to a brittle state. The dies themselves were heated, and the reduction progressed in stages, each of which reduced the wire's diameter by a thousandth of an inch.

Finally, Coolidge had been successful. Pressed tungsten bars measuring $\frac{1}{4} \times \frac{3}{8} \times 6$ inches were hammered and rolled into rods $\frac{1}{8}$ inch, or $\frac{125}{1000}$ inch, in diameter. The unit $\frac{1}{1000}$ inch is often called a "mil." These rods were then swaged to approximately 30 mil and then passed through dies to achieve the filament size of 25 mil or smaller, depending on the power output of the lamp in which the filament was to be used. Tungsten wires of 1 mil or smaller are now readily available.

IMPACT

Ductile tungsten wire filaments are superior in several respects to platinum, carbon, or sintered tungsten filaments. Ductile filament lamps can withstand more mechanical shock without breaking. This means that they can be used in, for example, automobile headlights, in which jarring frequently occurs. Ductile wire can also be coiled into compact cylinders within the lamp bulb, which makes for a more concentrated source of light and easier focusing. Ductile tungsten filament lamps require less electricity than do carbon filament lamps, and they also last longer. Because the size of the filament wire can be carefully controlled, the light output from lamps of the same power rating is more reproducible. One 60-watt bulb is therefore exactly like another in terms of light production.

Improved production techniques have greatly reduced the cost of manufacturing ductile tungsten filaments and of light-bulb manufacturing in general. The modern world is heavily dependent upon this reliable, inexpensive light source, which turns darkness into daylight.

See also Fluorescent lighting; Memory metal; Steelmaking process.

FURTHER READING

Baldwin, Neil. *Edison: Inventing the Century*. Chicago: University of Chicago Press, 2001.
Cramer, Carol. *Thomas Edison*. San Diego, Calif.: Greenhaven Press, 2001.
Israel, Paul. *Edison: A Life of Invention*. New York: John Wiley, 1998.
Liebhafsky, H. A. *William David Coolidge: A Centenarian and His Work*. New York: Wiley, 1974.
Miller, John A. *Yankee Scientist: William David Coolidge*. Schenectady, N.Y.: Mohawk Development Service, 1963.

TUPPERWARE

THE INVENTION: Trademarked food-storage products that changed the way Americans viewed plastic products and created a model for selling products in consumers' homes.

THE PEOPLE BEHIND THE INVENTION:
Earl S. Tupper (1907-1983), founder of Tupperware
Brownie Wise, the creator of the vast home sales network for Tupperware
Morison Cousins (1934-2001), a designer hired by Tupperware to modernize its products in the early 1990's

"THE WAVE OF THE FUTURE"?

Relying on a belief that plastic was the wave of the future and wanting to improve on the newest refrigeration technology, Earl S. Tupper, who called himself "a ham inventor and Yankee trader," created an empire of products that changed America's kitchens. Tupper, a self-taught chemical engineer, began working at Du Pont in the 1930's. This was a time of important developments in the field of polymers and the technology behind plastics. Wanting to experiment with this new material yet unable to purchase the needed supplies, Tupper went to his employer for help. Because of the limited availability of materials, major chemical companies had been receiving all the raw goods for plastic production. Although Du Pont would not part with raw materials, the company was willing to let Tupper have the slag.

Polyethylene slag was a black, rock-hard, malodorous waste product of oil refining. It was virtually unusable. Undaunted, Tupper developed methods to purify the slag. He then designed an injection molding machine to form bowls and other containers out of his "Poly-T." Tupper did not want to call the substance plastic because of a public distrust of that substance. In 1938, he founded the Tupper Plastics Company to pursue his dream. It was during those first years that he formulated the design for the famous Tupperware seal.

Refrigeration techniques had improved tremendously during the first part of the twentieth century. The iceboxes in use prior to the 1940's were inconsistent in their interior conditions and were usually damp inside because of melting of the ice. In addition, the metal, glass, or earthenware food storage containers used during the first half of the century did not seal tightly and allowed food to stay moist. Iceboxes allowed mixing of food odors, particularly evident with strong-smelling items such as onions and fish.

ELECTRIC REFRIGERATORS

In contrast to iceboxes, the electric refrigerators available starting in the 1940's maintained dry interiors and low temperatures. This change in environment resulted in food drying out and wilting. Tupper set out to alleviate this problem through his plastic containers. The key to Tupper's solution was his containers' seal. He took his design from paint can lids and inverted it. This tight seal created a partial vacuum that protected food from the dry refrigeration process and kept food odors sealed within containers.

In 1942, Tupper bought his first manufacturing plant, in Farnumsville, Massachusetts. There he continued to improve on his designs. In 1945, Tupper introduced Tupperware, selling it through hardware and department stores as well as through catalog sales. Tupperware products were made of flexible, translucent plastic. Available in frosted crystal and five pastel colors, the new containers were airtight and waterproof. In addition, they carried a lifetime warranty against chipping, cracking, peeling, and breaking in normal noncommercial use. Early supporters of Tupperware included the American Thermos Bottle Company, which purchased seven million nesting cups, and the Tek Corporation, which ordered fifty thousand tumblers to sell with toothbrushes.

Even though he benefited from this type of corporate support, Tupper wanted his products to be for home use. Marketing the new products proved to be difficult in the early years. Tupperware sat on hardware and department store shelves, and catalog sales were nearly nonexistent. The problem appeared to involve a basic distrust of plastic by consumers and an unfamiliarity with how to use the new products. The product did not come with instructions on how to seal the containers or descriptions of how the closed container protected the food within. Brownie Wise, an early direct seller and veteran distributor of Stanley Home Products, stated that it took her several days to understand the technology behind the seal and the now-famous Tupperware "burp," the sound made when air leaves the container as it seals.

Wise and two other direct sellers, Tom Damigella and Harvey Hollenbush, found the niche for selling Tupperware for daily use—home sales. Wise approached Tupper with a home party sales strategy and detailed how it provided a relaxed atmosphere in which to learn about the products and thus lowered sales resistance. In April, 1951, Tupper took his product off store shelves and hired Wise to create a new direct selling system under the name of Tupperware Home Parties, Inc.

IMPACT

Home sales had already proved to be successful for the Fuller Brush Company and numerous encyclopedia publishers, yet Brownie Wise wanted to expand the possibilities. Her first step was to found a campus-like headquarters in Kissimmee, Florida. There, Tupper and a design department worked to develop new products, and Tupperware Home Parties, Inc., under Wise's direction, worked to develop new incentives for Tupperware's direct sellers, called hostesses.

Wise added spark to the notion of home demonstrations. "Parties," as they were called, included games, recipes, giveaways, and other ideas designed to help housewives learn how to use Tupperware products. The marketing philosophy was to make parties appealing events at which women could get together while their children were in school. This fit into the suburban lifestyle of the 1950's. These parties offered a nonthreatening means for home sales representatives to attract audiences for their demonstrations and gave guests a chance to meet and socialize with their neighbors. Often compared to the barbecue parties of the 1950's, Tupperware parties were social, yet educational, affairs. While guests ate lunch or snacked on desserts, the Tupperware hostess educated them about the technology behind the bowls and their seals as well as suggesting a wide variety of uses for the products. For example, a party might include recipes for dinner parties, with information provided on how party leftovers could be stored efficiently and economically with Tupperware products.

While Tupperware products were changing the kitchens of America, they were also changing the women who sold them (almost all the hosts were women). Tupperware sales offered employment for women at a time when society disapproved of women working outside the home. Being a hostess, however, was not a nine-to-five position. The job allowed women freedom to tailor their schedules to meet family needs. Employment offered more than the economic incentive of 35 percent of gross sales. Hostesses also learned new skills and developed self-esteem. An acclaimed mentoring program for new and advancing employees provided motivational training. Managers came only from the ranks of hostesses; moving up the corporate ladder meant spending time selling Tupperware at home parties.

The opportunity to advance offered incentive. In addition, annual sales conventions were renowned for teaching new marketing strategies in fun-filled classes. These conventions also gave women an opportunity to network and establish contacts. These experiences proved to be invaluable as women entered the workforce in increasing numbers in later decades.

Expanding Home-Sales Business

The tremendous success of Tupperware's marketing philosophy helped to set the stage for other companies to enter home sales. These companies used home-based parties to educate potential customers in familiar surroundings, in their own homes or in the homes of friends. The Mary Kay Cosmetics Company, founded in 1963, used beauty makeovers in the home party setting as its chief marketing tool. Discovery Toys, founded in 1978, encouraged guests to get on the floor and play with the toys demonstrated at its home parties. Both companies extended the socialization aspects found in Tupperware parties.

In addition to setting the standard for home sales, Tupperware is also credited with starting the plastic revolution. Early plastics were of poor quality and cracked or broke easily. This created distrust of plastic products among consumers. Earl Tupper's demand for quality set the stage for the future of plastics. He started with high-quality resin and developed a process that kept the "Poly-T" from splitting. He then invented an injection molding machine that mass-produced his bowl and cup designs. His standards of quality from start to finish helped other companies expand into plastics. The 1950's saw a wide variety of products appear in the improved material, including furniture and toys. This shift from wood, glass, and metal to plastic continued for decades.

Maintaining the position of Tupperware within the housewares

EARL S. TUPPER

Born in 1907, Earl Silas Tupper came from a family of go-getters. His mother, Lulu Clark Tupper, kept a boardinghouse and took in laundry, while his father, Earnest, ran a small farm and greenhouse in New Hampshire. The elder Tupper was also a small-time inventor, patenting a device for stretching out chickens to make cleaning them easier. Earl absorbed the family's taste for invention and enterprise.

Fresh out of high school in 1925, Tupper vowed to turn himself into a millionaire by the time he was thirty. He started a landscaping and nursery business in 1928, but the Depression led his company, Tupper Tree, into bankruptcy in 1936. Tupper was undeterred. He hired on with Du Pont the next year. Du Pont taught him a great deal about the chemistry and manufacturing of plastics, but it did not give him scope to apply his ideas, so in 1938 he founded the Earl S. Tupper Company. He continued to work as a contractor for Du Pont to make the fledgling company profitable, and during World War II the company made plastic moldings for gas masks and Navy signal lamps. Finally, in the 1940's Tupper could devote himself to his dream—designing plastic food containers, cups, and such small household conveniences as cases for cigarette packs.

Thanks to aggressive, innovative direct marketing, Tupper's kitchenware, Tupperware, became synonymous with plastic containers during the 1950's. In 1958 Tupper sold his company to Rexall for $16 million, having finally realized his youthful ambition to make himself wealthy through Yankee wit and hard work. He died in 1983.

market meant keeping current. As more Americans were able to purchase the newest refrigerators, Tupperware expanded to meet their needs. The company added new products, improved marketing strategies, and changed or updated designs. Over the years, Tupperware added baking items, toys, and home storage containers for such items as photographs, sewing materials, and holiday ornaments. The 1980's and 1990's brought microwaveable products.

As women moved into the work force in great numbers, Tupperware moved with them. The company introduced lunchtime parties at the workplace and parties at daycare centers for busy working parents. Tupperware also started a fund-raising line, in special colors, that provided organizations with a means to bring in money while not necessitating full-fledged parties. New party themes developed around time-saving techniques and health concerns such as diet planning. Beginning in 1992, customers too busy to attend a party could call a toll-free number, request a catalog, and be put in contact with a "consultant," as "hostesses" now were called.

Another marketing strategy developed out of a public push for environmentally conscious products. Tupperware consultants stressed the value of buying food in bulk to create less trash as well as saving money. To store these increased purchases, the company developed a new line for kitchen staples called Modular Mates. These stackable containers came in a wide variety of shapes and sizes to hold everything from cereal to flour to pasta. They were made of see-through plastic, allowing the user to see if the contents needed replenishing. Some consultants tailored parties around ideas to better organize kitchen cabinets using the new line. Another environmentally conscious product idea was the Tupperware lunch kit. These kits did away with the need for throwaway products such as paper plates, plastic storage bags, and aluminum foil. Lunch kits marketed in other countries were developed to accommodate the countries' particular needs. For example, Japanese designs included chopsticks, while Latin American styles were designed to hold tortillas.

DESIGN CHANGES

Tupperware designs have been well received over the years. Early designs prompted a 1947 edition of *House Beautiful* to call the product "Fine Art for 39 cents." Fifteen of Tupper's earliest designs are housed in a permanent collection at the Museum of Modern Art in New York City. Other museums, such as the Metropolitan Museum of Art and the Brooklyn Museum, also house Tupperware designs. Tupperware established its own Museum of Historic Food Containers at its international headquarters in Florida. Despite this critical acclaim, the company faced a constant struggle to keep product lines competitive with more accessible products, such as those made by Rubbermaid, that could be found on the shelves of local grocery or department stores.

Some of the biggest design changes came with the hiring of Morison Cousins in the early 1990's. Cousins, an accomplished designer, set out to modernize the Tupperware line. He sought to return to simple, traditional styles while bringing in time-saving aspects. He changed lid designs to make them easier to clean and rounded the bottoms of bowls so that every portion could be scooped out. Cousins also added thumb handles to bowls.

Backed by a knowledgeable sales force and quality product, the company experienced tremendous growth. Tupperware sales reached $25 million in 1954. By 1958, the company had grown from seven distributorships to a vast system covering the United States and Canada. That same year, Brownie Wise left the company, and Tupper Plastics was sold to Rexall Drug Company for $9 million. Rexall Drug changed its name to Dart Industries, Inc., in 1969, then merged with Kraft, Inc., eleven years later to become Dart and Kraft, Inc. During this time of parent-company name changing, Tupperware continued to be an important subsidiary. Through the 1960's and 1970's, the company spread around the world, with sales in Western Europe, the Far East, and Latin America. In 1986, Dart and Kraft, Inc., split into Kraft, Inc., and Premark International, Inc., of which Dart (and therefore Tupperware) was a subsidiary. Premark International included other home product companies such as West Bend, Precor, and Florida Tile.

By the early 1990's, annual sales of Tupperware products reached $1.1 billion. Manufacturing plants in Halls, Tennessee, and Hemingway, South Carolina, worked to meet the high demand for Tupperware products in more than fifty countries. Foreign sales accounted for almost 75 percent of the company's business. By meeting the needs of consumers and keeping current with design changes, new sales techniques, and new products, Tupperware was able to reach 90 percent of America's homes.

See also Electric refrigerator; Food freezing; Freeze-drying; Microwave cooking; Plastic; Polystyrene; Pyrex glass; Teflon.

FURTHER READING

Brown, Patricia Leigh. "New Designs to Keep Tupperware Fresh." *New York Times* (June 10, 1993).
Clarke, Alison J. *Tupperware: The Promise of Plastic in 1950s America.* Washington, D.C.: Smithsonian Institution Press, 1999.
Gershman, Michael. *Getting It Right the Second Time.* Reading, Mass.: Addison-Wesley, 1990.
Martin, Douglas. "Morison S. Cousins, Sixty-six, Designer, Dies; Revamped Tupperware's Look with Flair." *New York Times* (February 18, 2001).
Sussman, Vic. "I Was the Only Virgin at the Party." *Sales and Marketing Management* 141 (September 1, 1989).

Turbojet

THE INVENTION: A jet engine with a turbine-driven compressor that uses its hot-gas exhaust to develop thrust.

THE PEOPLE BEHIND THE INVENTION:
Henry Harley Arnold (1886-1950), a chief of staff of the U.S. Army Air Corps
Gerry Sayer, a chief test pilot for Gloster Aircraft Limited
Hans Pabst von Ohain (1911-   ), a German engineer
Sir Frank Whittle (1907-1996), an English Royal Air Force officer and engineer

Developments in Aircraft Design

On the morning of May 15, 1941, some eleven months after France had fallen to Adolf Hitler's advancing German army, an experimental jet-propelled aircraft was successfully tested by pilot Gerry Sayer. The airplane had been developed in a little more than two years by the English company Gloster Aircraft under the supervision of Sir Frank Whittle, the inventor of England's first jet engine.

Like the jet engine that powered it, the plane had a number of predecessors. In fact, the May, 1941, flight was not the first jet-powered test flight: That flight occurred on August 27, 1939, when a Heinkel aircraft powered by a jet engine developed by Hans Pabst von Ohain completed a successful test flight in Germany. During this period, Italian airplane builders were also engaged in jet aircraft testing, with lesser degrees of success.

Without the knowledge that had been gained from Whittle's experience in experimental aviation, the test flight at the Royal Air Force's Cranwell airfield might never have been possible. Whittle's repeated efforts to develop turbojet propulsion engines had begun in 1928, when, as a twenty-one-year-old Royal Air Force (RAF) flight cadet at Cranwell Academy, he wrote a thesis entitled "Future Developments in Aircraft Design." One of the principles of Whittle's earliest research was that if aircraft were eventually to achieve very high speeds over long distances, they would have to fly at very high altitudes, benefiting from the reduced wind resistance encountered at such heights.

Whittle later stated that the speed he had in mind at that time was about 805 kilometers per hour—close to that of the first jet-powered aircraft. His earliest idea of the engines that would be necessary for such planes focused on rocket propulsion (that is, "jets" in which the fuel and oxygen required to produce the explosion needed to propel an air vehicle are entirely contained in the engine, or, alternatively, in gas turbines driving propellers at very high speeds). Later, it occurred to him that gas turbines could be used to provide forward thrust by what would become "ordinary" jet propulsion (that is, "thermal air" engines that take from the surrounding atmosphere the oxygen they need to ignite their fuel). Eventually, such ordinary jet engines would function according to one of four possible systems: the so-called athodyd, or continuous-firing duct; the pulsejet, or intermittent-firing duct; the turbojet, or gas-turbine jet; or the propjet, which uses a gas turbine jet to rotate a conventional propeller at very high speeds.

Passing the Test

The aircraft that was to be used to test the flight performance was completed by April, 1941. On April 7, tests were conducted on the ground at Gloster Aircraft's landing strip at Brockworth by chief test pilot Sayer. At this point, all parties concerned tried to determine whether the jet engine's capacity would be sufficient to push the aircraft forward with enough speed to make it airborne. Sayer dared to take the plane off the ground for a limited distance of between 183 meters and 273 meters, despite the technical staff's warnings against trying to fly in the first test flights.

On May 15, the first real test was conducted at Cranwell. During that test, Sayer flew the plane, now called the Pioneer, for seventeen minutes at altitudes exceeding 300 meters and at a conservative test speed exceeding 595 kilometers per hour, which was equivalent to the top speed then possible in the RAF's most versatile fighter plane, the Spitfire.

Once it was clear that the tests undertaken at Cranwell were not only successful but also highly promising in terms of even better performance, a second, more extensive test was set for May 21, 1941. It was this later demonstration that caused the Ministry of Air Production (MAP) to initiate the first steps to produce the Meteor jet fighter aircraft on a full industrial scale barely more than a year after the Cranwell test flight.

IMPACT

Since July, 1936, the Junkers engine and aircraft companies in Hitler's Germany had been a part of a new secret branch dedicated to the development of a turbojet-driven aircraft. In the same period, Junkers' rival in the German aircraft industry, Heinkel, Inc., approached von Ohain, who was far enough along in his work on the turbojet principle to have patented a device very similar to Whittle's in 1935. A later model of this jet engine would power a test aircraft in August, 1939.

In the meantime, the wider impact of the flight was the result of decisions made by General Henry Harley Arnold, chief of staff of the U.S. Army Air Corps. Even before learning of the successful flight in May, he made arrangements to have one of Whittle's engines shipped to the United States to be used by General Electric Company as a model for U.S. production. The engine arrived in October, 1941, and within one year, a General Electric-built engine powered a Bell Aircraft plane, the XP-59 A Airacomet, in its maiden flight.

The jet airplane was not perfected in time to have any significant impact on the outcome of World War II, but all of the wartime experimental jet aircraft developments that were either sparked by the flight in 1941 or preceded it prepared the way for the research and development projects that would leave a permanent revolutionary mark on aviation history in the early 1950's.

See also Airplane; Dirigible; Rocket; Rocket; Stealth aircraft; Supersonic passenger plane; V-2 rocket.

FURTHER READING

Adams, Robert. "Smithsonian Horizons." *Smithsonian* 18 (July, 1987).
Boyne, Walter J., Donald S. Lopez, and Anselm Franz. *The Jet Age: Forty Years of Jet Aviation.* Washington: National Air and Space Museum, 1979.
Constant, Edward W. *The Origins of the Turbojet Revolution.* Baltimore: Johns Hopkins University Press, 1980.
Launius, Roger D. *Innovation and the Development of Flight.* College Station: Texas A&M University Press, 1999.

Typhus Vaccine

THE INVENTION: the first effective vaccine against the virulent typhus disease.

THE PERSON BEHIND THE INVENTION:
Hans Zinsser (1878-1940), an American bacteriologist and immunologist

Studying Diseases

As a bacteriologist and immunologist, Hans Zinsser was interested in how infectious diseases spread. During an outbreak of typhus in Serbia in 1915, he traveled with a Red Cross team so that he could study the disease. He made similar trips to the Soviet Union in 1923, Mexico in 1931, and China in 1938. His research showed that, as had been suspected, typhus was caused by the rickettsia, an organism that had been identified in 1916 by Henrique da Rocha-Lima. The organism was known to be carried by a louse or a rat flea and transmitted to humans through a bite. Poverty, dirt, and overcrowding led to environments that helped the typhus disease to spread.

The rickettsia is a microorganism that is rod-shaped or spherical. Within the insect's body, it works its way into the cells that line the gut. Multiplying within this tissue, the rickettsia passes from the insect body with the feces. Since its internal cells are being destroyed, the insect dies within three weeks after it has been infected with the microorganism. As the infected flea or louse feeds on a human, it causes itching. When the bite is scratched, the skin may be opened, and the insect feces, carrying rickettsia, can then enter the body. Also, dried airborne feces can be inhaled.

Once inside the human, the rickettsia invades endothelial cells and causes an inflammation of the blood vessels. Cell death results, and this leads to tissue death. In a few days, the infected person may have a rash, a severe headache, a fever, dizziness, ringing in the ears, or deafness. Also, light may hurt the person's eyes, and the thinking processes become foggy and mixed up. (The word "typhus" comes from a Greek word meaning "cloudy" or "misty.") Without treatment, the victim dies within nine to eighteen days.

Medical science now recognizes three forms of typhus: the epidemic louse-borne, the Brill-Zinsser, and the murine (or rodent-related) form. The epidemic louse-borne (or "classical") form is the most severe. The Brill-Zinsser (or "endemic") form is similar but less severe. The murine form of typhus is also milder then the epidemic type.

In 1898, a researcher named Brill studied typhus among immigrants in New York City; the form of typhus he found was called "Brill's disease." In the late 1920's, Hermann Mooser proved that Brill's disease was carried by the rat flea.

When Zinsser began his work on typhus, he realized that what was known about the disease had never been properly organized. Zinsser and his coworkers, including Mooser and others, worked to identify the various types of typhus. In the 1930's, Zinsser suggested that the typhus studied by Brill in New York City had actually included two types: the rodent-associated form and Brill's disease. As a result of Zinsser's effort to identify the types of typhus disease, it was renamed Brill-Zinsser disease.

MAKING A VACCINE

Zinsser's studies had shown him that the disease-causing organism in typhus contained some kind of antigen, most likely a polysaccharide. In 1932, Zinsser would identify agglutinins, or antibodies, in the blood serum of patients who had the murine and classical forms of typhus. Zinsser believed that a vaccine could be developed to prevent the spread of typhus. He realized, however, that a large number of dead microorganisms was needed to help people develop an immunity.

Zinsser and his colleagues set out to develop a method of growing organisms in large quantities in tissue culture. The infected tissue was used to inoculate large quantities of normal chick tissue, and this tissue was then grown in flasks. In this way, Zinsser's team was able to produce the quantities of microorganisms they needed.

The type of immunization that Zinsser developed (in 1930) is known as "passive immunity." The infecting organisms carry antigens, which stimulate the production of antibodies. The antigens can elicit an immune reaction even if the cell is weak or dead.

"B" cells and macrophages, both of which are used in fighting disease organisms, recognize and respond to the antigen. The B cells produce antibodies that can destroy the invading organism directly or attract more macrophages to the area so that they can attack the organism. B cells also produce "memory cells," which remain in the blood and trigger a quick second response if there is a later infection. Since the vaccine contains weakened or dead organisms, the person who is vaccinated may have a mild reaction but does not actually come down with the disease.

IMPACT

Typhus is still common in many parts of the world, especially where there is poverty and overcrowding. Classical typhus is quite rare; the last report of this type of typhus in the United States was in 1921. Endemic and murine typhus are more common. In the United States, where children are vaccinated against the disease, only about fifty cases are now reported each year. Antibiotics such as tetracycline and chloramphenicol are effective in treating the disease, so few infected people now die of the disease in areas where medical care is available.

The work of Zinsser and his colleagues was very important in stopping the spread of typhus. Zinsser's classification of different types of the disease meant that it was better understood, and this led to the development of cures. The control of lice and rodents and improved cleanliness in living conditions helped bring typhus under control. Once Zinsser's vaccine was available, even people who lived in crowded inner cities could be protected against the disease.

Zinsser's research in growing the rickettsia in tissue culture also inspired further work. Other researchers modified and improved his technique so that the use of tissue culture is now standard in laboratories.

See also Antibacterial drugs; Birth control pill; Penicillin; Polio vaccine (Sabin); Polio vaccine (Salk); Salvarsan; Tuberculosis vaccine; Yellow fever vaccine.

FURTHER READING

DeJauregui, Ruth. *100 Medical Milestones That Shaped World History.* San Mateo, Calif.: Bluewood Books, 1998.

Gray, Michael W. "Rickettsia in Medicine and History." *Nature* 396, no. 6707 (November, 1998).

Hoff, Brent H., Carter Smith, and Charles H. Calisher. *Mapping Epidemics: A Historical Atlas of Disease.* New York: Franklin Watts, 2000.

Ultracentrifuge

THE INVENTION: A super-high-velocity centrifuge designed to separate colloidal or submicroscopic substances, the ultracentrifuge was used to measure the molecular weight of proteins and proved that proteins are large molecules.

THE PEOPLE BEHIND THE INVENTION:
Theodor Svedberg (1884-1971), a Swedish physical chemist and 1926 Nobel laureate in chemistry
Jesse W. Beams (1898-1977), an American physicist
Arne Tiselius (1902-1971), a Swedish physical biochemist and 1948 Nobel laureate in chemistry

SVEDBERG STUDIES COLLOIDS

Theodor "The" Svedberg became the principal founder of molecular biology when he invented the ultracentrifuge and used it to examine proteins in the mid-1920's. He began to study materials called "colloids" as a Swedish chemistry student at the University of Uppsala and continued to conduct experiments with colloidal systems when he joined the faculty in 1907. A colloid is a kind of mixture in which very tiny particles of one substance are mixed uniformly with a dispersing medium (often water) and remain suspended indefinitely. These colloidal dispersions play an important role in many chemical and biological systems.

The size of the colloid particles must fall within a certain range. The force of gravity will cause them to settle if they are too large. If they are too small, the properties of the mixture change, and a solution is formed. Some examples of colloidal systems include mayonnaise, soap foam, marshmallows, the mineral opal, fog, India ink, jelly, whipped cream, butter, paint, and milk. Svedberg wondered what such different materials could have in common. His early work helped to explain why colloids remain in suspension. Later, he developed the ultracentrifuge to measure the weight of colloid particles by causing them to settle in a controlled way.

SVEDBERG BUILDS AN ULTRACENTRIFUGE

Svedberg was a successful chemistry professor at the University of Uppsala in Sweden when he had the idea that colloids could be made to separate from suspension by means of centrifugal force. Centrifugal force is caused by circular motion and acts on matter much as gravity does. A person can feel this force by tying a ball to a rope and whirling it rapidly in a circle. The pull on the rope becomes stronger as the ball moves faster in its circular orbit. A centrifuge works the same way: It is a device that spins balanced containers of substances very rapidly.

Svedberg figured that it would take a centrifugal force thousands of times the force of gravity to cause colloid particles to settle. How fast they settle depends on their size and weight, so the ultracentrifuge can also provide a measure of these properties. Centrifuges were already used to separate cream from whole milk and blood corpuscles from plasma, but these centrifuges were too slow to cause the separation of colloids. An *ultra*centrifuge—one that could spin samples much faster—was needed, and Svedberg made plans to build one.

The opportunity came in 1923, when Svedberg spent eight months as visiting professor in the chemistry department of the University of Wisconsin at Madison and worked with J. Burton Nichols, one of the six graduate students assigned to assist him. Here, Svedberg announced encouraging results with an electrically driven centrifuge—not yet an ultracentrifuge—which attained a rotation equal to about 150 times the force of gravity. Svedberg returned to Sweden and, within a year, built a centrifuge capable of generating 7,000 times the force of gravity. He used it with Herman Rinde, a colleague at the University of Uppsala, to separate the suspended particles of colloidal gold. This was in 1924, which is generally accepted as the date of the first use of a true ultracentrifuge. From 1925 to 1926, Svedberg raised the funds to build an even more powerful ultracentrifuge. It would be driven by an oil turbine, a machine capable of producing more than 40,000 revolutions per minute to generate a force 100,000 times that of gravity.

Svedberg and Robin Fahraeus used the new ultracentrifuge to separate the protein hemoglobin from its colloidal suspension. Together with fats and carbohydrates, proteins are one of the most abundant organic constituents of living organisms. No protein had been isolated in pure form before Svedberg began this study, and it was uncertain whether proteins consisted of molecules of a single compound or mixtures of different substances working together in biological systems. The colloid particles of Svedberg's previous studies separated at different rates, some settling faster than others, showing that they had different sizes and weights. Colloid particles of the protein, however, separated together. The uniform separation observed for proteins, such as hemoglobin, demonstrated for the first time that each protein consists of identical well-defined molecules. More than one hundred proteins were studied by Svedberg and his coworkers, who extended their technique to carbohydrate polymers such as cellulose and starch.

IMPACT

Svedberg built more and more powerful centrifuges so that smaller and smaller molecules could be studied. In 1936, he built an ultracentrifuge that produced a centrifugal force of more than a half-million times the force of gravity. Jesse W. Beams was an American pioneer in ultracentrifuge design. He reduced the friction of an air-driven rotor by first housing it in a vacuum, in 1934, and later by supporting it with a magnetic field.

The ultracentrifuge was a central tool for providing a modern understanding of the molecular basis of living systems, and it is employed in thousands of laboratories for a variety of purposes. It is used to analyze the purity and the molecular properties of substances containing large molecules, from the natural products of the biosciences to the synthetic polymers of chemistry. The ultracentrifuge is also employed in medicine to analyze body fluids, and it is used in biology to isolate viruses and the components of fractured cells.

Svedberg, while at Wisconsin in 1923, invented a second, very different method to separate proteins in suspension using electric currents. It is called "electrophoresis," and it was later improved by his student, Arne Tiselius, for use in his famous study of the proteins in blood serum. The technique of electrophoresis is as widespread and important as is the ultracentrifuge.

See also Ultramicroscope; X-ray crystallography.

FURTHER READING

Lechner, M. D. *Ultracentrifugation*. New York: Springer, 1994.
Rickwood, David. *Preparative Centrifugation: A Practical Approach*. New York: IRL Press at Oxford University Press, 1992.
Schuster, Todd M. *Modern Analytical Ultracentrifugation: Acquisition and Interpretation of Data for Biological and Synthetic Polymer Systems*. Boston: Birkhäuser, 1994.
Svedberg, Theodor B., Kai Oluf Pedersen, and Johannes Henrik Bauer. *The Ultracentrifuge*. Oxford: Clarendon Press, 1940.

ULTRAMICROSCOPE

THE INVENTION: A microscope characterized by high-intensity illumination for the study of exceptionally small objects, such as colloidal substances.

THE PEOPLE BEHIND THE INVENTION:

Richard Zsigmondy (1865-1929), an Austrian-born German organic chemist who won the 1925 Nobel Prize in Chemistry

H. F. W. Siedentopf (1872-1940), a German physicist-optician

Max von Smouluchowski (1879-1961), a German organic chemist

ACCIDENTS OF ALCHEMY

Richard Zsigmondy's invention of the ultramicroscope grew out of his interest in colloidal substances. Colloids consist of tiny particles of a substance that are dispersed throughout a solution of another material or substance (for example, salt in water). Zsigmondy first became interested in colloids while working as an assistant to the physicist Adolf Kundt at the University of Berlin in 1892. Although originally trained as an organic chemist, in which discipline he took his Ph.D. at the University of Munich in 1890, Zsigmondy became particularly interested in colloidal substances containing fine particles of gold that produce lustrous colors when painted on porcelain. For this reason, he abandoned organic chemistry and devoted his career to the study of colloids.

Zsigmondy began intensive research into his new field of interest in 1893, when he returned to Austria to accept a post as lecturer at a technical school at Graz. Zsigmondy became especially interested in gold-ruby glass, the accidental invention of the seventeenth century alchemist Johann Kunckle. Kunckle, while pursuing the alchemist's pipe dream of transmuting base substances (such as lead) into gold, discovered instead a method of producing glass with a beautiful, deep red luster by suspending very fine particles of gold throughout the liquid glass before it was cooled. Zsigmondy also began studying a colloidal pigment called "purple of Cassius," the discovery of another seventeenth century alchemist, Andreas Cassius.

Zsigmondy soon discovered that purple of Cassius was a colloidal solution and not, as most chemists believed at the time, a chemical compound. This fact allowed him to develop techniques for glass and porcelain coloring with great commercial value, which led directly to his 1897 appointment to a research post with the Schott Glass Manufacturing Company in Jena, Germany. With the Schott Company, Zsigmondy concentrated on the commercial production of colored glass objects. His most notable achievement during this period was the invention of Jena milk glass, which is still prized by collectors throughout the world.

BRILLIANT PROOF

While studying colloids, Zsigmondy devised experiments that proved that purple of Cassius was colloidal. When he published the results of his research in professional journals, however, they were not widely accepted by the scientific community. Other scientists were not able to replicate Zsigmondy's experiments and consequently denounced them as flawed. The criticism of his work in technical literature stimulated Zsigmondy to make his greatest discovery, the ultramicroscope, which he developed to prove his theories regarding purple of Cassius.

The problem with proving the exact nature of purple of Cassius was that the scientific instruments available at the time were not sensitive enough for direct observation of the particles suspended in a colloidal substance. Using the facilities and assisted by the staff (especially H. F. W. Siedentopf, an expert in optical lens grinding) of the Zeiss Glass Manufacturing Company of Jena, Zsigmondy developed an ingenious device that permitted direct observation of individual colloidal particles.

This device, which its developers named the "ultramicroscope," made use of a principle that already existed. Sometimes called "dark-field illumination," this method consisted of shining a light (usually sunlight focused by mirrors) through the solution under the microscope at right angles to the observer, rather than shining the light directly from the observer into the solution. The resulting effect is similar to that obtained when a beam of sunlight is admitted to a closed room through a small window. If an observer stands back from and at

RICHARD ZSIGMONDY

Born in Vienna, Austria, in 1865, Richard Adolf Zsigmondy came from a talented, energetic family. His father, a celebrated dentist and inventor of medical equipment, inspired his children to study the sciences, while his mother urged them to spend time outdoors in strenuous exercise. Although his father died when Zsigmondy was fifteen, the teenager's interest in chemistry was already firmly established. He read advanced chemistry textbooks and worked on experiments in his own home laboratory.

After taking his doctorate at the University of Munich and teaching in Berlin and Graz, Austria, he became an industrial chemist at the glassworks in Jena, Germany. However, pure research was his love, and he returned to it, working entirely on his own after 1900. In 1907 he received an appointment as professor and director of the Institute of Inorganic Chemistry at the University of Göttingen, one of the scientific centers of the world. There he accomplished much of his ground-breaking work on colloids and Brownian motion, despite the severe shortages that hampered him during the economic depression in Germany following World War I. His 1925 Nobel Prize in Chemistry, especially the substantial money award, helped him overcome his supply problems. He retired in early 1929 and died seven months later.

right angles to such a beam, many dust particles suspended in the air will be observed that otherwise would not be visible.

Zsigmondy's device shines a very bright light through the substance or solution being studied. From the side, the microscope then focuses on the light shaft. This process enables the observer using the ultramicroscope to view colloidal particles that are ordinarily invisible even to the strongest conventional microscope. To a scientist viewing purple of Cassius, for example, colloidal gold particles as small as one ten-millionth of a millimeter in size become visible.

IMPACT

After Zsigmondy's invention of the ultramicroscope in 1902, the University of Göttingen appointed him professor of inorganic chemistry and director of its Institute for Inorganic Chemistry. Using the ultramicroscope, Zsigmondy and his associates quickly proved that purple of Cassius is indeed a colloidal substance.

That finding, however, was the least of the spectacular discoveries that resulted from Zsigmondy's invention. In the next decade, Zsigmondy and his associates found that color changes in colloidal gold solutions result from coagulation—that is, from changes in the size and number of gold particles in the solution caused by particles bonding together. Zsigmondy found that coagulation occurs when the negative electrical charge of the individual particles is removed by the addition of salts. Coagulation can be prevented or slowed by the addition of protective colloids.

These observations also made possible the determination of the speed at which coagulation takes place, as well as the number of particles in the colloidal substance being studied. With the assistance of the organic chemist Max von Smouluchowski, Zsigmondy worked out a complete mathematical formula of colloidal coagulation that is valid not only for gold colloidal solutions but also for all other colloids. Colloidal substances include blood and milk, which both coagulate, thus giving Zsigmondy's work relevance to the fields of medicine and agriculture. These observations and discoveries concerning colloids—in addition to the invention of the ultramicroscope—earned for Zsigmondy the 1925 Nobel Prize in Chemistry.

See also Scanning tunneling microscope; Ultracentrifuge; X-ray crystallography.

FURTHER READING

Zsigmondy, Richard, and Jerome Alexander. *Colloids and the Ultramicroscope.* New York: J. Wiley & Sons, 1909.
Zsigmondy, Richard, Ellwood Barker Spear, and John Foote Norton. *The Chemistry of Colloids.* New York: John Wiley & Sons, 1917.

ULTRASOUND

THE INVENTION: A medically safe alternative to X-ray examination, ultrasound uses sound waves to detect fetal problems in pregnant women.

THE PEOPLE BEHIND THE INVENTION:
Ian T. Donald (1910-1987), a British obstetrician
Paul Langévin (1872-1946), a French physicist
Marie Curie (1867-1946) and Pierre Curie (1859-1906), the French husband-and-wife team that researched and developed the field of radioactivity
Alice Stewart, a British researcher

AN UNDERWATER BEGINNING

In the early 1900's, two major events made it essential to develop an appropriate means for detecting unseen underwater objects. The first event was the *Titanic* disaster in 1912, which involved a largely submerged, unseen, and silent iceberg. This iceberg caused the sinking of the *Titanic* and resulted in the loss of many lives as well as valuable treasure. The second event was the threat to the Allied Powers from German U-boats during World War I (1914-1918). This threat persuaded the French and English Admiralties to form a joint committee in 1917. The Anti-Submarine Detection and Investigation Committee (ASDIC) found ways to counter the German naval developments. Paul Langévin, a former colleague of Pierre Curie and Marie Curie, applied techniques developed in the Curies' laboratories in 1880 to formulate a crude ultrasonic system to detect submarines. These techniques used beams of sound waves of very high frequency that were highly focused and directional.

The advent of World War II (1939-1945) made necessary the development of faster electronic detection technology to improve the efforts of ultrasound researchers. Langévin's crude invention evolved into the sophisticated system called "sonar" (*so*und *na*vigation *r*anging), which was important in the success of the Allied forces. Sonar was based on pulse echo principles and, like the system called "ra-

IAN DONALD

Ian Donald was born in Paisley, Scotland, in 1910 and educated in Edinburgh until he was twenty, when he moved to South Africa with his parents. He graduated with a bachelor of arts degree from Diocesan College, Cape Town, and then moved to London to study medicine, graduating from the University of London in 1937. During World War II he served as a medical officer in the Royal Air Force and received a medal for rescuing flyers from a burning airplane. After the war he began his long teaching career in medicine, first at St. Thomas Hospital Medical School and then as the Regius Professor of Midwifery at Glasgow University. His specialties were obstetrics and gynecology.

While at Glasgow he accomplished his pioneering work with diagnostic ultrasound technology, but he also championed laparoscopy, breast feeding, and the preservation of membranes during the delivery of babies. In addition to his teaching duties and medical practice he wrote a widely used textbook, oversaw the building of the Queen Mother's Hospital in Glasgow, and campaigned against England's 1967 Abortion Act.

His expertise with ultrasound came to his own rescue after he had cardiac surgery in the 1960's. He diagnosed himself as having internal bleeding from a broken blood vessel. The cardiologists taking care of him were skeptical until an ultrasound proved him right. Widely honored among physicians, he died in England in 1987.

dar" (*r*adio *d*etecting *a*nd *r*anging), had military implications. This vital technology was classified as a military secret and was kept hidden until after the war.

AN ALTERNATIVE TO X RAYS

Ian Donald's interest in engineering and the principles of sound waves began when he was a schoolboy. Later, while he was in the British Royal Air Force, he continued and maintained his enthusiasm by observing the development of the anti-U-boat warfare efforts. He went to medical school after World War II and began a career in obstetrics. By the early 1950's, Donald had em-

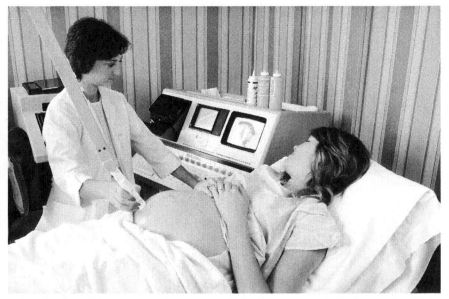

Safe and not requiring surgery, ultrasonography has become the principal means for obtaining information about fetal structures. (Digital Stock)

barked on a study of how to apply sonar technology in medicine. He moved to Glasgow, Scotland, a major engineering center in Europe that presented a fertile environment for interdisciplinary research. There Donald collaborated with engineers and technicians in his medical ultrasound research. They used inanimate and tissue materials in many trials. Donald hoped to apply ultrasound technology to medicine, especially to gynecology and obstetrics, his specialty.

His efforts led to new pathways and new discoveries. He was interested in adapting a certain type of ultrasound technology method (used to probe metal structures and welds for cracks and flaws) to medicine. Kelvin Hughes, the engineering manufacturing company that produced the flaw detector apparatus, gave advice, expertise, and equipment to Donald and his associates, who were then able to devise water tanks with flexible latex bottoms. These were coated with a film of grease and placed into contact with the abdomens of pregnant women.

The use of diagnostic radiography (such as X rays) became controversial when it was evident that it caused potential leukemias and other injuries to the fetus. It was realized from the earliest days of radiology that radiation could cause tumors, particularly of the skin. The aftereffects of radiological studies were recognized much later and confirmed by studies of atomic bomb survivors and of patients receiving therapeutic irradiation. The use of radiation in obstetrics posed several major threats to the developing fetus, most notably the production of tumors later in life, genetic damage, and developmental anomalies in the unborn fetus.

In 1958, bolstered by earlier clinical reports and animal research findings, Alice Stewart and her colleagues presented a major case study of more than thirteen hundred children in England and Wales who had died of cancer before the age of ten between 1953 and 1958. There was a 91 percent increase in leukemias in children who were exposed to intrauterine radiation, as well as a higher percentage of fetal death. Although controversial, this report led to a reduction in the exposure of pregnant women to X rays, with subsequent reductions in fetal abnormalities and death.

These reports came at a very opportune time for Donald: The development of ultrasonography would provide useful information about the unborn fetus without the adverse effects of radiation. Stewart's findings and Donald's experiments convinced others of the need for ultrasonography in obstetrics.

CONSEQUENCES

Diagnostic ultrasound first gained clinical acceptance in obstetrics, and its major contributions have been in the assessment of fetal size and growth. In combination with amniocentesis (the study of fluid taken from the womb), ultrasound is an invaluable tool in operative procedures necessary to improve the outcomes of pregnancies.

As can be expected, safety has been a concern, especially for a developing, vulnerable fetus that is exposed to high-frequency sound. Research has not been able to document any harmful effect of ultrasonography on the developing fetus. The procedure produces neither heat nor cold. It has not been shown to produce any toxic or destructive effect on the auditory or balancing organs of the developing fetus. Chromosomal abnormalities have not been reported in any of the studies conducted.

Ultrasonography, because it is safe and does not require surgery, has become the principal means for obtaining information about fetal structures. With this procedure, the contents of the uterus—as well as the internal structure of the placenta, fetus, and fetal organs—can be evaluated at any time during pregnancy. The use of ultrasonography remains a most valued tool in medicine, especially obstetrics, because of Donald's work.

See also Amniocentesis; Birth control pill; CAT scanner; Electrocardiogram; Electroencephalogram; Mammography; Nuclear magnetic resonance; Pap test; Sonar; Syphilis test; X-ray image intensifier.

FURTHER READING

Danforth, David N., and James R. Scott. *Danforth's Obstetrics and Gynecology*. 7th ed. Philadelphia: Lippincott, 1994.
DeJauregui, Ruth. *100 Medical Milestones That Shaped World History*. San Mateo, Calif.: Bluewood Books, 1998.
Rozycki, Grace S. *Surgeon-Performed Ultrasound: Its Use in Clinical Practice*. Philadelphia: W. B. Saunders, 1998.
Wolbarst, Anthony B. *Looking Within: How X-ray, CT, MRI, Ultrasound, and Other Medical Images Are Created, and How They Help Physicians Save Lives.* Berkeley: University of California Press, 1999.

UNIVAC COMPUTER

THE INVENTION: The first commercially successful computer system.

THE PEOPLE BEHIND THE INVENTION:

John Presper Eckert (1919-1995), an American electrical engineer
John W. Mauchly (1907-1980), an American physicist
John von Neumann (1903-1957), a Hungarian American
mathematician
Howard Aiken (1900-1973), an American physicist
George Stibitz (1904-1995), a scientist at Bell Labs

THE ORIGINS OF COMPUTING

On March 31, 1951, the U.S. Census Bureau accepted delivery of the first Universal Automatic Computer (UNIVAC). This powerful electronic computer, far surpassing anything then available in technological features and capability, ushered in the first computer generation and pioneered the commercialization of what had previously been the domain of academia and the interest of the military. The fanfare that surrounded this historic occasion, however, masked the turbulence of the previous five years for the young upstart Eckert-Mauchly Computer Corporation (EMCC), which by this time was a wholly owned subsidiary of Remington Rand Corporation.

John Presper Eckert and John W. Mauchly met in the summer of 1941 at the University of Pennsylvania. A short time later, Mauchly, then a physics professor at Ursinus College, joined the Moore School of Engineering at the University of Pennsylvania and embarked on a crusade to convince others of the feasibility of creating electronic digital computers. Up to this time, the only computers available were called "differential analyzers," which were used to solve complex mathematical equations known as "differential equations." These slow machines were good only for solving a relatively narrow range of mathematical problems.

Eckert and Mauchly landed a contract that eventually resulted in the development and construction of the world's first operational general-purpose electronic computer, the Electronic Numerical Integrator and Calculator (ENIAC). This computer, used eventually by the Army for the calculation of ballistics tables, was deficient in many obvious areas, but this was caused by economic rather than engineering constraints. One major deficiency was the lack of automatic program control; the ENIAC did not have stored program memory. This was addressed in the development of the Electronic Discrete Variable Automatic Computer (EDVAC), the successor to the ENIAC.

Fighting the Establishment

A symbiotic relationship had developed between Eckert and Mauchly that worked to their advantage on technical matters. They worked well with each other, and this contributed to their success in spite of external obstacles. They both were interested in the commercial applications of computers and envisioned uses for these machines far beyond the narrow applications required by the military.

This interest brought them into conflict with the administration at the Moore School of Engineering as well as with the noted mathematician John von Neumann, who "joined" the ENIAC/EDVAC development team in 1945. Von Neumann made significant contributions and added credibility to the Moore School group, which often had to fight against the conservative scientific establishment characterized by Howard Aiken at Harvard University and George Stibitz at Bell Labs. Philosophical differences between von Neumann and Eckert and Mauchly, as well as patent issue disputes with the Moore School administration, eventually caused the resignation of Eckert and Mauchly on March 31, 1946.

Eckert and Mauchly, along with some of their engineering colleagues at the University of Pennsylvania, formed the Electronic Control Company and proceeded to interest potential customers (including the Census Bureau) in an "EDVAC-type" machine. On May 24, 1947, the EDVAC-type machine became the UNIVAC. This new computer would overcome the shortcomings of the ENIAC and the EDVAC (which was eventually completed by the Moore School in 1951). It would be a stored-program computer and would allow input to and output from the computer via magnetic tape. The prior method of input/output used punched paper cards that were extremely slow compared to the speed at which data in the computer could be processed.

A series of poor business decisions and other unfortunate circumstances forced the newly renamed Eckert-Mauchly Computer Corporation to look for a buyer. They found one in Remington Rand in 1950. Remington Rand built tabulating equipment and was a competitor of International Business Machines Corporation (IBM). IBM was approached about buying EMCC, but the negotiations fell apart. EMCC became a division of Remington Rand and had access to the resources necessary to finish the UNIVAC.

CONSEQUENCES

Eckert and Mauchly made a significant contribution to the advent of the computer age with the introduction of the UNIVAC I. The words "computer" and "UNIVAC" entered the popular vocabulary as synonyms. The efforts of these two visionaries were rewarded quickly as contracts started to pour in, taking IBM by surprise and propelling the inventors into the national spotlight.

This spotlight shone brightest, perhaps, on the eve of the national presidential election of 1952, which pitted war hero General Dwight D. Eisenhower against statesman Adlai Stevenson. At the suggestion of Remington Rand, CBS was invited to use UNIVAC to predict the outcome of the election. Millions of television viewers watched as CBS anchorman Walter Cronkite "asked" UNIVAC for its predictions. A program had been written to analyze the results of thousands of voting districts in the elections of 1944 and 1948. Based on only 7 percent of the votes coming in, UNIVAC had Eisenhower winning by a landslide, in contrast with all the prior human forecasts of a close election. Surprised by this answer and not willing to suffer the embarrassment of being wrong, the programmers quickly directed the program to provide an answer that was closer to the perceived situation. The outcome of the election, however, matched UNIVAC's original answer. This prompted CBS commentator Edward R. Murrow's famous quote, "The trouble with machines is people."

The development of the UNIVAC I produced many technical innovations. Primary among these is the use of magnetic tape for input and output. All machines that preceded the UNIVAC (with one exception) used either paper tape or cards for input and cards for output. These methods were very slow and created a bottleneck of information. The great advantage of magnetic tape was the ability to store the equivalent of thousands of cards of data on one 30-centimeter reel of tape. Another advantage was its speed.

See also Apple II computer; BINAC computer; Colossus computer; ENIAC computer; IBM Model 1401 computer; Personal computer; Supercomputer.

FURTHER READING

Metropolis, Nicholas, Jack Howlett, and Gian Carlo Rota. *A History of Computing in the Twentieth Century: A Collection of Essays.* New York: Academic Press, 1980.

Slater, Robert. *Portraits in Silicon.* Cambridge, Mass.: MIT Press, 1987.

Stern, Nancy B. *From ENIAC to UNIVAC: An Appraisal of the Eckert-Mauchly Computers.* Bedford, Mass.: Digital Press, 1981.

Vacuum cleaner

The invention: The first portable domestic vacuum cleaner successfully adapted to electricity, the original machine helped begin the electrification of domestic appliances in the early twentieth century.

The people behind the invention:
H. Cecil Booth (1871-1955), a British civil engineer
Melville R. Bissell (1843-1889), the inventor and marketer of the Bissell carpet sweeper in 1876
William Henry Hoover (1849-1932), an American industrialist
James Murray Spangler (1848-1915), an American inventor

From Brooms to Bissells

During most of the nineteenth century, the floors of homes were cleaned primarily with brooms. Carpets were periodically dragged out of the home by the boys and men of the family, stretched over rope lines or fences, and given a thorough beating to remove dust and dirt. In the second half of the century, carpet sweepers, perhaps inspired by the success of street-sweeping machines, began to appear. Although there were many models, nearly all were based upon the idea of a revolving brush within an outer casing that moved on rollers or wheels when pushed by a long handle.

Melville Bissell's sweeper, patented in 1876, featured a knob for adjusting the brushes to the surface. The Bissell Carpet Company, also formed in 1876, became the most successful maker of carpet sweepers and dominated the market well into the twentieth century. Electric vacuum cleaners were not feasible until homes were wired for electricity and the small electric motor was invented. Thomas Edison's success with an incandescent lighting system in the 1880's and Nikola Tesla's invention of a small electric motor that was used in 1889 to drive a Westinghouse Electric Corporation fan opened the way for the application of electricity to household technologies.

CLEANING WITH ELECTRICITY

In 1901, H. Cecil Booth, a British civil engineer, observed a London demonstration of an American carpet cleaner that blew compressed air at the fabric. Booth was convinced that the process should be reversed so that dirt would be sucked out of the carpet. In developing this idea, Booth invented the first successful suction vacuum sweeper.

Booth's machines, which were powered by gasoline or electricity, worked without brushes. Dust was extracted by means of a suction action through flexible tubes with slot-shaped nozzles. Some machines were permanently installed in buildings that had wall sockets for the tubes in every room. Booth's British Vacuum Cleaner Company also employed horse-drawn mobile units from which white-uniformed men unrolled long tubes that they passed into buildings through windows and doors. His company's commercial triumph came when it cleaned Westminster Abbey for the coronation of Edward VII in 1902. Booth's company also manufactured a 1904 domestic model that had a direct-current electric motor and a vacuum pump mounted on a wheeled carriage. Dust was sucked into the nozzle of a long tube and deposited into a metal container. Booth's vacuum cleaner used electricity from overhead light sockets.

The portable electric vacuum cleaner was invented in 1907 in the United States by James Murray Spangler. When Spangler was a janitor in a department store in Canton, Ohio, his asthmatic condition was worsened by the dust he raised with a large Bissell carpet sweeper. Spangler's modifications of the Bissell sweeper led to his own invention. On June 2, 1908, he received a patent for his Electric Suction Sweeper. The device consisted of a cylindrical brush in the front of the machine, a vertical-shaft electric motor above a fan in the main body, and a pillowcase attached to a broom handle behind the main body. The brush dislodged the dirt, which was sucked into the pillowcase by the movement of air caused by a fan powered by the electric motor. Although Spangler's initial attempt to manufacture and sell his machines failed, Spangler had, luckily for him, sold one of his machines to a cousin, Susan Troxel Hoover, the wife of William Henry Hoover.

The Hoover family was involved in the production of leather goods, with an emphasis on horse saddles and harnesses. William Henry Hoover, president of the Hoover Company, recognizing that the adoption of the automobile was having a serious impact on the family business, was open to investigating another area of production. In addition, Mrs. Hoover liked the Spangler machine that she had been using for a couple of months, and she encouraged her husband to enter into an agreement with Spangler. An agreement made on August 5, 1908, allowed Spangler, as production manager, to manufacture his machine with a small work force in a section of Hoover's plant. As sales of vacuum cleaners increased, what began as a sideline for the Hoover Company became the company's main line of production.

Few American homes were wired for electricity when Spangler and Hoover joined forces; not until 1920 did 35 percent of American homes have electric power. In addition to this inauspicious fact, the first Spangler-Hoover machine, the Model O, carried the relatively high price of seventy-five dollars. Yet a full-page ad for the Model O in the December, 1908, issue of the *Saturday Evening Post* brought a deluge of requests. American women had heard of the excellent performance of commercial vacuum cleaners, and they hoped that the Hoover domestic model would do as well in the home.

IMPACT

As more and more homes in the United States and abroad became wired for electric lighting, a clean and accessible power source became available for household technologies. Whereas electric lighting was needed only in the evening, the electrification of household technologies made it necessary to use electricity during the day. The electrification of domestic technologies therefore matched the needs of the utility companies, which sought to maximize the use of their facilities. They became key promoters of electric appliances. In the first decades of the twentieth century, many household technologies became electrified. In addition to fans and vacuum cleaners, clothes-washing machines, irons, toasters, dishwashing machines, refrigerators, and kitchen ranges were being powered by electricity.

The application of electricity to household technologies came as large numbers of women entered the work force. During and after World War I, women found new employment opportunities in industrial manufacturing, department stores, and offices. The employment of women outside the home continued to increase throughout the twentieth century. Electrical appliances provided the means by which families could maintain the same standards of living in the home while both parents worked outside the home.

It is significant that Bissell was motivated by an allergy to dust and Spangler by an asthmatic condition. The employment of the carpet sweeper, and especially the electric vacuum cleaner, not only

---

H. CECIL BOOTH

Although Hubert Cecil Booth (1871-1955), an English civil engineer, designed battleship engines, factories, and bridges, he was not above working on homier problems when they intrigued him. That happened in 1900 when he watched the demonstration of a device that used forced air to blow the dirt out of railway cars. It worked poorly, and the reason, it seemed to Booth, was that blowing just stirred up the dirt. Sucking it into a receptacle, he thought, would work better. He tested his idea by placing a wet cloth over furniture upholstery and sucking through it. The grime that collected on the side of the cloth facing the upholstery proved him right.

He built his first vacuum cleaner—a term that he coined—in 1901. It cleaned houses, but only with considerable effort. Measuring 54 inches by 42 inches by 10 inches, it had to be carried in a horse-driven van to the cleaning site. A team of workmen from Booth's Vacuum Cleaner Company then did the cleaning with hoses that reached inside the house through windows and doors. Moreover, the machine cost the equivalent of more than fifteen hundred dollars. It was beyond the finances and physical powers of home owners.

Booth marketed the first successful British one-person vacuum cleaner, the Trolley-Vac, in 1906. Weighing one hundred pounds, it was still difficult to wrestle into position, but it came with hoses and attachments that made possible the cleaning of different types of surfaces and hard-to-reach areas.

made house cleaning more efficient and less physical but also led to a healthier home environment. Whereas sweeping with a broom tended only to move dust to a different location, the carpet sweeper and the electric vacuum cleaner removed the dirt from the house.

See also Disposable razor; Electric refrigerator; Microwave cooking; Robot (household); Washing machine.

FURTHER READING

Jailer-Chamberlain, Mildred. "This Is the Way We Cleaned Our Floors." *Antiques & Collecting Magazine* 101, no. 4 (June, 1996).
Kirkpatrick, David D. "The Ultimate Victory of Vacuum Cleaners." *New York Times* (April 14, 2001).
Shapiro, Laura. "Household Appliances." *Newsweek* 130, no. 24A (Winter, 1997/1998).

Vacuum tube

THE INVENTION: A sealed glass tube from which air and gas have been removed to permit electrons to move more freely, the vacuum tube was the heart of electronic systems until it was displaced by transistors.

THE PEOPLE BEHIND THE INVENTION:

Sir John Ambrose Fleming (1849-1945), an English physicist and professor of electrical engineering
Thomas Alva Edison (1847-1931), an American inventor
Lee de Forest (1873-1961), an American scientist and inventor
Arthur Wehnelt (1871-1944), a German inventor

A SOLUTION IN SEARCH OF A PROBLEM

The vacuum tube is a sealed tube or container from which almost all the air has been pumped out, thus creating a near vacuum. When the tube is in operation, currents of electricity are made to travel through it. The most widely used vacuum tubes are cathode-ray tubes (television picture tubes).

The most important discovery leading to the invention of the vacuum tube was the Edison effect by Thomas Alva Edison in 1884. While studying why the inner glass surface of light bulbs blackened, Edison inserted a metal plate near the filament of one of his light bulbs. He discovered that electricity would flow from the positive side of the filament to the plate, but not from the negative side to the plate. Edison offered no explanation for the effect.

Edison had, in fact, invented the first vacuum tube, which was later termed the *diode*; at that time there was no use for this device. Therefore, the discovery was not recognized for its true significance. A diode converts electricity that alternates in direction (alternating current) to electricity that flows in the same direction (direct current). Since Edison was more concerned with producing direct current in generators, and not household electric lamps, he essentially ignored this aspect of his discovery. Like many other inventions or discoveries that were ahead of their time—such as the laser—for a number of years, the Edison effect was "a solution in search of a problem."

The explanation for why this phenomenon occurred would not come until after the discovery of the electron in 1897 by Sir Joseph John Thomson, an English physicist. In retrospect, the Edison effect can be identified as one of the first observations of *thermionic emission*, the freeing up of electrons by the application of heat. Electrons were attracted to the positive charges and would collect on the positively charged plate, thus providing current; but they were repelled from the plate when it was made negative, meaning that no current was produced. Since the diode permitted the electrical current to flow in only one direction, it was compared to a valve that allowed a liquid to flow in only one direction. This analogy is popular since the behavior of water has often been used as an analogy for electricity, and this is the reason that the term *valves* became popular for vacuum tubes.

Same Device, Different Application

Sir John Ambrose Fleming, acting as adviser to the Edison Electric Light Company, had studied the light bulb and the Edison effect starting in the early 1880's, before the days of radio. Many years later, he came up with an application for the Edison effect as a radio detector when he was a consultant for the Marconi Wireless Telegraph Company. Detectors (devices that conduct electricity in one direction only, just as the diode does, but at higher frequencies) were required to make the high-frequency radio waves audible by converting them from alternating current to direct current. Fleming was able to detect radio waves quite effectively by using the Edison effect. Fleming used essentially the same device that Edison had created, but for a different purpose. Fleming applied for a patent on his detector on November 16, 1904.

In 1906, Lee de Forest refined Fleming's invention by adding a zigzag piece of wire between the metal plate and the filament of the vacuum tube. The zigzag piece of wire was later replaced by a screen called a "grid." The grid allowed a small voltage to control a larger voltage between the filament and plate. It was the first com-

John Ambrose Fleming

John Ambrose Fleming had a remarkably long and fruitful scientific career. He was born in Lancaster, England, in 1849, the eldest son of a minister. When he was a boy, the family moved to London, which remained his home for the rest of his life. An outstanding student, Fleming matriculated at University College, London, graduating in 1870 with honors. Scholarships took him to other colleges until his skill with electrical experiments earned him a job as a lab instructor at Cambridge University in 1880. In 1885, he returned to University College, London, as professor of electrical technology. He taught there for the following forty-one years, occasionally taking time off to serve as a consultant for such electronics industry leaders as Thomas Edison and Guglielmo Marconi.

Fleming's passion was electricity and electronics, and he was sought after as a teacher with a knack for memorable explanations. For instance, he thought up the "right-hand" rule (also called Fleming's rule) to illustrate the relation of electromagnetic forces during induction: When the thumb, index finger, and middle finger of a human hand are held at right angles to one another so that the thumb points in the direction of motion through a magnetic field—which is indicated by the index finger—then the middle finger shows the direction of induced current. During his extensive research, Fleming investigated transformers, high-voltage transmitters, electrical conduction, cryogenic electrical effects, radio, and television, and also invented the vacuum tube.

Advanced age hardly slowed him down. He wrote three books and more than one hundred articles and remarried at eighty-four. He also delivered public lectures—to audiences at the Royal Institution and the Royal Society among other venues—until he was ninety. He died in 1945, ninety-five years old, having helped give birth to telecommunications.

plete vacuum tube and the first device ever constructed capable of amplifying a signal—that is, taking a small-voltage signal and making it much larger. He named it the "audion" and was granted a U.S. patent in 1907.

In 1907-1908, the American Navy carried radios equipped with de Forest's audion in its goodwill tour around the world. While useful as an amplifier of the weak radio signals, it was not useful at this point for the more powerful signals of the telephone. Other developments were made quickly as the importance of the emerging fields of radio and telephony were realized.

IMPACT

With many industrial laboratories working on vacuum tubes, improvements came quickly. For example, tantalum and tungsten filaments quickly replaced the early carbon filaments. In 1904, Arthur Wehnelt, a German inventor, discovered that if metals were coated with certain materials such as metal oxides, they emitted far more electrons at a given temperature. These materials enabled electrons to escape the surface of the metal oxides more easily. Thermionic emission and, therefore, tube efficiencies were greatly improved by this method.

Another important improvement in the vacuum tube came with the work of the American chemist Irving Langmuir of the General Electric Research Laboratory, starting in 1909, and Harold D. Arnold of Bell Telephone Laboratories. They used new devices such as the mercury diffusion pump to achieve higher vacuums. Working independently, Langmuir and Arnold discovered that very high vacuum used with higher voltages increased the power these tubes could handle from small fractions of a watt to hundreds of watts. The de Forest tube was now useful for the higher-power audio signals of the telephone. This resulted in the introduction of the first transamerican speech transmission in 1914, followed by the first transatlantic communication in 1915.

The invention of the transistor in 1948 by the American physicists William Shockley, Walter H. Brattain, and John Bardeen ultimately led to the downfall of the tube. With the exception of the cathode-ray tube, transistors could accomplish the jobs of nearly all vacuum tubes much more efficiently. Also, the development of the integrated circuit allowed the creation of small, efficient, highly complex devices that would be impossible with radio tubes. By 1977, the major producers of the vacuum tube had stopped making it.

See also Color television; FM radio; Radar; Radio; Radio crystal sets; Television; Transistor; Transistor radio.

FURTHER READING

Baldwin, Neil. *Edison: Inventing the Century.* Chicago: University of Chicago Press, 2001.

Fleming, John Ambrose. *Memories of a Scientific Life.* London: Marshall, Morgan & Scott, 1934.

Hijiya, James A. *Lee de Forest and the Fatherhood of Radio.* Bethlehem, Pa.: Lehigh University Press, 1992.

Read, Oliver, and Walter L. Welch. *From Tin Foil to Stereo: Evolution of the Phonograph.* 2d ed. Indianapolis: H. W. Sams, 1976.

Vat Dye

The invention: The culmination of centuries of efforts to mimic the brilliant colors displayed in nature in dyes that can be used in many products.

The people behind the invention:
Sir William Henry Perkin (1838-1907), an English student in Hofmann's laboratory
René Bohn (1862-1922), a synthetic organic chemist
Karl Heumann (1850-1894), a German chemist who taught Bohn
Roland Scholl (1865-1945), a Swiss chemist who established the correct structure of Bohn's dye
August Wilhelm von Hofmann (1818-1892), an organic chemist

Synthesizing the Compounds of Life

From prehistoric times until the mid-nineteenth century, all dyes were derived from natural sources, primarily plants. Among the most lasting of these dyes were the red and blue dyes derived from alizarin and indigo.

The process of making dyes took a great leap forward with the advent of modern organic chemistry in the early years of the nineteenth century. At the outset, this branch of chemistry, dealing with the compounds of the element carbon and associated with living matter, hardly existed, and synthesis of carbon compounds was not attempted. Considerable data had accumulated showing that organic, or living, matter was basically different from the compounds of the nonliving mineral world. It was widely believed that although one could work with various types of organic matter in physical ways and even analyze their composition, they could be produced only in a living organism.

Yet, in 1828, the German chemist Friedrich Wöhler found that it was possible to synthesize the organic compound urea from mineral compounds. As more chemists reported the successful preparation of compounds previously isolated only from plants or animals, the theory that organic compounds could be produced only in a living organism faded.

One field ripe for exploration was that committed to exploiting the uses of coal tar. Here, August Wilhelm von Hofmann was an active worker. He and his students made careful studies of this complex mixture. The high-quality stills they designed allowed for the isolation of pure samples of important compounds for further study.

Of greater importance was the collection of able students Hofmann attracted. Among them was Sir William Henry Perkin, who is regarded as the founder of the dyestuffs industry. In 1856, Perkin undertook the task of synthesizing quinine (a bitter crystalline alkaloid used in medicine) from a nitrogen-containing coal tar material called toluidine. Luck played a decisive role in the outcome of his experiment. The sticky compound Perkin obtained contained no quinine, so he decided to investigate the simpler related compound aniline. A small amount of the impurity toluidine in his aniline gave Perkin the first synthetic dye, Mauveine.

SEARCHING FOR STRUCTURE

From this beginning, the great dye industries of Europe, particularly Germany, grew. The trial-and-error methods gave way to more systematic searches as the structural theory of organic chemistry was formulated.

As the twentieth century began, great progress had been made, and German firms dominated the industry. Badische Anilin- und Soda-Fabrik (BASF) was incorporated at Ludwigshafen in 1865 and undertook extensive explorations of both alizarin and indigo. A chemist, René Bohn, had made important discoveries in 1888, which helped the company recover lost ground in the alizarin field. In 1901, he undertook the synthesis of a dye he hoped would combine the desirable attributes of both alizarin and indigo.

As so often happens in science, nothing like the expected occurred. Bohn realized that the beautiful blue crystals that resulted from his synthesis represented a far more important product. Not only was this the first synthetic vat dye, Indanthrene, ever prepared, but also, by studying the reaction at higher temperature, a useful yellow dye, Flavanthrone, could be produced.

The term *vat dye* is used to describe a method of applying the dye, but it also serves to characterize the structure of the dye, because all

WILLIAM HENRY PERKIN

Born in England in 1838, William Henry Perkin saw a chemical experiment for the first time when he was a small boy. He found his calling there and then, much to the dismay of his father, who wanted him to be a builder and architect like himself.

Perkin studied chemistry every chance he found as a teenager and was only seventeen when he won an appointment as the assistant to the German chemist August Wilhelm von Hofmann. A year later, while trying to synthesize quinine at Hofmann's suggestion, Perkin discovered a deep purple dye—now known as aniline purple or Mauveine, but popularly called mauve. In 1857 he opened a small dyeworks by the Grand Union Canal in West London, hoping to make his fortune by manufacturing the dye.

He succeeded brilliantly. His ambitions were helped along royally when Queen Victoria wore a silk gown dyed with Mauveine to the Royal Exhibition of 1862. In 1869, he perfected a method for producing another new dye, alizarin, which is red. A wealthy man, he sold his business in 1874 when he was just thirty-six years old and devoted himself to research, which included isolation of the first synthetic perfume, coumarin, from coal tar.

Perkin died in 1907, a year after receiving a knighthood, one of his many awards and honors for starting the artificial dye industry. His son William Henry Perkin, Jr. (1860-1927) also became a well-known researcher in organic chemistry.

currently useful vat dyes share a common unit. One fundamental problem in dyeing relates to the extent to which the dye is water-soluble. A beautifully colored molecule that is easily soluble in water might seem attractive given the ease with which it binds with the fiber; however, this same solubility will lead to the dye's rapid loss in daily use.

Vat dyes are designed to solve this problem by producing molecules that can be made water-soluble, but only during the dyeing or vatting process. This involves altering the chemical structure of the dye so that it retains its color throughout the life of the cloth.

By 1907, Roland Scholl had showed unambiguously that the

chemical structure proposed by Bohn for Indanthrene was correct, and a major new area of theoretical and practical importance was opened for organic chemists.

IMPACT

Bohn's discovery led to the development of many new and useful dyes. The list of patents issued in his name fills several pages in *Chemical Abstracts* indexes.

The true importance of this work is to be found in a consideration of all synthetic chemistry, which may perhaps be represented by this particular event. More than two hundred dyes related to Indanthrene are in commercial use. The colors represented by these substances are a rainbow making nature's finest hues available to all. The dozen or so natural dyes have been synthesized into more than seven thousand superior products through the creativity of the chemist.

Despite these desirable outcomes, there is doubt whether there is any real benefit to society from the development of new dyes. This doubt is the result of having to deal with limited natural resources. With so many urgent problems to be solved, scientists are not sure whether to search for greater luxury. If the field of dye synthesis reveals a single theme, however, it must be to expect the unexpected. Time after time, the search for one goal has led to something quite different—and useful.

See also Buna rubber; Color film; Neoprene.

FURTHER READING

Clark, Robin J. H., et al. "Indigo, Woad, and Tyrian Purple: Important Vat Dyes from Antiquity to the Present." *Endeavour* 17, no. 4 (December, 1993).

Farber, Eduard. *The Evolution of Chemistry: A History of Its Ideas, Methods, and Materials*. 2d ed. New York: Ronald Press, 1969.

Partington, J. R. *A History of Chemistry*. Staten Island, N.Y.: Martino, 1996.

Schatz, Paul F. "Anniversaries: 2001." *Journal of Chemical Education* 78, no. 1 (January, 2001).

Velcro

The invention: A material comprising millions of tiny hooks and loops that work together to create powerful and easy-to-use fasteners for a wide range of applications.

The person behind the invention:
Georges de Mestral (1904-1990), a Swiss engineer and inventor

From Cockleburs to Fasteners

Since prehistoric times, people have walked through weedy fields and arrived at home with cockleburs all over their clothing. In 1948, a Swiss engineer and inventor, Georges de Mestral, found his clothing full of cockleburs after walking in the Swiss Alps near Geneva. Wondering why cockleburs stuck to clothing, he began to examine them under a microscope. De Mestral's initial examination showed that each of the thousands of fibrous ends of the cockleburs was tipped with a tiny hook; it was the hooks that made the cockleburs stick to fabric. This observation, combined with much subsequent work, led de Mestral to invent velcro, which was patented in 1957 in the form of two strips of nylon material. One of the strips contained millions of tiny hooks, while the other contained a similar number of tiny loops. When the two strips were pushed together, the hooks were inserted into the loops, joining the two strips of nylon very firmly. This design makes velcro extremely useful as a material for fasteners that is used in applications ranging from sneaker fasteners to fasteners used to join heart valves during surgery.

Making Velcro Practical

Velcro is not the only invention credited to de Mestral, who also invented such items as a toy airplane and an asparagus peeler, but it was his greatest achievement. It is said that his idea for the material was partly the result of a problem his wife had with a jammed dress zipper just before an important social engagement. De Mestral's idea was to design a sort of locking tape that used the hook-and-loop principle that he had observed under the microscope. Such a tape, he believed, would never jam. He also believed that the tape would do away with such annoyances as buttons that popped open unexpectedly and knots in shoelaces that refused to be untied. The design of the material envisioned by de Mestral took seven years of painstaking effort. When it was finished, de Mestral named it "velcro" (a contraction of the French phrase *velvet crochet*, meaning velvet hook), patented it, and opened a factory to manufacture it. Velcro's design required that de Mestral identify the optimal number of hooks and loops to be used. He eventually found that using approximately three hundred per square inch worked best. In addition, his studies showed that nylon was an excellent material for his purposes, although it had to be stiffened somewhat to work well. Much additional experimentation showed that the most effective way of producing the necessary stiffening was to subject the velcro to infrared light after manufacturing it.

Other researchers have demonstrated that velcrolike materials need not be made of nylon. For example, a new micromechanical velcrolike material (microvelcro) that medical researchers believe will soon be used to hold together blood vessels after surgery is made of minute silicon loops and hooks. This material is thought to be superior to other materials for such applications because it will not be redissolved prematurely by the body. Other uses for microvelcro may be to hold together tiny electronic components in miniaturized computers without the use of glue or other adhesives. A major advantage of the use of microvelcro in such situations is that it is resistant to changes of temperature as well as to most chemicals that destroy glue and other adhesives.

IMPACT

In 1957, when velcro was patented, there were four main ways to hold things together. These involved the use of buttons, laces, snaps, and zippers (which had been invented by Chicagoan Whitcomb L. Judson in 1892). All these devices had drawbacks; zippers can jam, buttons can come open at embarrassing times, and shoelaces can form knots that are difficult to unfasten. Almost immediately after velcro was introduced, its use became widespread; velcro fasteners can be found on or in clothing, shoes, watchbands, wallets, back-

---

GEORGES DE MESTRAL

Georges de Mestral got his idea for Velcro in part during a hunting trip on his estates and in part before an important formal social function. These contexts are evidence of the high standing in Swiss society held by de Mestral, an engineer and manufacturer. In fact, de Mestral, who was born in 1904, came from a illustrious line of noble landowners. Their prize possession was one of Switzerland's famous residences, the castle of Saint Saphorin on Morges.

Built on the site of yet older fortifications, the castle was completed by François-Louis de Pesme in 1710. An enemy of King Louis XIV, de Pesme served in the military forces of Austria, Holland, and England, rising to the rank of lieutenant general, but he is best known for driving off a Turkish invasion fleet on the Danube in 1695. Other forebears include the diplomat Armand- François Louis de Mestral (1738-1805) and his father, Albert-Georges-Constantin de Mestral (1878-1966), an agricultural engineer.

The castle passed to the father's four sons and eventually into the care of the inventor. It in turn was inherited by Georges de Mestral's sons Henri and François when he died in 1990 in Genolier, Switzerland.

---

packs, bookbags, motor vehicles, space suits, blood-pressure cuffs, and in many other places. There is even a "wall jumping" game incorporating velcro in which a wall is covered with a well-supported piece of velcro. People who want to play put on jackets made of velcro and jump as high as they can. Wherever they land on the wall, the velcro will join together, making them stick.

Wall jumping, silly though it may be, demonstrates the tremendous holding power of velcro; a velcro jacket can keep a two-hundred-pound person suspended from a wall. This great strength is used in a more serious way in the design of the items used to anchor astronauts to space shuttles and to buckle on parachutes. In addition, velcro is washable, comes in many colors, and will not jam. No doubt many more uses for this innovative product will be found.

See also Artificial heart.

FURTHER READING

"George De Mestral: Inventor of Velcro Fastener." *Los Angeles Times* (February 13, 1990).

LaFavre Yorks, Cindy. "Hidden Helpers Velcro Fasteners, Pull-On Loops and Other Extras Make Dressing Easier for People with Disabilities." *Los Angeles Times* (November 1, 1991).

Roberts, Royston M., and Jeanie Roberts. *Lucky Science: Accidental Discoveries from Gravity to Velcro, with Experiments.* New York: John Wiley, 1994.

Stone, Judith. "Stuck on Velcro!" *Reader's Digest* (September, 1988).

"Velcro-wrapped Armor Saves Lives in Bosnia." *Design News* 52, no. 7 (April 7, 1997).

VENDING MACHINE SLUG REJECTOR

THE INVENTION: A device that separates real coins from counterfeits, the slug rejector made it possible for coin-operated vending machines to become an important marketing tool for many products

THE PEOPLE BEHIND THE INVENTION:

Thomas Adams, the founder of Adams Gum Company

Frederick C. Lynde, an Englishman awarded the first American patent on a vending machine

Nathaniel Leverone (1884-1969), a founder of the Automatic Canteen Company of America

Louis E. Leverone (1880-1957), a founder, with his brother, of the Automatic Canteen Company of America

THE GROWTH OF VENDING MACHINES

One of the most imposing phenomena to occur in the United States economy following World War II was the growth of vending machines. Following the 1930's invention and perfection of the slug rejector, vending machines became commonplace as a means of marketing gum and candy. By the 1960's, almost every building had soft drink and coffee machines. Street corners featured machines that dispensed newspapers, and post offices even used vending machines to sell stamps. Occasionally someone fishing in the backwoods could find a vending machine next to a favorite fishing hole that would dispense a can of fishing worms upon deposit of the correct amount of money. The primary advantage offered by vending machines is their convenience. Unlike people, machines can provide goods and services around the clock, with no charge for the "labor" of standing duty.

The decade of the 1950's brought not only an increase in the number of vending machines but also an increase in the types of goods that were marketed through them. Before World War II, the major products had been cigarettes, candy, gum, and soft drinks. The 1950's brought far more products into the vending machine market.

The first recognized vending machine in history was invented in the third century B.C.E. by the mathematician Hero. This first machine was a coin-activated device that dispensed sacrificial water in an Egyptian temple. It was not until the year 1615 that another vending machine was recorded. In that year, snuff and tobacco vending boxes began appearing in English pubs and taverns. These tobacco boxes were less sophisticated machines than was Hero's, since they left much to the honesty of the customer. Insertion of a coin opened the box; once it was open, the customer could take out as much tobacco as desired. One of the first United States patents on a machine was issued in 1886 to Frederick C. Lynde. That machine was used to vend postcards.

If any one person can be considered the father of vending machines in the United States, it would probably be Thomas Adams, the founder of Adams Gum Company. Adams began the first successful vending operation in America in 1888 when he placed gum machines on train platforms in New York City.

Other early vending machines included scales (which vended a service rather than a product), photograph machines, strength testers, beer machines, and hot water vendors (to supply poor people who had no other source of hot water). These were followed, around 1900, by complete automatic restaurants in Germany, cigar vending machines in Chicago, perfume machines in Paris, and an automatic divorce machine in Utah.

Also around 1900 came the introduction of coin-operated gambling machines. These "slot machines" are differentiated from normal vending machines. The vending machine industry does not consider gambling machines to be a part of the vending industry since they do not vend merchandise. The primary importance of the gambling machines was that they induced the industry to do research into slug rejection. Early machines allowed coins to be retrieved by the use of strings tied to them and accepted counterfeit lead coins, called slugs. It was not until the 1930's that the slug rejector was perfected. Invention of the slug rejection device gave rise to the tremendous growth in the vending machine industry in the 1930's by giving vendors more confidence that they would be paid for their products or services.

Soft drink machines got their start just prior to the beginning of the twentieth century. By 1906, improved models of these machines could dispense up to ten different flavors of soda pop. The drinks were dispensed into a drinking glass or tin cup that was placed near the machine (there was usually only one glass or cup to a machine, since paper cups had not been invented). Public health officials became concerned that everyone was drinking from the same cup. At that point, someone came up with the idea of setting a bucket of water next to the machine so that each customer could rinse off the cup before drinking from it. The year 1909 witnessed one of the monumental inventions in the history of vending machines, the pay toilet.

IMPACT

The 1930's witnessed improved vending machines. Slug rejectors were the most important introduction. In addition, change-making machines were instituted, and a few machines would even say "thank you" after a coin was deposited. These improved machines led many marketers to experiment with automatic vending. Coin-operated washing machines were one of the new applications of the 1930's. During the Depression, many appliance dealers attached coin metering devices to washing machines, allowing the user to accumulate money to make the monthly payments by using the appliance. This was a form of forced saving. It was not long before some enterprising appliance dealer got the idea of placing washing machines in apartment house basements. This idea was soon followed by stores full of coin-operated laundry machines, giving rise to a new kind of automatic vending business.

Following World War II, there was a surge of innovation in the vending machine industry. Much of that surge resulted from the discovery of vending machines by industrial management. Prior to the war, the managements of most factories had been tolerant of vending machines. Following the war, managers discovered that the machines could be an inexpensive means of keeping workers happy. They became aware that worker productivity could be increased by access to candy bars or soft drinks. As a result, the demand for machines exceeded the supply offered by the industry during the late 1940's.

Vending machines have had a surprising effect on the total retail sales of the U.S. economy. In 1946, sales through vending machines totaled $600 million. By 1960, that figure had increased to $2.5 billion; by 1970, it exceeded $6 billion. The decade of the 1950's began with individual machines that would dispense cigarettes, candy, gum, coffee, and soft drinks. By the end of that decade, it was much more common to see vending machines in groups. The combination of machines in a group could, in many cases, meet the requirements to assemble a complete meal.

Convenience is the key to the popularity of vending machines. Their ability to sell around the clock has probably been the major impetus to vending machine sales as opposed to more conventional marketing. Lower labor costs have also played a role in their popularity, and their location in areas of dense pedestrian traffic prompts impulse purchases.

Despite the advances made by the vending machine industry during the 1950's, there was still one major limitation to growth, to be solved during the early 1960's. That problem was that vending machines were effectively limited to low-priced items, since the machines would accept nothing but coins. The inconvenience of inserting many coins kept machine operators from trying to market expensive items; as they expected consumer reluctance. The early 1960's witnessed the invention of vending machines that would accept and make change for $1, $5, and $10 bills. This invention paved the way for expansion into lines of grocery items and tickets.

The first use of vending machines to issue tickets was at an Illinois race track, where pari-mutuel tickets were dispensed upon deposit of $2. Penn Central Railroad was one of the first transportation companies to sell tickets by means of vending machines. These machines, used in high-traffic areas on the East Coast, permitted passengers to deal directly with a computer when buying reserved-seat train tickets. The machines would accept $1 bills and $5 bills as well as coins.

LIMITATIONS TO VENDING MACHINES

There are limitations to the use of vending machines. Primary among these are mechanical failure and vandalism of machines. Another limitation often mentioned is that not every product can be sold by machine. There are several factors that make some goods more vendable than others. National advertising and wide consumer acceptance help. Product must have a high turnover in order to justify the cost of a machine and the cost of servicing it. A third factor in measuring the potential success of an item is where it will be consumed or used. The most successful products are used within a short distance of the machine; consumers must be made willing to pay the usually higher prices of machine-bought products by the convenience of machine location.

The automatic vending of merchandise plays the largest role in the vending machine industry, but the vending of services also plays a role. The largest percentage of service vending comes from coin laundries. Other types of services are vended by weighing machines, parcel lockers, and pay toilets. By depositing a coin, a person can even get shoes shined. Some motel beds offer a "massage." Even the lowly parking meter is an example of a vending machine that dispenses services. Coin-operated photocopy machines account for a large portion of service vending.

A later advance in the vending machine industry is the use of credit. The cashless society began to make strides with vending machines as well as conventional vendors. As of the early 1990's, credit cards could be used to operate only a few types of vending machines, primarily those that dispense transportation tickets. Vending machines operated by banks dispense money upon deposit of a credit card. Credit-card gasoline pumps reduced labor requirements at gasoline stations, pushing the concept of self-service a step further. As credit card transactions become more common in general and as the cost of making them falls, use of credit cards for vending machines will increase.

Thousands of items have been marketed through vending machines, and firms must continue to evaluate the use of automatic retailing as a marketing channel. Many products are not conducive to automatic vending, but before dismissing that option for a particular product, a marketer should consider the range of products sold through vending machines. The producers of Band-Aid flexible plastic bandages saw the possibilities in the vending field. The only product modification necessary was to put Band-Aids in a package the size of a candy bar, able to be sold from renovated candy machines.

The next problem was to determine areas where there would be a high turnover of Band-Aids. Bowling alleys were an obvious answer, since many bowlers suffered from abrasions on their fingers.

The United States is not alone in the development of vending machines; in fact, it is not as advanced as some nations of the world. In Japan, machines operated by credit cards have been used widely since the mid-1960's, and the range of products offered has been larger than in the United States. Western Europe is probably the most advanced area of the world in terms of vending machine technology. Germany of the early 1990's probably had the largest selection of vending machines of any European country. Many gasoline stations in Germany featured beer dispensing machines. In rural areas of the country, vending machines hung from utility poles. These rural machines provided candy and gum, among other products, to farmers who did not often travel into town.

Most vending machine business in Europe was done not in individual machines but in automated vending shops. The machines offered a creative solution to obstacles created by regulations and laws. Some countries had laws stating that conventional retail stores could not be open at night or on Sundays. To increase sales and satisfy consumer needs, stores built vending operations that could be used by customers during off hours. The machines, or combinations of them, often stocked a tremendous variety of items. At one German location, consumers could choose among nearly a thousand grocery items.

The Future

The future will see a broadening of product lines offered in vending machines as marketers come to recognize the opportunities that exist in automatic retailing. In the United States, vending machines of the early 1990's primarily dispensed products for immediate consumption. If labor costs increase, it will become economically feasible to sell more items from vending machines. Grocery items and tickets offered the most potential for expansion.

Vending machines offer convenience to the consumer. Virtually any company that produces for the retail market must consider vending machines as a marketing channel. Machines offer an alternative to conventional stores that cannot be ignored as the range of products offered through machines increases.

Vending machines appear to be a permanent fixture and have only scratched the surface of the market. Although machines have a long history, their popularization came from innovations of the 1930's, particularly the slug rejector. Marketing managers came to recognize that vending machine sales are more than a sideline. Increasingly, firms established separate departments to handle sales through vending machines. Successful companies make the best use of all channels of distribution, and vending machines had become an important marketing channel.

See also Geiger counter; Sonar; Radio interferometer.

FURTHER READING

Ho, Rodney. "Vending Machines Make Change—Now They Sell Movie Soundtracks, Underwear—Even Art." *Wall Street Journal* (July 7, 1999).

Rosen, Cheryl. "Vending Machines Get a High-Tech Makeover." *Informationweek* 822 (January 29, 2001).

Ryan, James. "In Vending Machine, Brains That Tell Good Money from Bad." *New York Times* (April 8, 1999).

Tagliabue, John. "Vending Machines Face an Upheaval of Change." *New York Times* (February 16, 1999).

Videocassette recorder

The invention: A device for recording and playing back movies and television programs, the videocassette recorder (VCR) revolutionized the home entertainment industry in the late 1970's.

The company behind the invention:
Philips Corporation, a Dutch Company

Videotape Recording

Although television sets first came on the market before World War II, video recording on magnetic tape was not developed until the 1950's. Ampex marketed the first practical videotape recorder in 1956. Unlike television, which manufacturers aimed at retail consumers from its inception, videotape recording was never expected to be attractive to the individual consumer. The first videotape recorders were meant for use within the television industry.

Developed not long after the invention of magnetic tape recording of audio signals, the early videotape recorders were large machines that employed an open reel-to-reel tape drive similar to that of a conventional audiotape recorder. Recording and playback heads scanned the tape longitudinally (lengthwise). Because video signals have a much wider frequency ("frequency" is the distance between the tops and the bottoms of the signal waves) than audio signals do, this scanning technique meant that the amount of recording time available on one reel of tape was extremely limited. In addition, open reels were large and awkward, and the magnetic tape itself was quite expensive.

Still, within the limited application area of commercial television, videotape recording had its uses. It made it possible to play back recorded material immediately rather than having to wait for film to be processed in a laboratory. As television became more popular and production schedules became more hectic, with more material being produced in shorter and shorter periods of time, videotape solved some significant problems.

HELICAL SCANNING BREAKTHROUGH

Engineers in the television industry continued to search for innovations and improvements in videotape recording following Ampex's marketing of the first practical videotape recorder in the 1950's. It took more than ten years, however, for the next major breakthrough to occur. The innovation that proved to be the key to reducing the size and awkwardness of video recording equipment came in 1967 with the invention by the Philips Corporation of helical scanning.

All videocassette recorders eventually employed multiple-head helical scanning systems. In a helical scanning system, the record and playback heads are attached to a spinning drum or head that rotates at exactly 1,800 revolutions per minute, or 30 revolutions per second. This is the number of video frames per second used in the NTSC-TV broadcasts in the United States and Canada. The heads are mounted in pairs 180 degrees apart on the drum. Two fields on the tape are scanned for each revolution of the drum. Perhaps the easiest way to understand the helical scanning system is to visualize the spiral path followed by the stripes on a barber's pole.

Helical scanning deviated sharply from designs based on audio recording systems. In an audiotape recorder, the tape passes over stationary playback and record heads; in a videocassette recorder, both the heads and the tape move. Helical scanning is, however, one of the few things that competing models and formats of videocassette recorders have in common. Different models employ different tape delivery systems and, in the case of competing formats such as Beta and VHS, there may be differences in the composition of the video signal to be recorded. Beta uses a 688-kilohertz (kHz) frequency, while VHS employs a frequency of 629 kHz. This difference in frequency is what allows Beta videocassette recorders (VCRs) to provide more lines of resolution and thus a superior picture quality; VHS provides 240 lines of resolution, while Beta has 400. (For this reason, it is perhaps unfortunate that the VHS format eventually dominated the market.)

In addition to helical scanning, Philips introduced another innovation: the videocassette. Existing videotape recorders employed a reel-to-reel tape drive, as do videocassettes, but videocassettes enclose the tape reels in a protective case. The case prevents the tape from being damaged in handling.

The first VCRs were large and awkward compared to later models. Industry analysts still thought that the commercial television and film industries would be the primary markets for VCRs. The first videocassettes employed wide—¾-inch or 1-inch—videotapes, and the machines themselves were cumbersome. Although Philips introduced a VCR in 1970, it took until 1972 before the machines actually became available for purchase, and it would be another ten years before VCRs became common appliances in homes.

CONSEQUENCES

Following the introduction of the VCR in 1970, the home entertainment industry changed radically. Although the industry did not originally anticipate that the VCR would have great commercial potential as a home entertainment device, it quickly became obvious that it did. By the late 1970's, the size of the cassette had been reduced and the length of recording time available per cassette had been increased from one hour to six. VCRs became so widespread that advertisers on television became concerned with a phenomenon known as "timeshifting," which refers to viewers setting the VCR to record programs for later viewing. Jokes about the complexity of programming VCRs appeared in the popular culture, and an inability to cope with the VCR came to be seen as evidence of technological illiteracy.

Consumer demand for VCRs was so great that, by the late 1980's, compact portable video cameras became widely available. The same technology—helical scanning with multiple heads—was successfully miniaturized, and "camcorders" were developed that were not much larger than a paperback book. By the early 1990's, "reality television"—that is, television shows based on actual events—began relying on video recordings supplied by viewers rather than material produced by professionals. The video recorder had completed a circle: It began as a tool intended for use in the television studio, and it returned there four decades later. Along the way, it had an effect no one could have predicted; passive viewers in the audience had evolved into active participants in the production process.

See also Cassette recording; Color television; Compact disc; Dolby noise reduction; Television; Walkman cassette player.

FURTHER READING

Gilder, George. *Life After Television*. New York: W. W. Norton, 1992.
Lardner, James. *Fast Forward: Hollywood, the Japanese, and the Onslaught of the VCR*. New York: Norton, 1987.
Luther, Arch C. *Digital Video in the PC Environment*. New York: McGraw-Hill, 1989.
Wassser, Frederick. *Veni, Vidi, Video: The Hollywood Empire and the VCR*. Austin: University of Texas Press, 2001.

Virtual Machine

The invention: The first computer to swap storage space between its random access memory (RAM) and hard disk to create a larger "virtual" memory that enabled it to increase its power.

The people behind the invention:
International Business Machines (IBM) Corporation, an American data processing firm
Massachusetts Institute of Technology (MIT), an American university
Bell Labs, the research and development arm of the American Telephone and Telegraph Company

A Shortage of Memory

During the late 1950's and the 1960's, computers generally used two types of data storage areas. The first type, called "magnetic disk storage," was slow and large, but its storage space was relatively cheap and abundant. The second type, called "main memory" (also often called "random access memory," or RAM), was much faster. Computation and program execution occurred primarily in the "central processing unit" (CPU), which is the "brain" of the computer. The CPU accessed RAM as an area in which to perform intermediate computations, store data, and store program instructions.

To run programs, users went through a lengthy process. At that time, keyboards with monitors that allowed on-line editing and program storage were very rare. Instead, most users used typewriter-like devices to type their programs or text on paper cards. Holding decks of such cards, users waited in lines to use card readers. The cards were read and returned to the user, and the programs were scheduled to run later. Hours later or even overnight, the output of each program was printed in some predetermined order, after which all the outputs were placed in user bins. It might take as long as several days to make any program corrections that were necessary.

Because CPUs were expensive, many users had to share a single CPU. If a computer had a monitor that could be used for editing or could run more than one program at a time, more memory was required. RAM was extremely expensive, and even multimillion-dollar computers had small memories. In addition, this primitive RAM was extremely bulky.

VIRTUALLY UNLIMITED MEMORY

The solution to the problem of creating affordable, convenient memory came in a revolutionary reformulation of the relationship between main memory and disk space. Since disk space was large and cheap, it could be treated as an extended "scratch pad," or temporary-use area, for main memory. While a program ran, only small parts of it (called pages or segments), normally the parts in use at that moment, would be kept in the main memory. If only a few pages of each program were kept in memory at any time, more programs could coexist in memory. When pages lay idle, they would be sent from RAM to the disk, as newly requested pages were loaded from the disk to the RAM. Each user and program "thought" it had essentially unlimited memory (limited only by disk space), hence the term "virtual memory."

The system did, however, have its drawbacks. The swapping and paging processes reduced the speed at which the computer could process information. Coordinating these activities also required more circuitry. Integrating each program and the amount of virtual memory space it required was critical. To keep the system operating accurately, stably, and fairly among users, all computers have an "operating system." Operating systems that support virtual memory are more complex than the older varieties are.

Many years of research, design, simulations, and prototype testing were required to develop virtual memory. CPUs and operating systems were designed by large teams, not individuals. Therefore, the exact original discovery of virtual memory is difficult to trace. Many people contributed at each stage.

The first rudimentary implementation of virtual memory concepts was on the Atlas computer, which was constructed in the early 1960's in England, at the University of Manchester. It coupled RAM with a device that read a magnetizable cylinder, or drum, which meant that it was a two-part storage system.

In the late 1960's, the Massachusetts Institute of Technology (MIT), Bell Telephone Labs, and the General Electric Company (later Honeywell) jointly designed a high-level operating system called MULTICS, which had virtual memory.

During the 1960's, IBM worked on virtual memory, and the IBM 360 series supported the new memory system. With the evolution of engineering concepts such as circuit integration, IBM produced a new line of computers called the IBM 370 series. The IBM 370 supported several advances in hardware (equipment) and software (program instructions), including full virtual memory capabilities. It was a platform for a new and powerful "environment," or set of conditions, in which software could be run; IBM called this environment the VM/370. The VM/370 went far beyond virtual memory, using virtual memory to create virtual machines. In a virtual machine environment, each user can select a separate and complete operating system. This means that separate copies of operating systems such as OS/360, CMS, DOS/360, and UNIX can all run in separate "compartments" on a single computer. In effect, each operating system has its own machine. Reliability and security were also increased. This was a major breakthrough, a second computer revolution.

Another measure of the significance of the IBM 370 was the commercial success and rapid, widespread distribution of the system. The large customer base for the older IBM 360 also appreciated the IBM 370's compatibility with that machine. The essentials of the IBM 370 virtual memory model were retained even in the 1990's generation of large, powerful mainframe computers. Furthermore, its success carried over to the design decisions of other computers in the 1970's.

The second-largest computer manufacturer, Digital Equipment Corporation (DEC), followed suit; its popular VAX minicomputers had virtual memory in the late 1970's. The celebrated UNIX operating system also added virtual memory. IBM's success had led to industry-wide acceptance.

CONSEQUENCES

The impact of virtual memory extends beyond large computers and the 1970's. During the late 1970's and early 1980's, the computer world took a giant step backward. Small, single-user computers called personal computers (PCs) became very popular. Because they were single-user models and were relatively cheap, they were sold with weak CPUs and deplorable operating systems that did not support virtual memory. Only one program could run at a time. Larger and more powerful programs required more memory than was physically installed. These computers crashed often.

Virtual memory raises PC user productivity. With virtual memory space, during data transmissions or long calculations, users can simultaneously edit files if physical memory runs out. Most major PCs now have improved CPUs and operating systems, and these advances support virtual memory. Popular virtual memory systems such as OS/2, Windows/DOS, and MAC-OS are available. Even old virtual memory UNIX has been used in PCs.

The concept of a virtual machine has been revived, in a weak form, on PCs that have dual operating systems (such as UNIX and DOS, OS/2 and DOS, MAC and DOS, and UNIX and DOS combinations).

Most powerful programs benefit from virtual memory. Many dazzling graphics programs require massive RAM but run safely in virtual memory. Scientific visualization, high-speed animation, and virtual reality all benefit from it. Artificial intelligence and computer reasoning are also part of a "virtual" future.

See also Colossus computer; Differential analyzer; ENIAC computer; IBM Model 1401 computer; Personal computer; Robot (industrial); SAINT; Virtual reality.

FURTHER READING

Bashe, Charles J. *IBM's Early Computers*. Cambridge, Mass.: MIT Press, 1986.
Ceruzzi, Paul E. *A History of Modern Computing*. Cambridge, Mass.: MIT Press, 2000.

Chposky, James, and Ted Leonsis. *Blue Magic: The People, Power, and Politics Behind the IBM Personal Computer.* New York: Facts on File, 1988.

Seitz, Frederick, and Norman G. Einspruch. *Electronic Genie: The Tangled History of Silicon.* Urbana: University of Illinois Press, 1998.

VIRTUAL REALITY

THE INVENTION: The creation of highly interactive, computer-based multimedia environments in which the user becomes a participant with the computer in a "virtually real" world.

THE PEOPLE BEHIND THE INVENTION:
Ivan Sutherland (1938-   ), an American computer scientist
Myron W. Krueger (1942-   ), an American computer scientist
Fred P. Brooks (1931-   ), an American computer scientist

HUMAN/COMPUTER INTERFACE

In the early 1960's, the encounter between humans and computers was considered to be the central event of the time. The computer was evolving more rapidly than any technology in history; humans seemed not to be evolving at all. The "user interface" (the devices and language with which a person communicates with a computer) was a veneer that had been applied to the computer to make it slightly easier to use, but it seemed obvious that the ultimate interface would be connecting the human body and senses directly to the computer.

Against this background, Ivan Sutherland of the University of Utah identified the next logical step in the development of computer graphics. He implemented a head-mounted display that allowed a person to look around in a graphically created "room" simply by turning his or her head. Two small cathode-ray tubes, or CRTs (which are the basis of television screens and computer monitors), driven by vector graphics generators (mathematical image-creating devices) provided the appropriate view for each eye, and thus, stereo vision.

In the early 1970's, Fred P. Brooks of the University of North Carolina created a system that allowed a person to handle graphic objects by using a mechanical manipulator. When the user moved the physical manipulator, a graphic manipulator moved accordingly. If a graphic block was picked up, the user felt its weight and its resistance to his or her fingers closing around it.

A New Reality

Beginning in 1969, Myron W. Krueger of the University of Wisconsin created a series of interactive environments that emphasized unencumbered, full-body, multisensory participation in computer events. In one demonstration, a sensory floor detected participants' movements around a room. A symbol representing each participant moved through a projected graphic maze that changed in playful ways if participants tried to cheat. In another demonstration, participants could use the image of a finger to draw on the projection screen. In yet another, participants' views of a projected three-dimensional room changed appropriately as they moved around the physical space.

It was interesting that people naturally accepted these projected experiences as reality. They expected their bodies to influence graphic objects and were delighted when they did. They regarded their electronic images as extensions of themselves. What happened to their images also happened to them; they felt what touched their images. These observations led to the creation of the Videoplace, a graphic world that people could enter from different places to interact with each other and with graphic creatures. Videoplace is an installation at the Connecticut Museum of Natural History in Storrs, Connecticut. Videoplace visitors in separate rooms can fingerpaint together, perform free-fall gymnastics, tickle each other, and experience additional interactive events.

The computer combines and alters inputs from separate cameras trained on each person, each of whom responds in turn to the computer's output, playing games in the world created by Videoplace software. Since participants' live video images can be manipulated (moved, scaled, or rotated) in real time, the world that is created is not bound by the laws of physics. In fact, the result is a virtual reality in which new laws of cause and effect are created, and can be changed, from moment to moment. Indeed, the term "virtual reality" describes the type of experience that can be created with Videoplace or with the technology invented by Ivan Sutherland.

Virtual realities are part of certain ongoing trends. Most obvious are the trend from interaction to participation in computer events and the trend from passive to active art forms. In addition, artificial

---

IVAN SUTHERLAND

Ivan Sutherland was born in Hastings, Nebraska, in 1938. His father was an engineer, and from an early age Sutherland considered engineering his own destiny, too. He earned a bachelor's degree from the Carnegie Institute of Technology in 1959, a master's degree from the California Institute of Technology in 1960, and a doctorate from the Massachusetts Institute of Technology (MIT) in 1963.

His adviser at MIT was Claude Shannon, creator of information theory, who directed Sutherland to find ways to simplify the interface between people and computers. Out of this research came Sketchpad. It was software that allowed people to draw designs on a computer terminal with a light pen, an early form of computer-assisted design (CAD). The U.S. Defense Department's Advanced Research Projects Center became interested in Sutherland's work and hired him to direct its Information Processing Techniques Office in 1964. In 1966 he left to become an associate professor of electrical engineering at Harvard University, moving to the University of Utah in 1968, and then to Caltech in 1975. During his academic career he developed the graphic interface for virtual reality, first announced in his ground-breaking 1968 article "A Head-Mounted Three-Dimensional Display."

In 1980 Sutherland left academia for industry. He already had business experience as cofounder of Evans & Sutherland in Salt Lake City. The new firm, Sutherland, Sproull, and Associates, which provided consulting services and venture capital, later became part of Sun Microsystems, Inc. Sutherland remained as a research fellow and vice president. A member of the National Academy of Engineering and National Academy of Sciences, in 1988 Sutherland was awarded the AM Turing Award, the highest honor in information technology.

---

experiences are taking on increasing significance. Businesspersons like to talk about "doing it right the first time." This can now be done in many cases, not because fewer mistakes are being made by people but because those mistakes are being made in simulated environments.

Most important is that virtual realities provide means of expressing and experiencing, as well as new ways for people to interact. Entertainment uses of virtual reality will be as economically significant as more practical uses, since entertainment is the United States' number-two export. Vicarious experience through theater, novels, movies, and television represents a significant percentage of people's lives in developed countries. The addition of a radically new form of physically involving, interactive experience is a major cultural event that may shape human consciousness as much as earlier forms of experience have.

CONSEQUENCES

Most religions offer their believers an escape from this world, but few technologies have been able to do likewise. Not so with virtual reality, the fledgling technology in which people explore a simulated three-dimensional environment generated by a computer. Using this technology, people can not only escape from this world but also design the world in which they want to live.

In most virtual reality systems, many of which are still experimental, one watches the scene, or alternative reality, through three dimensional goggles in a headset. Sound and tactile sensations enhance the illusion of reality. Because of the wide variety of actual and potential applications of virtual reality, from three-dimensional video games and simulators to remotely operated "telepresence" systems for the nuclear and undersea industries, interest in the field is intense.

The term "virtual reality" describes the computer-generated simulation of reality with physical, tactile, and visual dimensions. The interactive technology is used by science and engineering researchers as well as by the entertainment industry, especially in the form of video games. Virtual reality systems can, for example, simulate a walk-through of a building in an architectural graphics program. Virtual reality technology in which the artificial world overlaps with reality will have major social and psychological implications.

See also Personal computer; Virtual machine.

FURTHER READING

Earnshaw, Rae A., M. A. Gigante, and H. Jones. *Virtual Reality Systems*. San Diego: Academic Press, 1993.

Moody, Fred. *The Visionary Position: The Inside Story of the Digital Dreamers Who Are Making Virtual Reality a Reality*. New York: Times Business, 1999.

Sutherland, Ivan Edward. *Sketchpad: A Man-Machine Graphical Communication System*. New York: Garland, 1980.

V-2 ROCKET

THE INVENTION: The first first long-range, liquid-fueled rocket, the V-2 was developed by Germany to carry bombs during World War II.

THE PEOPLE BEHIND THE INVENTION:
Wernher von Braun (1912-1977), the chief engineer and prime motivator of rocket research in Germany during the 1930's and 1940's
Walter Robert Dornberger (1895-1980), the former commander of the Peenemünde Rocket Research Institute
Ing Fritz Gosslau, the head of the V-1 development team
Paul Schmidt, the designer of the impulse jet motor

THE "BUZZ BOMB"

On May 26, 1943, in the middle of World War II, key German military officials were briefed by two teams of scientists, one representing the air force and the other representing the army. Each team had launched its own experimental aerial war craft. The military chiefs were to decide which project merited further funding and development. Each experimental craft had both advantages and disadvantages, and each counterbalanced the other. Therefore, it was decided that both craft were to be developed. They were to become the V-1 and the V-2 aircraft.

The impulse jet motor used in the V-1 craft was designed by Munich engineer Paul Schmidt. On April 30, 1941, the motor had been used to assist power on a biplane trainer. The development team for the V-1 was headed by Ing Fritz Gosslau; the aircraft was designed by Robert Lusser.

The V-1, or "buzz bomb," was capable of delivering a one-ton warhead payload. While still in a late developmental stage, it was launched, under Adolf Hitler's orders, to terrorize inhabited areas of London in retaliation for the damage that had been wreaked on Germany during the war. More than one hundred V-1's were launched daily between June 13 and early September, 1944. Because the V-1 flew in a straight line and at a constant speed, Allied aircraft were able to intercept it more easily than they could the V-2.

Two innovative systems made the V-1 unique: the drive operation and the guidance system. In the motor, oxygen entered the grid valves through many small flaps. Fuel oil was introduced and the mixture of fuel and oxygen was ignited. After ignition, the expanded gases produced the reaction propulsion. When the expanded gases had vacated, the reduced internal pressure allowed the valve flaps to reopen, admitting more air for the next cycle.

The guidance system included a small propeller connected to a revolution counter that was preset based on the distance to the target. The number of propeller revolutions that it would take to reach the target was calculated before launch and punched into the counter. During flight, after the counter had measured off the selected number of revolutions, the aircraft's elevator flaps became activated, causing the craft to dive at the target. Understandably, the accuracy was not what the engineers had hoped.

VENGEANCE WEAPON 2

According to the Treaty of Versailles (1919), world military forces were restricted to 100,000 men and a certain level of weaponry. The German military powers realized very early, however, that the treaty had neglected to restrict rocket-powered weaponry, which did not exist at the end of World War I (1914-1918). Wernher von Braun was hired as chief engineer for developing the V-2 rocket.

The V-2 had a lift-off thrust of 11,550.5 newtons and was propelled by the combustion of liquid oxygen and alcohol. The propellants were pumped into the combustion chamber by a steam-powered turboprop. The steam was generated by the decomposition of hydrogen peroxide, using sodium permanganate as a catalyst. One innovative feature of the V-2 that is still used was regenerative cooling, which used alcohol to cool the double-walled combustion chamber.

The guidance system included two phases: powered and ballistic. Four seconds after launch, a preprogrammed tilt to 17 degrees was begun, then acceleration was continued to achieve the desired trajectory. At the desired velocity, the engine power was cut off via one of two systems. In the automatic system, a device shut off the engine at the velocity desired; this method, however, was inaccurate. The second system sent a radio signal to the rocket's receiver, which cut off the power. This was a far more accurate method, but the extra equipment required at the launch site was an attractive target for Allied bombers. This system was more often employed toward the end of the war.

Even the 907-kilogram warhead of the V-2 was a carefully tested device. The detonators had to be able to withstand 6 g's of force during lift-off and reentry, as well as the vibrations inherent in a rocket flight. Yet they also had to be sensitive enough to ignite the bomb upon impact and before the explosive became buried in the target and lost power through diffusion of force.

The V-2's first successful test was in October of 1942, but it continued to be developed until August of 1944. During the next eight months, more than three thousand V-2's were launched against England and the Continent, causing immense devastation and living up to its name: *Vergeltungswaffe zwei* (vengeance weapon 2). Unfortunately for Hitler's regime, the weapon that took fourteen years of research and testing to perfect entered the war too late to make an impact upon the outcome.

IMPACT

The V-1 and V-2 had a tremendous impact on the history and development of space technology. Even during the war, captured V-2's were studied by Allied scientists. American rocket scientists were especially interested in the technology, since they too were working to develop liquid-fueled rockets.

After the war, German military personnel were sent to the United States, where they signed contracts to work with the U.S. Army in a program known as "Operation Paperclip." Testing of the captured V-2's was undertaken at White Sands Missile Range near Alamogordo, New Mexico. The JB-2 Loon Navy jet-propelled bomb was developed following the study of the captured German craft.

The Soviet Union also benefited from captured V-2's and from the German V-2 factories that were dismantled following the war. With these resources, the Soviet Union developed its own rocket technology, which culminated in the launch of *Sputnik 1*, the world's first artificial satellite, on October 4, 1957. The United States was not far behind. It launched its first satellite, Explorer 1, on January 31, 1958. On April 12, 1961, the world's first human space traveler, Soviet cosmonaut Yuri A. Gagarin, was launched into Earth orbit.

See also Airplane; Cruise missile; Hydrogen bomb; Radar; Rocket; Stealth aircraft.

FURTHER READING

Bergaust, Erik. *Wernher von Braun: The Authoritative and Definitive Biographical Profile of the Father of Modern Space Flight.* Washington: National Space Institute, 1976.
De Maeseneer, Guido. *Peenemünde: The Extraordinary Story of Hitler's Secret Weapons V-1 and V-2.* Vancouver: AJ Publishing, 2001.
Piszkiewicz, Dennis. *Wernher von Braun: The Man Who Sold the Moon.* Westport, Conn.: Praeger, 1998.

Walkman cassette player

The invention: Inexpensive portable device for listening to stereo cassettes that was the most successful audio product of the 1980's and the forerunner of other portable electronic devices.

The people behind the invention:

Masaru Ibuka (1908-1997), a Japanese engineer who cofounded Sony

Akio Morita (1921-1999), a Japanese physicist and engineer, cofounder of Sony

Norio Ohga (1930-   ), a Japanese opera singer and businessman who ran Sony's tape recorder division before becoming president of the company in 1982

Convergence of Two Technologies

The Sony Walkman was the result of the convergence of two technologies: the transistor, which enabled miniaturization of electronic components, and the compact cassette, a worldwide standard for magnetic recording tape. As the smallest tape player devised, the Walkman was based on a systems approach that made use of advances in several unrelated areas, including improved loudspeaker design and reduced battery size. The Sony company brought them together in an innovative product that found a mass market in a remarkably short time.

Tokyo Telecommunications Engineering, which became Sony, was one of many small entrepreneurial companies that made audio products in the years following World War II. It was formed in the ruins of Tokyo, Japan, in 1946, and got its start manufacturing components for inexpensive radios and record players. They were the ideal products for a company with some expertise in electrical engineering and a limited manufacturing capability.

Akio Morita and Masaru Ibuka formed Tokyo Telecommunications Engineering to make a variety of electrical testing devices and instruments, but their real interests were in sound, and they decided to concentrate on audio products. They introduced a reel-to-reel tape recorder in 1946. Its success ensured that the company would remain in the audio field. The trade name of the magnetic tape they manufactured was "Soni," this was the origin of the company's new name, adopted in 1957. The 1953 acquisition of a license to use Bell Laboratories' transistor technology was a turning point in the fortunes of Sony, for it led the company to the highly popular transistor radio and started it along the path to reducing the size of consumer products. In the 1960's, Sony led the way to smaller and cheaper radios, tape recorders, and television sets, all using transistors instead of vacuum tubes.

THE CONSUMER MARKET

The original marketing strategy for manufacturers of mechanical entertainment devices had been to put one into every home. This was the goal for Edison's phonograph, the player piano, the Victrola, and the radio receiver. Sony and other Japanese manufacturers found out that if a product were small enough and cheap enough, two or three might be purchased for home use, or even for outdoor use. This was the marketing lesson of the transistor radio.

The unparalleled sales of transistor radios indicated that consumer durables intended for entertainment were not exclusively used in the home. The appeal of the transistor radio was that it made entertainment portable. Sony applied this concept to televisions and tape recorders, developing small portable units powered by batteries. Sony was first to produce a "personal" television set, with a five-inch screen. To the surprise of many manufacturers who said there would never be a market for such a novelty item, it sold well.

It was impossible to reduce tape recorders to the size of transistor radios because of the problems of handling very small reels of tape and the high power required to turn them. Portable tape recorders required several large flashlight batteries. Although tape had the advantage of recording capability, it could not challenge the popularity of the microgroove 45 revolution-per-minute (rpm) disc because the tape player was much more difficult to operate. In the 1960's, several types of tape cartridge were introduced to overcome this problem, including the eight-track tape cartridge and the Philips compact cassette. Sony and Matsushita were two of the leading Japanese manufacturers that quickly incorporated the compact cassette into their audio products, producing the first cassette players available in the United States.

The portable cassette players of the 1960's and 1970's were based on the transistor radio concept: small loudspeaker, transistorized amplifier, and flashlight batteries all enclosed in a plastic case. The size of transistorized components was being reduced constantly, and new types of batteries, notably the nickel cadmium combination, offered higher power output in smaller sizes. The problem of reducing the size of the loudspeaker without serious deterioration of sound quality blocked the path to very small cassette players. Sony's engineers solved the problem with a very small loudspeaker device using plastic diaphragms and new, lighter materials for the magnets. These devices were incorporated into tiny stereo headphones that set new standards of fidelity.

The first "walkman" was made by Sony engineers for the personal use of Masaru Ibuka. He wanted to be able to listen to high fidelity recorded sound wherever he went, and the tiny player was small enough to fit inside a pocket. Sony was experienced in reducing the size of machines. At the same time the walkman was being made up, Sony engineers were struggling to produce a video recording cassette that was also small enough to fit into Ibuka's pocket.

Although the portable stereo was part of a long line of successful miniaturized consumer products, it was not immediately recognized as a commercial technology. There were already plenty of cassette players in home units, in automobiles, and in portable players. Marketing experts questioned the need for a tiny version. The board of directors of Sony had to be convinced by Morita that the new product had commercial potential. The Sony Soundabout portable cassette player was introduced to the market in 1979.

IMPACT

The Soundabout was initially treated as a novelty in the audio equipment industry. At a price of $200, it could not be considered as a product for the mass market. Although it sold well in Japan, where people were used to listening to music on headphones, sales in the United States were not encouraging. Sony's engineers, working under the direction of Kozo Ohsone, reduced the size and cost of the machine. In 1981, the Walkman II was introduced. It was 25 percent smaller than the original version and had 50 percent fewer moving parts. Its price was considerably lower and continued to fall.

The Walkman opened a huge market for audio equipment that nobody knew existed. Sony had again confounded the marketing experts who doubted the appeal of a new consumer electronics product. It took about two years for Sony's Japanese competitors, including Matsushita, Toshiba, and Aiwa, to bring out portable personal stereos. Such was the popularity of the device that any miniature cassette player was called a "walkman," irrespective of the manufacturer. Sony kept ahead of the competition by constant innovation: Dolby noise reduction circuits were added in 1982, and a rechargeable battery feature was introduced in 1985. The machine became smaller, until it was barely larger than the audio cassette it played.

Sony developed a whole line of personal stereos. Waterproofed Walkmans were marketed to customers who wanted musical accompaniment to water sports. There were special models for tennis players and joggers. The line grew to encompass about forty different types of portable cassette players, priced from about $30 to $500 for a high-fidelity model.

In the ten years following the introduction of the Walkman, Sony sold fifty million units, including twenty-five million in the United States. Its competitors sold millions more. They were manufactured all over the Far East and came in a broad range of sizes and prices, with the cheapest models about $20. Increased competition in the portable tape player market continually forced down prices. Sony had to respond to the huge numbers of cheap copies by redesigning the Walkman to bring down its cost and by automating its production. The playing mechanism became part of the integrated circuit that provided amplification, allowing manufacturing as one unit.

The Walkman did more than revive sales of audio equipment in the sagging market of the late 1970's. It stimulated demand for cassette tapes and helped make the compact cassette the worldwide standard for magnetic tape. At the time the Walkman was introduced, the major form of prerecorded sound was the vinyl micro-

---

MASARU IBUKA

Nicknamed "genius inventor" in college, Masaru Ibuka developed into a visionary corporate leader and business philosopher. Born in Nikko City, Japan, in 1908, he took a degree in engineering from Waseda University in 1933 and went to work at Photo-Chemical Laboratory, which developed movie film. Changing to naval research during World War II, he met Akio Morita, another engineer. After the war they opened an electronics shop together, calling it the Tokyo Telecommunications Engineering Corporation, and began experimenting with tape recorders.

Their first model was a modest success, and the business grew under Ibuka, who was president and later chairman He thought up a new, less daunting name for his company, Sony, in the 1950's, when it rapidly became a leader in consumer electronics. His goal was to make existing technology useful to people in everyday life. "He sowed the seeds of a deep conviction that our products must bring joy and fun to users," one of his successors as president, Nobuyuki Idei, said in 1997.

While American companies were studying military applications for the newly developed transistor in the 1950's, Ibuka and Morita put it to use in an affordable transistor radio and then found ways to shrink its size and power it with batteries so that it could be taken anywhere. In a similar fashion, they made tape recorders and players (such as the Walkman), video players, compact disc players, and televisions ever cheaper, more reliable, and more efficiently designed.

A hero in the Japanese business world, Ibuka retired as Sony chairman in 1976 but continued to help out as a consultant until his death in 1997.

---

groove record. In 1983, the ratio of vinyl to cassette sales was 3:2. By the end of the decade, the audio cassette was the bestselling format for recorded sound, outselling vinyl records and compact discs combined by a ratio of 2:1. The compatibility of the audio cassette used in personal players with the home stereo ensured that it would be the most popular tape recording medium.

The market for portable personal players in the United States during the decade of the 1990's was estimated to be more than twenty million units each year. Sony accounted for half of the 1991 American market of fifteen million units selling at an average price of $50. It appeared that there would be more than one in every home. In some parts of Western Europe, there were more cassette players than people, reflecting the level of market penetration achieved by the Walkman.

The ubiquitous Walkman had a noticeable effect on the way that people listen to music. The sound from the headphones of a portable player is more intimate and immediate than the sound coming from the loudspeakers of a home stereo. The listener can hear a wider range of frequencies and more of the lower amplitudes of music, while the reverberation caused by sound bouncing off walls is reduced. The listening public has become accustomed to the Walkman sound and expects it to be duplicated on commercial recordings. Recording studios that once mixed their master recordings to suit the reproduction characteristics of car or transistor radios began to mix them for Walkman headphones. Personal stereos also enable the listener to experience more of the volume of recorded sound because it is injected directly into the ear.

The Walkman established a market for portable tape players that exerted an influence on all subsequent audio products. The introduction of the compact disc (CD) in 1983 marked a completely new technology of recording based on digital transformation of sound. It was jointly developed by the Sony and Philips companies. Despite the enormous technical difficulties of reducing the size of the laser reader and making it portable, Sony's engineers devised the Discman portable compact disc player, which was unveiled in 1984. It followed the Walkman concept exactly and offered higher fidelity than the cassette tape version. The Discman sold for about $300 when it was introduced, but its price soon dropped to less than $100. It did not achieve the volume of sales of the audio cassette version because fewer CDs than audio cassettes were in use. The slow acceptance of the compact disc hindered sales growth. The Discman could not match the portability of the Walkman because vibrations caused the laser reader to skip tracks.

In the competitive market for consumer electronics products, a company must innovate to survive. Sony had watched cheap competition erode the sales of many of its most successful products, particularly the transistor radio and personal television, and was committed to both product improvement and new entertainment technologies. It knew that the personal cassette player had a limited sales potential in the advanced industrial countries, especially after the introduction of digital recording in the 1980's. It therefore sought new technology to apply to the Walkman concept. Throughout the 1980's, Sony and its many competitors searched for a new version of the Walkman.

The next generation of personal players was likely to be based on digital recording. Sony introduced its digital audio tape (DAT) system in 1990. This used the same digital technology as the compact disc but came in tape form. It was incorporated into expensive home players; naturally, Sony engineered a portable version. The tiny DAT Walkman offered unsurpassed fidelity of reproduction, but its incompatibility with any other tape format and its high price limited its sales to professional musicians and recording engineers.

After the failure of DAT, Sony refocused its digital technology into a format more similar to the Walkman. Its Mini Disc (MD) used the same technology as the compact disc but had the advantage of a recording capability. The 2.5-inch disc was smaller than the CD, and the player was smaller than the Walkman. The play-only version fit in the palm of a hand. A special feature prevented the skipping of tracks that caused problems with the Discman. The Mini Disc followed the path blazed by the Walkman and represented the most advanced technology applied to personal stereo players. At a price of about $500 in 1993, it was still too expensive to compete in the audio cassette Walkman market, but the history of similar products illustrates that rapid reduction of price could be achieved even with a complex technology.

The Walkman had a powerful influence on the development of other digital and optical technologies. The laser readers of compact disc players can access visual and textual information in addition to sound. Sony introduced the Data Discman, a handheld device that displayed text and pictures on a tiny screen. Several other manufacturers marketed electronic books. Whatever the shape of future entertainment and information technologies, the legacy of the Walkman will put a high premium on portability, small size, and the interaction of machine and user.

See also Cassette recording; Compact disc; Dolby noise reduction; Electronic synthesizer; Laser; Transistor; Videocassette recorder.

FURTHER READING

Bull, Michael. *Sounding Out the City: Personal Stereos and the Management of Everyday Life.* New York: Berg, 2000.
Lyons, Nick. *The Sony Vision.* New York: Crown Publishers, 1976.
Morita, Akio, with Edwin M. Reingold, and Mitsuko Shimomura. *Made in Japan: Akio Morita and Sony.* London: HarperCollins, 1994.
Nathan, John. *Sony: The Private Life.* London: HarperCollins Business, 2001.
Schlender, Brenton R. "How Sony Keeps the Magic Going." *Fortune* 125 (February 24, 1992).

Washing machine

The invention: Electrical-powered machines that replaced hand-operated washing tubs and wringers, making the job of washing clothes much easier.

The people behind the invention:

O. B. Woodrow, a bank clerk who claimed to be the first to adapt electricity to a remodeled hand-operated washing machine

Alva J. Fisher (1862-1947), the founder of the Hurley Machine Company, who designed the Thor electric washing machine, claiming that it was the first successful electric washer

Howard Snyder, the mechanical genius of the Maytag Company

Hand Washing

Until the development of the electric washing machine in the twentieth century, washing clothes was a tiring and time-consuming process. With the development of the washboard, dirt was loosened by rubbing. Clothes and tubs had to be carried to the water, or the water had to be carried to the tubs and clothes. After washing and rinsing, clothes were hand-wrung, hang-dried, and ironed with heavy, heated irons. In nineteenth century America, the laundering process became more arduous with the greater use of cotton fabrics. In addition, the invention of the sewing machine resulted in the mass production of inexpensive ready-to-wear cotton clothing. With more clothing, there was more washing.

One solution was hand-operated washing machines. The first American patent for a hand-operated washing machine was issued in 1805. By 1857, more than 140 patents had been issued; by 1880, between 4,000 and 5,000 patents had been granted. While most of these machines were never produced, they show how much the public wanted to find a mechanical means of washing clothes. Nearly all the early types prior to the Civil War (1861-1865) were modeled after the washboard.

Washing machines based upon the rubbing principle had two limitations: They washed only one item at a time, and the rubbing was hard on clothes. The major conceptual breakthrough was to move away from rubbing and to design machines that would clean by forcing water through a number of clothes at the same time.

An early suction machine used a plunger to force water through clothes. Later electric machines would have between two and four suction cups, similar to plungers, attached to arms that went up and down and rotated on a vertical shaft. Another hand-operated washing machine was used to rock a tub on a frame back and forth. An electric motor was later substituted for the hand lever that rocked the tub. A third hand-operated washing machine was the dolly type. The dolly, which looked like an upside-down three-legged milking stool, was attached to the inside of the tub cover and was turned by a two-handled lever on top of the enclosed tub.

MACHINE WASHING

The hand-operated machines that would later dominate the market as electric machines were the horizontal rotary cylinder and the underwater agitator types. In 1851, James King patented a machine of the first type that utilized two concentric half-full cylinders. Water in the outer cylinder was heated by a fire beneath it; a hand crank turned the perforated inner cylinder that contained clothing and soap. The inner-ribbed design of the rotating cylinder raised the clothes as the cylinder turned. Once the clothes reached the top of the cylinder, they dropped back down into the soapy water.

The first underwater agitator-type machine, the second type, was patented in 1869. In this machine, four blades at the bottom of the tub were attached to a central vertical shaft that was turned by a hand crank on the outside. The agitation created by the blades washed the clothes by driving water through the fabric. It was not until 1922, when Howard Snyder of the Maytag Company developed an underwater agitator with reversible motion, that this type of machine was able to compete with the other machines. Without reversible action, clothes would soon wrap around the blades and not be washed.

Claims for inventing the first electric washing machine came from O. B. Woodrow, who founded the Automatic Electric Washer Company, and Alva J. Fisher, who developed the Thor electric washing machine for the Hurley Machine Corporation. Both Woodrow and Fisher made their innovations in 1907 by adapting electric power to modified hand-operated, dolly-type machines. Since only 8 percent of American homes were wired for electricity in 1907, the early machines were advertised as adaptable to electric or gasoline power but could be hand-operated if the power source failed. Soon, electric power was being applied to the rotary cylinder, oscillating, and suction-type machines. In 1910, a number of companies introduced washing machines with attached wringers that could be operated by electricity. The introduction of automatic washers in 1937 meant that washing machines could change phases without the action of the operator.

IMPACT

By 1907 (the year electricity was adapted to washing machines), electric power was already being used to operate fans, ranges, coffee percolators, flatirons, and sewing machines. By 1920, nearly 35 percent of American residences had been wired for electricity; by 1941, nearly 80 percent had been wired. The majority of American homes had washing machines by 1941; by 1958, this had risen to an estimated 90 percent.

The growth of electric appliances, especially washing machines, is directly related to the decline in the number of domestic servants in the United States. The development of the electric washing machine was, in part, a response to a decline in servants, especially laundresses. Also, rather than easing the work of laundresses with technology, American families replaced their laundresses with washing machines.

Commercial laundries were also affected by the growth of electric washing machines. At the end of the nineteenth century, they were in every major city and were used widely. Observers noted that just as spinning, weaving, and baking had once been done in the home but now were done in commercial establishments, laundry work had now begun its move out of the home. After World

War II (1939-1945), however, although commercial laundries continued to grow, their business was centered more and more on institutional laundry, rather than residential laundry, which they had lost to the home washing machine.

Some scholars have argued that, on one hand, the return of laundry to the home resulted from marketing strategies that developed the image of the American woman as one who is home operating her appliances. On the other hand, it was probably because the electric washing machine made the task much easier that American women, still primarily responsible for the family laundry, were able to pursue careers outside the home.

See also Electric refrigerator; Microwave cooking; Robot (household); Vacuum cleaner; Vending machine slug rejector.

FURTHER READING

Ierley, Merritt. *Comforts of Home: The American House and the Evolution of Modern Convenience*. New York: C. Potter, 1999.
"Maytag Heritage Embraces Innovation, Dependable Products." *Machine Design* 71, no. 18 (September, 1999).
Shapiro, Laura. "Household Appliances." *Newsweek* 130, no. 24A (Winter, 1997/1998).

Weather Satellite

The invention: A series of cloud-cover meteorological satellites that pioneered the reconnaissance of large-scale weather systems and led to vast improvements in weather forecasting.

The person behind the invention:
Harry Wexler (1911-1962), director of National Weather Bureau meteorological research

Cameras in Space

The first experimental weather satellite, Tiros 1, was launched from Cape Canaveral on April 1, 1960. Tiros's orbit was angled to cover the area from Montreal, Canada, to Santa Cruz, Argentina, in the Western Hemisphere. Tiros completed an orbit every ninety-nine minutes and, when launched, was expected to survive at least three months in space, returning thousands of images of large-scale weather systems.

Tiros 1 was equipped with a pair of vidicon scanner television cameras, one equipped with a wide-angle lens and the other with a narrow-angle lens. Both cameras created pictures with five hundred lines per frame at a shutter speed of 1.5 milliseconds. Each television camera's imaging data were stored on magnetic tape for downloading to ground stations when Tiros 1 was in range. The wide-angle lens provided a low-resolution view of an area covering 2,048 square kilometers. The narrow-angle lens had a resolution of half a kilometer within a viewing area of 205 square kilometers.

Tiros transmitted its data to ground stations, which displayed the data on television screens. Photographs of these displays were then made for permanent records. Tiros weather data were sent to the Naval Photographic Interpretation Center for detailed meteorological analysis. Next, the photographs were passed along to the National Weather Bureau for further study.

Tiros caused some controversy because it was able to image large areas of the communist world: the Soviet Union, Cuba, and Mongolia. The weather satellite's imaging system was not, however, partic-

Hurricane off the coast of Florida photographed from space. (PhotoDisc)

ularly useful as a spy satellite, and only large-scale surface features were visible in the images. Nevertheless, the National Aeronautics and Space Administration (NASA) skirted adverse international reactions by carefully scrutinizing Tiros's images for evidence of sensitive surface features before releasing them publicly.

A STARTLING DISCOVERY

Tiros 1 was not in orbit very long before it made a significant and startling discovery. It was the first satellite to document that large storms have vortex patterns that resemble whirling pinwheels. Within its lifetime, Tiros photographed more than forty northern mid-latitude storm systems, and each one had a vortex at its center. These storms were in various stages of development and were between 800 and 1,600 kilometers in diameter. The storm vortex in most of these was located inside a 560-kilometer-diameter circle around the center of the storm's low-pressure zone. Nevertheless, Tiros's images did not reveal at what stage in a storm's development the vortex pattern formed.

This was typical of Tiros's data. The satellite was truly an experiment, and, as is the case with most initial experiments, various new phenomena were uncovered but were not fully understood. The data showed clearly that weather systems could be investigated from orbit and that future weather satellites could be outfitted with sensors that would lead to better understanding of meteorology on a global scale.

Tiros 1 did suffer from a few difficulties during its lifetime in orbit. Low contrast in the television imaging system often made it difficult to distinguish between cloud cover and snow cover. The magnetic tape system for the high-resolution camera failed at an early stage. Also, Earth's magnetic field tended to move Tiros 1 away from an advantageous Earth observation attitude. Experience with Tiros 1 led to improvements in later Tiros satellites and many other weather-related satellites.

CONSEQUENCES

Prior to Tiros 1, weather monitoring required networks of ground-based instrumentation centers, airborne balloons, and instrumented aircraft. Brief high-altitude rocket flights provided limited coverage of cloud systems from above. Tiros 1 was the first step in the development of the permanent monitoring of weather systems. The resulting early detection and accurate tracking of hurricanes alone have resulted in savings in both property and human life.

As a result of the Tiros 1 experiment, meteorologists were not ready to discard ground-based and airborne weather systems in favor of satellites alone. Such systems could not provide data about pressure, humidity, and temperature, for example. Tiros 1 did, however, introduce weather satellites as a necessary supplement to ground-based and airborne systems for large-scale monitoring of weather systems and storms. Satellites could provide more reliable and expansive coverage at a far lower cost than a large contingent of aircraft. Tiros 1, which was followed by nine similar spacecraft, paved the way for modern weather satellite systems.

See also Artificial satellite; Communications satellite; Cruise missile; Radio interferometer; Rocket.

FURTHER READING

Fishman, Jack, and Robert Kalish. *The Weather Revolution: Innovations and Imminent Breakthroughs in Accurate Forecasting.* New York: Plenum Press, 1994.

Kahl, Jonathan D. *Weather Watch: Forecasting the Weather.* Minneapolis, Minn.: Lerner, 1996.

Rao, Krishna P. *Weather Satellites: Systems, Data, and Environmental Applications.* Boston: American Meteorological Society, 1990.

Artist's depiction of a weather satellite. (PhotoDisc)

Xerography

THE INVENTION: Process that makes identical copies of documents with a system of lenses, mirrors, electricity, chemicals that conduct electricity in bright light, and dry inks (toners) that fuse to paper by means of heat.

THE PEOPLE BEHIND THE INVENTION:

Chester F. Carlson (1906-1968), an American inventor

Otto Kornei (1903-   ), a German physicist and engineer

XEROGRAPHY, XEROGRAPHY, EVERYWHERE

The term *xerography* is derived from the Greek for "dry writing." The process of xerography was invented by an American, Chester F. Carlson, who made the first xerographic copy of a document in 1938. Before the development of xerography, the preparation of copies of documents was often difficult and tedious. Most often, unclear carbon copies of typed documents were the only available medium of information transfer.

The development of xerography led to the birth of the giant Xerox Corporation, and the term *xerographic* was soon shortened to *Xerox*. The process of xerography makes identical copies of a document by using lens systems, mirrors, electricity, chemicals that conduct electricity in bright light ("semiconductors"), and dry inks called "toners" that are fused to copy paper by means of heat. The process makes it easy to produce identical copies of a document quickly and cheaply. In addition, xerography has led to huge advances in information transfer, the increased use of written documents, and rapid decision-making in all areas of society. Xeroxing can produce both color and black-and-white copies.

FROM THE FIRST XEROX COPY TO MODERN PHOTOCOPIES

On October 22, 1938, after years of effort, Chester F. Carlson produced the first Xerox copy. Reportedly, his efforts grew out of his 1930's job in the patent department of the New York firm P. R.

Mallory and Company. He was looking for a quick, inexpensive method for making copies of patent diagrams and other patent specifications. Much of Carlson's original work was conducted in the kitchen of his New York City apartment or in a room behind a beauty parlor in Astoria, Long Island. It was in Astoria that Carlson, with the help of Otto Kornei, produced the first Xerox copy (of the inscription "10-22-38 Astoria") on waxed paper.

The first practical method of xerography used the element selenium, a substance that conducts electricity only when it is exposed to light. The prototype Xerox copying machines were developed as a result of the often frustrating, nerve-wracking, fifteen-year collaboration of Carlson, scientists and engineers at the Battelle Memorial Institute in Columbus, Ohio, and the Haloid Company of Rochester, New York. The Haloid Company financed the effort after 1947, based on an evaluation made by an executive, John H. Dessauer. In return, the company obtained the right to manufacture and market Xerox machines. The company, which was originally a manufacturer of photographic paper, evolved into the giant Xerox Corporation. Carlson became very wealthy as a result of the royalties and dividends paid to him by the company.

Early xerographic machines operated in several stages. First, the document to be copied was positioned above a mirror so that its image, lit by a flash lamp and projected by a lens, was reflected onto a drum coated with electrically charged selenium. Wherever dark sections of the document's image were reflected, the selenium coating retained its positive charge. Where the image was light, the charge of the selenium was lost, because of the photoactive properties of the selenium.

Next, the drum was dusted with a thin layer of a negatively charged black powder called a "toner." Toner particles stuck to positively charged dark areas of the drum and produced a visible image on the drum. Then, Xerox copy paper, itself positively charged, was put in contact with the drum, where it picked up negatively charged toner. Finally, an infrared lamp heated the paper and the toner, fusing the toner to the paper and completing the copying process.

In ensuing years, the Xerox Corporation engineered many changes in the materials and mechanics of Xerox copiers. For example, the semiconductors and toners were changed, which increased both the

CHESTER F. CARLSON

The copying machine changed Chester Floyd Carlson's life even before he invented it. While he was experimenting with photochemicals in his apartment, the building owner's daughter came by to complain about the stench Carlson was creating. However, she found Carlson himself more compelling than her complaints and married him not long afterward. Soon Carlson transferred his laboratory to a room behind his mother-in-law's beauty parlor, where he devoted ten dollars a month from his meager wages to spend on research.

Born in Seattle, Washington, in 1906, Carlson learned early to husband his resources, set his goals high, and never give up. Both his father and mother were sickly, and so after he was fourteen, Carlson was the family's main breadwinner. His relentless drive and native intelligence got him through high school and into a community college, where an impressed teacher inspired him to go even further—into the California Institute of Technology. After he graduated, he worked for General Electric but lost his job during the layoffs caused by the Great Depression. In 1933 he hired on with P. R. Mallory Company, an electrical component manufacturer, which, although not interested in his invention, at least paid him enough in wages to keep going.

His thirteen-year crusade to invent a copier and then find a manufacturer to build it ended just as Carlson was nearly broke. In 1946 Haloid Corporation licensed the rights to Carlson's copying machine, but even then the invention did not become an important part of American communications culture until the company marketed the Xerox 914 in 1960. The earnings for Xerox Corporation (as it was called after 1961) leapt from $33 million to more than $500 million in the next six years, and Carlson became enormously wealthy. He won the Inventor of the Year Award in 1964 and the Horatio Alger Award in 1966. Before he died in 1968, he remembered the hardships of his youth by donating $100 million to research organizations and charitable foundations.

quality of copies and the safety of the copying process. In addition, auxiliary lenses of varying focal length were added, along with other features, which made it possible to produce enlarged or reduced copies. Furthermore, modification of the mechanical and chemical properties of the components of the system made it possible to produce thousands of copies per hour, sort them, and staple them.

The next development was color Xerox copying. Color systems use the same process steps that the black-and-white systems use, but the document exposure and toning operations are repeated three times to yield the three overlaid colored layers (yellow, magenta, and cyan) that are used to produce multicolored images in any color printing process. To accomplish this, blue, green, and red filters are rotated in front of the copier's lens system. This action produces three different semiconductor images on three separate rollers. Next, yellow, magenta, and cyan toners are used—each on its own roller—to yield three images. Finally, all three images are transferred to one sheet of paper, which is heated to produce the multicolored copy. The complex color procedure is slower and much more expensive than the black-and-white process.

IMPACT

The quick, inexpensive copying of documents is commonly performed worldwide. Memoranda that must be distributed to hundreds of business employees can now be copied in moments, whereas in the past such a process might have occupied typists for days and cost hundreds of dollars. Xerox copying also has the advantage that each copy is an exact replica of the original; no new errors can be introduced, as was the case when documents had to be retyped. Xerographic techniques are also used to reproduce X rays and many other types of medical and scientific data, and the facsimile (fax) machines that are now used to send documents from one place to another over telephone lines are a variation of the Xerox process.

All this convenience is not without some problems: The ease of photocopying has made it possible to reproduce copyrighted publications. Few students at libraries, for example, think twice about copying portions of books, since it is easy and inexpensive to do so. However, doing so can be similar to stealing, according to the law. With the advent of color photocopying, an even more alarming problem has arisen: Thieves are now able to use this technology to create counterfeit money and checks. Researchers will soon find a way to make such important documents impossible to copy.

See also Fax machine; Instant photography; Laser-diode recording process.

FURTHER READING

Kelley, Neil D. "Xerography: The Greeks Had a Word for It." *Infosystems* 24, no. 1 (January, 1977).
McClain, Dylan L. "Duplicate Efforts." *New York Times* (November 30, 1998).
Mort, J. *The Anatomy of Xerography: Its Invention and Evolution.* Jefferson, N.C.: McFarland, 1989.

X-RAY CRYSTALLOGRAPHY

THE INVENTION: Technique for using X rays to determine the crystal structures of many substances.

THE PEOPLE BEHIND THE INVENTION:

Sir William Lawrence Bragg (1890-1971), the son of Sir William Henry Bragg and cowinner of the 1915 Nobel Prize in Physics

Sir William Henry Bragg (1862-1942), an English mathematician and physicist and cowinner of the 1915 Nobel Prize in Physics

Max von Laue (1879-1960), a German physicist who won the 1914 Nobel Prize in Physics

Wilhelm Conrad Röntgen (1845-1923), a German physicist who won the 1901 Nobel Prize in Physics

René-Just Haüy (1743-1822), a French mathematician and mineralogist

Auguste Bravais (1811-1863), a French physicist

The Elusive Crystal

A crystal is a body that is formed once a chemical substance has solidified. It is uniformly shaped, with angles and flat surfaces that form a network based on the internal structure of the crystal's atoms. Determining what these internal crystal structures look like is the goal of the science of X-ray crystallography. To do this, it studies the precise arrangements into which the atoms are assembled.

Central to this study is the principle of X-ray diffraction. This technique involves the deliberate scattering of X rays as they are shot through a crystal, an act that interferes with their normal path of movement. The way in which the atoms are spaced and arranged in the crystal determines how these X rays are reflected off them while passing through the material. The light waves thus reflected form a telltale interference pattern. By studying this pattern, scientists can discover variations in the crystal structure.

The development of X-ray crystallography in the early twentieth century helped to answer two major scientific questions: What are X rays? and What are crystals? It gave birth to a new technology for the identification and classification of crystalline substances.

From studies of large, natural crystals, chemists and geologists had established the elements of symmetry through which one could classify, describe, and distinguish various crystal shapes. René-Just Haüy, about a century before, had demonstrated that diverse shapes of crystals could be produced by the repetitive stacking of tiny solid cubes.

Auguste Bravais later showed, through mathematics, that all crystal forms could be built from a repetitive stacking of three-dimensional arrangements of points (lattice points) into "space lattices," but no one had ever been able to prove that matter really was arranged in space lattices. Scientists did not know if the tiny building blocks modeled by space lattices actually were solid matter throughout, like Haüy's cubes, or if they were mostly empty space, with solid matter located only at the lattice points described by Bravais.

With the disclosure of the atomic model of Danish physicist Niels Bohr in 1913, determining the nature of the building blocks of crystals took on a special importance. If crystal structure could be shown to consist of atoms at lattice points, then the Bohr model would be supported, and science then could abandon the theory that matter was totally solid.

X Rays Explain Crystal Structure

In 1912, Max von Laue first used X rays to study crystalline matter. Laue had the idea that irradiating a crystal with X rays might cause diffraction. He tested this idea and found that X rays were scattered by the crystals in various directions, revealing on a photographic plate a pattern of spots that depended on the orientation and the symmetry of the crystal.

The experiment confirmed in one stroke that crystals were not solid and that their matter consisted of atoms occupying lattice sites with substantial space in between. Further, the atomic arrangements of crystals could serve to diffract light rays. Laue received the 1914 Nobel Prize in Physics for his discovery of the diffraction of X rays in crystals.

SIR WILLIAM HENRY BRAGG
AND SIR WILLIAM LAWRENCE BRAGG

William Henry Bragg, senior member of one of the most illustrious father-son scientific teams in history, was born in Cumberland, England, in 1862. Talented at mathematics, he studied that field at Trinity College, Cambridge, and physics at the Cavendish Laboratory, then moved into a professorship at the University of Adelaide in Australia. Despite an underequipped laboratory, he proved that the atom is not a solid body, and his work with X rays attracted the attention of Ernest Rutherford in England, who helped him win a professorship at the University of Leeds in 1908.

(Library of Congess)

William Henry Bragg

By then his eldest son, William Lawrence Bragg, was showing considerable scientific abilities of his own. Born in Adelaide in 1890, he also attended Trinity College, Cambridge, and performed research at the Cavendish. It was while there that father and son worked together to establish the specialty of X-ray crystallography. When they shared the 1915 Nobel Prize in Physics for their work, the son was only twenty-five years old—the youngest person ever to receive a Nobel Prize in any field.

The younger Bragg was also an artillery officer in France during World War I. Meanwhile, his father worked for the Royal Admiralty. The hydrophone he invented to detect submarines underwater earned him a knighthood in 1920. The father moved to University College, London, and became director of the Royal Institution. His popular lectures about the latest scientific developments made him famous among the public, while his elevation to president of the Royal Society in 1935 placed him among the most influential scientists in the world. He died in 1942.

The son taught at the University of Manchester in 1919 and then in 1938 became director of the National Physics Laboratory and professor of physics at the Cavendish. Following the father's example, he became an administrator and professor at the Royal Institution, where he also distinguished himself with his popular lectures. He encouraged research using X-ray crystallography, including the work that unlocked the structure of deoxyribonucleic acid (DNA). Knighted in 1941, he became a royal Companion of Honor in 1967. He died in 1971.

Still, the diffraction of X rays was not yet a proved scientific fact. Sir William Henry Bragg contributed the final proof by passing one of the diffracted beams through a gas and achieving ionization of the gas, the same effect that true X rays would have caused. He also used the spectrometer he built for this purpose to detect and measure specific wavelengths of X rays and to note which orientations of crystals produced the strongest reflections. He noted that X rays, like visible light, occupy a definite part of the electromagnetic spectrum. Yet most of Bragg's work focused on actually using X rays to deduce crystal structures.

Sir Lawrence Bragg was also deeply interested in this new phenomenon. In 1912, he had the idea that the pattern of spots was an indication that the X rays were being reflected from the planes of atoms in the crystal. If that were true, Laue pictures could be used to obtain information about the structures of crystals. Bragg developed an equation that described the angles at which X rays would be most effectively diffracted by a crystal. This was the start of the X-ray analysis of crystals.

Henry Bragg had at first used his spectrometer to try to determine whether X rays had a particulate nature. It soon became evident, however, that the device was a far more powerful way of analyzing crystals than the Laue photograph method had been. Not long afterward, father and son joined forces and founded the new science of X-ray crystallography. By experimenting with this technique, Lawrence Bragg came to believe that if the lattice models of Bravais applied to actual crystals, a crystal structure could be viewed as being composed of atoms arranged in a pattern consisting of a few sets of flat, regularly spaced, parallel planes.

Diffraction became the means by which the Braggs deduced the detailed structures of many crystals. Based on these findings, they built three-dimensional scale models out of wire and spheres that made it possible for the nature of crystal structures to be visualized clearly even by nonscientists. Their results were published in the book *X-Rays and Crystal Structure* (1915).

IMPACT

The Braggs founded an entirely new discipline, X-ray crystallography, which continues to grow in scope and application. Of particular importance was the early discovery that atoms, rather than molecules, determine the nature of crystals. X-ray spectrometers of the type developed by the Braggs were used by other scientists to gain insights into the nature of the atom, particularly the innermost electron shells. The tool made possible the timely validation of some of Bohr's major concepts about the atom.

X-ray diffraction became a cornerstone of the science of mineralogy. The Braggs, chemists such as Linus Pauling, and a number of mineralogists used the tool to do pioneering work in deducing the structures of all major mineral groups. X-ray diffraction became the definitive method of identifying crystalline materials.

Metallurgy progressed from a technology to a science as metallurgists became able, for the first time, to deduce the structural order of various alloys at the atomic level. Diffracted X rays were applied in the field of biology, particularly at the Cavendish Laboratory under the direction of Lawrence Bragg. The tool proved to be essential for deducing the structures of hemoglobin, proteins, viruses, and eventually the double-helix structure of deoxyribonucleic acid (DNA).

See also Field ion microscope; Geiger counter; Holography; Mass spectrograph; Neutrino detector; Scanning tunneling microscope; Thermal cracking process; Ultramicroscope.

FURTHER READING

Achilladelis, Basil, and Mary Ellen Bowden. *Structures of Life*. Philadelphia: The Center, 1989.
Bragg, William Lawrence. *The Development of X-Ray Analysis*. New York: Hafner Press, 1975.
Thomas, John Meurig. "Architecture of the Invisible." *Nature* 364 (August 5, 1993).

X-RAY IMAGE INTENSIFIER

THE INVENTION: A complex electronic device that increases the intensity of the light in X-ray beams exiting patients, thereby making it possible to read finer details.

THE PEOPLE BEHIND THE INVENTION:
Wilhelm Conrad Röntgen (1845-1923), a German physicist
Thomas Alva Edison (1847-1931), an American inventor
W. Edward Chamberlain, an American physician
Thomson Electron Tubes, a French company

RADIOLOGISTS NEED DARK ADAPTATION

Thomas Alva Edison invented the fluoroscope in 1896, only one year after Wilhelm Conrad Röntgen's discovery of X rays. The primary function of the fluoroscope is to create images of the internal structures and fluids in the human body. During fluoroscopy, the radiologist who performs the procedure views a continuous image of the motion of the internal structures.

Although much progress was made during the first half of the twentieth century in recording X-ray images on plates and film, fluoroscopy lagged behind. In conventional fluoroscopy, a radiologist observed an image on a dim fluoroscopic screen. In the same way that it is more difficult to read a telephone book in dim illumination than in bright light, it is much harder to interpret a dim fluoroscopic image than a bright one. In the early years of fluoroscopy, the radiologist's eyes had to be accustomed to dim illumination for at least fifteen minutes before performing fluoroscopy. "Dark adaptation" was the process of wearing red goggles under normal illumination so that the amount of light entering the eye was reduced.

The human retina contains two kinds of light-sensitive elements: rods and cones. The dim light emitted by the screen of the fluoroscope, even under the best conditions, required the radiologist to see only with the rods, and vision is much less accurate in such circumstances. For normal rod-and-cone vision, the brightness of the screen might have to be increased a thousandfold. Such an increase was impossible; even if an X-ray tube could have been built that was capable of emitting a beam of sufficient intensity, its rays would have been fatal to the patient in less than a minute.

FLUOROSCOPY IN AN UNDARKENED ROOM

In a classic paper delivered at the December, 1941, meeting of the Radiological Society of North America, Dr. W. Edward Chamberlain of Temple University Medical School proposed applying to fluoroscopy the techniques of image amplification (also known as image intensification) that had already been adapted for use in the electron microscope and in television. The idea was not original with him. Four or five years earlier, Irving Langmuir of General Electric Company had applied for a patent for a device that would intensify a fluoroscopic image. "It is a little hard to understand the delay in the creation of a practical device," Chamberlain noted. "Perhaps what is needed is a realization by the physicists and the engineers of the great need for brighter fluoroscopic images and the great advantage to humanity which their arrival would entail."

Chamberlain's brilliant analysis provided precisely that awareness. World War II delayed the introduction of fluoroscopic image intensification, but during the 1950's, a number of image intensifiers based on the principles Chamberlain had outlined came on the market.

The image-intensifier tube is a complex electronic device that receives the X-ray beam exiting the patient, converts it into light, and increases the intensity of that light. The tube is usually contained in a glass envelope that provides some structural support and maintains a vacuum. The X rays, after passing through the patient, impinge on the face of a screen and trigger the ejection of electrons, which are then speeded up and focused within the tube by means of electrical fields. When the speeded-up electrons strike the phosphor at the output end of the tube, they trigger the emission of light photons that re-create the desired image, which is several thousand times brighter than is the case with the conventional fluoroscopic screen. The output of the image intensifier can be viewed in an undarkened room without prior dark adaptation, thus saving the radiologist much valuable time.

Moving pictures can be taken of the output phosphor of the intensifying tube or of the television receiver image, and they can be stored on motion picture film or on magnetic tape. This permanently records the changing image and makes it possible to reduce further the dose of radiation that a patient must receive. Instead of prolonging the radiation exposure while examining various parts of the image or checking for various factors, the radiologist can record a relatively short exposure and then rerun the motion picture film or tape as often as necessary to analyze the information that it contains. The radiation dosage that is administered to the patient can be reduced to a tenth or even a hundredth of what it had been previously, and the same amount of diagnostic information or more can be obtained. The radiation dose that the radiologist receives is reduced to zero or almost zero. In addition, the combination of the brighter image and the lower radiation dosage administered to the patient has made it possible for radiologists to develop a number of important new diagnostic procedures that could not have been accomplished at all without image intensification.

IMPACT

The image intensifier that was developed by the French company Thomson Electron Tubes in 1959 had an input-phosphor diameter, or field, of four inches. Later on, image intensifiers with field sizes of up to twenty-two inches became available, making it possible to create images of much larger portions of the human anatomy.

The most important contribution made by image intensifiers was to increase fluoroscopic screen illumination to the level required for cone vision. These devices have made dark adaptation a thing of the past. They have also brought the television camera into the fluoroscopic room and opened up a whole new world of fluoroscopy.

See also Amniocentesis; CAT scanner; Electrocardiogram; Electroencephalogram; Mammography; Nuclear magnetic resonance; Ultrasound.

FURTHER READING

Glasser, Otto. *Dr. W. C. Röntgen*. 2d ed. Springfield, Ill.: Charles C. Thomas, 1972.

Isherwood, Ian, Adrian Thomas, and Peter Neil Temple Wells. *The Invisible Light: One Hundred Years of Medical Radiology*. Cambridge, Mass.: Blackwell Science, 1995.

Lewis, Ricki. "Radiation Continuing Concern with Fluoroscopy." *FDA Consumer* 27 (November, 1993).

YELLOW FEVER VACCINE

THE INVENTION: The first safe vaccine agaisnt the virulent yellow fever virus, which caused some of the deadliest epidemics of the nineteenth and early twentieth centuries.

THE PEOPLE BEHIND THE INVENTION:
Max Theiler (1899-1972), a South African microbiologist
Wilbur Augustus Sawyer (1879-1951), an American physician
Hugh Smith (1902-1995), an American physician

A YELLOW FLAG

Yellow fever, caused by a virus and transmitted by mosquitoes, infects humans and monkeys. After the bite of the infecting mosquito, it takes several days before symptoms appear. The onset of symptoms is abrupt, with headache, nausea, and vomiting. Because the virus destroys liver cells, yellowing of the skin and eyes is common. Approximately 10 to 15 percent of patients die after exhibiting terrifying signs and symptoms. Death occurs usually from liver necrosis (decay) and liver shutdown. Those that survive recover completely and are immunized.

At the beginning of the twentieth century, there was no cure for yellow fever. The best that medical authorities could do was to quarantine the afflicted. Those quarantines usually waved the warning yellow flag, which gave the disease its colloquial name, "yellow jack."

After the Aëdes aegypti mosquito was clearly identified as the carrier of the disease in 1900, efforts were made to combat the disease by wiping out the mosquito. Most famous in these efforts were the American army surgeon Walter Reed and the Cuban physician Carlos J. Finlay. This strategy was successful in Panama and Cuba and made possible the construction of the Panama Canal. Still, the yellow fever virus persisted in the tropics, and the opening of the Panama Canal increased the danger of its spreading aboard the ships using this new route.

Moreover, the disease, which was thought to be limited to the jungles of South and Central America, had begun to spread arounds the world to wherever the mosquito *Aëdes aegypti* could carry the virus. Mosquito larvae traveled well in casks of water aboard trading vessels and spread the disease to North America and Europe.

IMMUNIZATION BY MUTATION

Max Theiler received his medical education in London. Following that, he completed a four-month course at the London School of Hygiene and Tropical Medicine, after which he was invited to come to the United States to work in the department of tropical medicine at Harvard University.

While there, Theiler started working to identify the yellow fever organism. The first problem he faced was finding a suitable laboratory animal that could be infected with yellow fever. Until that time, the only animal successfully infected with yellow fever was the rhesus monkey, which was expensive and difficult to care for under laboratory conditions. Theiler succeeded in infecting laboratory mice with the disease by injecting the virus directly into their brains.

Laboratory work for investigators and assistants coming in contact with the yellow fever virus was extremely dangerous. At least six of the scientists at the Yellow Fever Laboratory at the Rockefeller Institute died of the disease, and many other workers were infected. In 1929, Theiler was infected with yellow fever; fortunately, the attack was so mild that he recovered quickly and resumed his work.

During one set of experiments, Theiler produced successive generations of the virus. First, he took virus from a monkey that had died of yellow fever and used it to infect a mouse. Next, he extracted the virus from that mouse and injected it into a second mouse, repeating the same procedure using a third mouse. All of them died of encephalitis (inflammation of the brain). The virus from the third mouse was then used to infect a monkey. Although the monkey showed signs of yellow fever, it recovered completely. When Theiler passed the virus through more mice and then into the abdomen of another monkey, the monkey showed no symptoms of the disease. The results of these experiments were published by Theiler in the journal *Science*.

This article caught the attention of Wilbur Augustus Sawyer, director of the Yellow Fever Laboratory at the Rockefeller Foundation International Health Division in New York. Sawyer, who was working on a yellow fever vaccine, offered Theiler a job at the Rockefeller Foundation, which Theiler accepted. Theiler's mouse-adapted, "attenuated" virus was given to the laboratory workers, along with human immune serum, to protect them against the yellow fever virus. This type of vaccination, however, carried the risk of transferring other diseases, such as hepatitis, in the human serum.

In 1930, Theiler worked with Eugen Haagen, a German bacteriologist, at the Rockefeller Foundation. The strategy of the Rockefeller laboratory was a cautious, slow, and steady effort to culture a strain of the virus so mild as to be harmless to a human but strong enough to confer a long-lasting immunity. (To "culture" something—tissue cells, microorganisms, or other living matter—is to grow it in a specially prepared medium under laboratory conditions.) They started with a new strain of yellow fever harvested from a twenty-eight-year-old West African named Asibi; it was later known as the "Asibi strain." It was a highly virulent strain that in four to seven days killed almost all the monkeys that were infected with it. From time to time, Theiler or his assistant would test the culture on a monkey and note the speed with which it died.

It was not until April, 1936, that Hugh Smith, Theiler's assistant, called to his attention an odd development as noted in the laboratory records of strain 17D. In its 176th culture, 17D had failed to kill the test mice. Some had been paralyzed, but even these eventually recovered. Two monkeys who had received a dose of 17D in their brains survived a mild attack of encephalitis, but those who had taken the infection in the abdomen showed no ill effects whatever. Oddly, subsequent subcultures of the strain killed monkeys and mice at the usual rate. The only explanation possible was that a mutation had occurred unnoticed.

The batch of strain 17D was tried over and over again on monkeys with no harmful effects. Instead, the animals were immunized effectively. Then it was tried on the laboratory staff, including Theiler and his wife, Lillian. The batch injected into humans had the same immunizing effect. Neither Theiler nor anyone else could explain how the mutation of the virus had resulted. Attempts to duplicate the experiment, using the same Asibi virus, failed. Still, this was the first safe vaccine for yellow fever. In June, 1937, Theiler reported this crucial finding in the *Journal of Experimental Medicine*.

IMPACT

Following the discovery of the vaccine, Theiler's laboratory became a production plant for the 17D virus. Before World War II (1939-1945), more than one million vaccination doses were sent to Brazil and other South American countries. After the United States entered the war, eight million soldiers were given the vaccine before being shipped to tropical war zones. In all, approximately fifty million people were vaccinated in the war years.

Yet although the vaccine, combined with effective mosquito control, eradicated the disease from urban centers, yellow fever is still present in large regions of South and Central America and of Africa. The most severe outbreak of yellow fever ever known occurred from 1960 to 1962 in Ethiopia; out of one hundred thousand people infected, thirty thousand died.

The 17D yellow fever vaccine prepared by Theiler in 1937 continues to be the only vaccine used by the World Health Organization, more than fifty years after its discovery. There is a continuous effort by that organization to prevent infection by immunizing the people living in tropical zones.

See also Antibacterial drugs; Penicillin; Polio vaccine (Sabin); Polio vaccine (Salk); Salvarsan; Tuberculosis vaccine; Typhus vaccine.

FURTHER READING

DeJauregui, Ruth. *One Hundred Medical Milestones That Shaped World History*. San Mateo, Calif.: Bluewood Books, 1998.

Delaporte, François. *The History of Yellow Fever: An Essay on the Birth of Tropical Medicine*. Cambridge, Mass.: MIT Press, 1991.

Theiler, Max, and Wilbur G. Downs. *The Arthropod-borne Viruses of Vertebrates: An Account of the Rockefeller Foundation Virus Program, 1951-1970*. New Haven, Conn.: Yale University Press, 1973.

Williams, Greer. *Virus Hunters*. London: Hutchinson, 1960.

Time Line

Date	Invention
c. 1900	Electrocardiogram
1900	Brownie camera
1900	Dirigible
1901	Artificial insemination
1901	Vat dye
1901-1904	Silicones
1902	Ultramicroscope
1903	Airplane
1903	Disposable razor
1903-1909	Laminated glass
1904	Alkaline storage battery
1904	Photoelectric cell
1904	Vacuum tube
1905	Blood transfusion
1905-1907	Plastic
1906	Gyrocompass
1906	Radio
1906-1911	Tungsten filament
1907	Autochrome plate
1908	Ammonia
1908	Geiger counter
1908	Interchangeable parts
1908	Oil-well drill bit
1908	Vacuum cleaner
1910	Radio crystal sets
1910	Salvarsan
1910	Washing machine
1910-1939	Electric refrigerator
1912	Color film
1912	Diesel locomotive
1912-1913	Solar thermal engine
1912-1914	Artificial kidney
1912-1915	X-ray crystallography

DATE	INVENTION
1913	Assembly line
1913	Geothermal power
1913	Mammography
1913	Thermal cracking process
1915	Long-distance telephone
1915	Propeller-coordinated machine gun
1915	Pyrex glass
1915	Long-distance radiotelephony
1916-1922	Internal combustion engine
1917	Food freezing
1917	Sonar
1919	Mass spectrograph
1921	Tuberculosis vaccine
1923	Rotary dial telephone
1923	Television
1923 and 1951	Syphilis test
1924	Ultracentrifuge
1925-1930	Differential analyzer
1926	Buna rubber
1926	Rocket
1926	Talking motion pictures
1927	Heat pump
1928	Pap test
1929	Electric clock
1929	Electroencephalogram
1929	Iron lung
1930's	Contact lenses
1930's	Vending machine slug rejector
1930	Refrigerant gas
1930	Typhus vaccine
1930-1935	FM Radio
1931	Cyclotron
1931	Electron microscope
1931	Neoprene
1932	Fuel cell
1932-1935	Antibacterial drugs

DATE	INVENTION
1933-1954	Freeze-drying
1934	Bathysphere
1935	Nylon
1935	Radar
1935	Richter scale
1936	Fluorescent lighting
1937	Yellow fever vaccine
1938	Polystyrene
1938	Teflon
1938	Xerography
1940's	Carbon dating
1940	Color television
1940	Penicillin
1940-1955	Microwave cooking
1941	Polyester
1941	Touch-tone telephone
1941	Turbojet
1942	Infrared photography
1942-1950	Orlon
1943	Aqualung
1943	Colossus computer
1943	Nuclear reactor
1943-1946	ENIAC computer
1944	Mark I calculator
1944	V-2 rocket
1945	Atomic bomb
1945	Tupperware
1946	Cloud seeding
1946	Synchrocyclotron
1947	Holography
1948	Atomic clock
1948	Broadcaster guitar
1948	Instant photography
1948-1960	Bathyscaphe
1949	BINAC computer
1949	Community antenna television

DATE	INVENTION
1950	Cyclamate
1950-1964	In vitro plant culture
1951	Breeder reactor
1951	UNIVAC computer
1951-1952	Hydrogen bomb
1952	Amniocentesis
1952	Hearing aid
1952	Polio vaccine (Salk)
1952	Reserpine
1952	Steelmaking process
1952-1956	Field ion microscope
1953	Artificial hormone
1953	Heart-lung machine
1953	Polyethylene
1953	Synthetic amino acid
1953	Transistor
1953-1959	Hovercraft
mid-1950's	Synthetic RNA
1954	Photovoltaic cell
1955	Radio interferometer
1955-1957	FORTRAN programming language
1956	Birth control pill
1957	Artificial satellite
1957	Nuclear power plant
1957	Polio vaccine (Sabin)
1957	Transistor radio
1957	Velcro
1957-1972	Pacemaker
1958	Ultrasound
1959	Atomic-powered ship
1959	COBOL computer language
1959	IBM Model 1401 computer
1959	X-ray image intensifier
1960's	Rice and wheat strains
1960's	Virtual machine
1960	Laser

DATE	INVENTION
1960	Memory metal
1960	Telephone switching
1960	Weather satellite
1961	SAINT
1962	Communications satellite
1962	Laser eye surgery
1962	Robot (industrial)
1963	Cassette recording
1964	Bullet train
1964	Electronic synthesizer
1964-1965	BASIC programming language
1966	Tidal power plant
1967	Coronary artery bypass surgery
1967	Dolby noise reduction
1967	Neutrino detector
1967	Synthetic DNA
1969	Bubble memory
1969	The Internet
1969-1983	Optical disk
1970	Floppy disk
1970	Videocassette recorder
1970-1980	Virtual reality
1972	CAT scanner
1972	Pocket calculator
1975-1979	Laser-diode recording process
1975-1990	Fax machine
1976	Supercomputer
1976	Supersonic passenger plane
1976-1988	Stealth aircraft
1977	Apple II computer
1977	Fiber-optics
1977-1985	Cruise missile
1978	Cell phone
1978	Compressed-air-accumulating power plant
1978	Nuclear magnetic resonance
1978-1981	Scanning tunneling microscope

DATE	INVENTION
1979	Artificial blood
1979	Walkman cassette player
1980's	CAD/CAM
1981	Personal computer
1982	Abortion pill
1982	Artificial heart
1982	Genetically engineered insulin
1982	Robot (household)
1983	Artificial chromosome
1983	Aspartame
1983	Compact disc
1983	Hard disk
1983	Laser vaporization
1985	Genetic "fingerprinting"
1985	Tevatron accelerator
1997	Cloning
2000	Gas-electric car

Topics by Category

Radar
Radio
Radio crystal sets
Rotary dial telephone
Sonar
Talking motion pictures
Telephone switching
Television
Touch-tone telephone
Transistor radio
Vacuum tube
Videocassette recorder
Xerography

COMPUTER SCIENCE

Apple II computer
BASIC programming language
BINAC computer
Bubble memory
COBOL computer language
Colossus computer
Computer chips
Differential analyzer
ENIAC computer
Floppy disk
FORTRAN programming
    language
Hard disk
IBM Model 1401 computer
Internet
Mark I calculator
Optical disk
Personal computer
Pocket calculator
SAINT
Supercomputer
UNIVAC computer

Virtual machine
Virtual reality

CONSUMER PRODUCTS

Apple II computer
Aspartame
Birth control pill
Broadcaster guitar
Brownie camera
Cassette recording
Cell phone
Color film
Color television
Compact disc
Cyclamate
Disposable razor
Electric refrigerator
FM radio
Gas-electric car
Hearing aid
Instant photography
Internet
Nylon
Orlon
Personal computer
Pocket calculator
Polyester
Pyrex glass
Radio
Rotary dial telephone
Teflon
Television
Touch-tone telephone
Transistor radio
Tupperware
Vacuum cleaner
Velcro

Videocassette recorder
Walkman cassette player
Washing machine

DRUGS AND VACCINES

Abortion pill
Antibacterial drugs
Artificial hormone
Birth control pill
Genetically engineered insulin
Penicillin
Polio vaccine (Sabin)
Polio vaccine (Salk)
Reserpine
Salvarsan
Tuberculosis vaccine
Typhus vaccine
Yellow fever vaccine

EARTH SCIENCE

Aqualung
Bathyscaphe
Bathysphere
Cloud seeding
Richter scale
X-ray crystallography

ELECTRONICS

Cassette recording
Cell phone
Color television
Communications satellite
Compact disc
Dolby noise reduction
Electronic synthesizer

Fax machine
Fiber-optics
FM radio
Hearing aid
Laser-diode recording process
Long-distance radiotelephony
Long-distance telephone
Radar
Radio
Radio crystal sets
Rotary dial telephone
Sonar
Telephone switching
Television
Touch-tone telephone
Transistor
Transistor radio
Vacuum tube
Videocassette recorder
Walkman cassette player
Xerography

ENERGY

Alkaline storage battery
Breeder reactor
Compressed-air-accumulating
    power plant
Fluorescent lighting
Fuel cell
Gas-electric car
Geothermal power
Heat pump
Nuclear power plant
Nuclear reactor
Oil-well drill bit
Photoelectric cell
Photovoltaic cell

Solar thermal engine
Tidal power plant
Vacuum tube

ENGINEERING

Airplane
Assembly line
Bullet train
CAD/CAM
Differential analyzer
Dirigible
ENIAC computer
Gas-electric car
Internal combustion engine
Oil-well drill bit
Robot (household)
Robot (industrial)
Steelmaking process
Tidal power plant
Vacuum cleaner
Washing machine

EXPLORATION

Carbon dating
Aqualung
Bathyscaphe
Bathysphere
Neutrino detector
Radar
Radio interferometer
Sonar

FOOD SCIENCE

Aspartame
Cyclamate

Electric refrigerator
Food freezing
Freeze-drying
Genetically engineered insulin
In vitro plant culture
Microwave cooking
Polystyrene
Refrigerant gas
Rice and wheat strains
Teflon
Tupperware

GENETIC ENGINEERING

Amniocentesis
Artificial chromosome
Artificial insemination
Cloning
Genetic "fingerprinting"
Genetically engineered insulin
In vitro plant culture
Rice and wheat strains
Synthetic amino acid
Synthetic DNA
Synthetic RNA

HOME PRODUCTS

Cell phone
Color television
Community antenna television
Disposable razor
Electric refrigerator
Fluorescent lighting
FM radio
Microwave cooking
Radio
Refrigerant gas

Robot (household)
Rotary dial telephone
Television
Touch-tone Telephone
Transistor radio
Tungsten filament
Tupperware
Vacuum cleaner
Videocassette recorder
Washing machine

MANUFACTURING

Assembly line
CAD/CAM
Interchangeable parts
Memory metal
Polystyrene
Steelmaking process

MATERIALS

Buna rubber
Contact lenses
Disposable razor
Laminated glass
Memory metal
Neoprene
Nylon
Orlon
Plastic
Polyester
Polyethylene
Polystyrene
Pyrex glass
Silicones
Steelmaking process
Teflon

Tungsten filament
Velcro

MEASUREMENT AND DETECTION

Amniocentesis
Atomic clock
Carbon dating
CAT scanner
Cyclotron
Electric clock
Electrocardiogram
Electroencephalogram
Electron microscope
Geiger counter
Gyrocompass
Mass spectrograph
Neutrino detector
Radar
Sonar
Radio interferometer
Richter scale
Scanning tunneling microscope
Synchrocyclotron
Tevatron accelerator
Ultracentrifuge
Ultramicroscope
Vending machine slug rejector
X-ray crystallography

MEDICAL PROCEDURES

Amniocentesis
Blood transfusion
CAT scanner
Cloning
Coronary artery bypass surgery
Electrocardiogram

Electroencephalogram
Heart-lung machine
Iron lung
Laser eye surgery
Laser vaporization
Mammography
Nuclear magnetic resonance
Pap test
Syphilis test
Ultrasound
X-ray image intensifier

Penicillin
Polio vaccine (Sabin)
Polio vaccine (Salk)
Reserpine
Salvarsan
Syphilis test
Tuberculosis vaccine
Typhus vaccine
Ultrasound
X-ray image intensifier
Yellow fever vaccine

MEDICINE

Abortion pill
Amniocentesis
Antibacterial drugs
Artificial blood
Artificial heart
Artificial hormone
Artificial kidney
Birth control pill
Blood transfusion
CAT scanner
Contact lenses
Coronary artery bypass surgery
Electrocardiogram
Electroencephalogram
Genetically engineered insulin
Hearing aid
Heart-lung machine
Iron lung
Laser eye surgery
Laser vaporization
Mammography
Nuclear magnetic resonance
Pacemaker
Pap test

MUSIC

Broadcaster guitar
Cassette recording
Dolby noise reduction
Electronic synthesizer
FM Radio
Radio
Transistor radio

PHOTOGRAPHY

Autochrome plate
Brownie camera
Color film
Electrocardiogram
Electron microscope
Fax machine
Holography
Infrared photography
Instant photography
Mammography
Mass spectrograph
Optical disk
Talking motion pictures
Weather satellite

Xerography
X-ray crystallography

PHYSICS

Atomic bomb
Cyclotron
Electron microscope
Field ion microscope
Geiger counter
Hydrogen bomb
Holography
Laser
Mass spectrograph
Scanning tunneling microscope
Synchrocyclotron
Tevatron accelerator
X-ray crystallography

SYNTHETICS

Artificial blood
Artificial chromosome
Artificial heart
Artificial hormone
Artificial insemination
Artificial kidney
Artificial satellite
Aspartame
Buna rubber
Cyclamate
Electronic synthesizer
Genetically engineered insulin
Neoprene

Synthetic amino acid
Synthetic DNA
Synthetic RNA
Vat dye

TRANSPORTATION

Airplane
Atomic-powered ship
Bullet train
Diesel locomotive
Dirigible
Gas-electric car
Gyrocompass
Hovercraft
Internal combustion engine
Supersonic passenger plane
Turbojet

WEAPONS TECHNOLOGY

Airplane
Atomic bomb
Cruise missile
Dirigible
Hydrogen bomb
Propeller-coordinated machine
    gun
Radar
Rocket
Sonar
Stealth aircraft
V-2 rocket

INDEX